MW01277520

Economics of Advanced Manufacturing Systems

Economics of Advanced Manufacturing Systems

Edited by

Hamid R. Parsaei
Center for Computer-Aided Engineering
University of Louisville, USA

and

Anil Mital
Ergonomics and Engineering Controls Research Laboratory
University of Cincinnati, USA

CHAPMAN & HALL
London · Glasgow · New York · Tokyo · Melbourne · Madras

Published by Chapman & Hall, 2–6 Boundary Row, London SE1 8HN

Chapman & Hall, 2–6 Boundary Row, London SE1 8HN, UK

Blackie Academic & Professional, Westercleddens Road, Bishopbriggs, Glasgow G64 2NZ, UK

Van Nostrand Reinhold Inc., 115 5th Avenue, New York NY10003, USA

Chapman & Hall Japan, Thomson Publishing Japan, Hirakawacho Nemoto Building, 7F, 1-7-11 Hirakawa-cho, Chiyoda-ku, Tokyo 102, Japan

Chapman & Hall Australia, Thomas Nelson Australia, 102 Dodds Street, South Melbourne, Victoria 3205, Australia

Chapman & Hall India, R. Seshadri, 32 Second Main Road, CIT East, Madras 600 035, India

First edition 1992

Typeset in 10/12 pt Times by Best-set Typesetter Ltd, Hong Kong
Printed in Great Britain by St Edmundsbury Press, Bury St Edmunds, Suffolk

ISBN 0 412 43350 8
0 442 31516 3

A catalogue record for this book is available from the British Library

Library of Congress Cataloging-in-Publication data
Economics of advanced manufacturing systems /
edited by Hamid R. Parsaei and Anil Mital.
p. cm.
Includes bibliographical references.
ISBN 0-442-31516-3
1. Manufactures—Technological innovations. 2. Manufacturing processes—Automation—Economic aspects. 3. Production engineering.
I. Parsaei, H. R. II. Mital, Anil.
HD9720.5.E27 1992 91-23613
338′.064—dc20 CIP

*To our wives, Farah and Chetna, and children,
Boback, Shadi, Aashi and Anubhav, who
suffered through our countless hours of neglect
and absence during the preparation of this book.*

Contents

Contributors

Giovannie Azzone
Dipartmento di Economia E Produzione, Politecnico di Milano, Piazza Leonardo da Vinci 32, Milan, Italy.

Adedeji B. Badiru
School of Industrial Engineering, University of Oklahoma, Oklahoma, USA.

Robert G. Batson
Department of Industrial Engineering, The University of Alabama, Alabama, USA.

Umberto Bertele'
Dipartmento di Economia E Produzione, Politecnico di Milano, Piazza Leonardo da Vinci 32, Milan, Italy.

Dr James A. Bontadelli
Department of Industrial Engineering, University of Tennessee-Knoxville, Knoxville, Tennessee, USA.

Hector R. Carrasco
Department of Industrial Engineering, Florida International University, Florida, USA.

M. Jeya Chandra
Department of Industrial Engineering, Pennsylvania State University, University Park, Pennsylvania, USA.

Subrato Chandra
Florida Solar Energy Center, University of Central Florida, Cape Canaveral, Florida, USA.

Adel S. Elmaghraby
Department of Engineering mathematics and computer science, University of Louisville, Louisville, Kentucky 40292, USA.

Ahmed K. Elshennawy
Department of Industrial Engineering, University of Central Florida, Orlando, Florida, USA.

Brian Goldiez
Institute for Simulation and Training, University of Central Florida, Orlando, Florida, USA.

David J. Hamblin
Putteridge Bury Management Centre, Luton, Bedfordshire, UK.

Catherine M. Harmonosky
Department of Industrial and Management Systems Engineering, Pennsylvania State University, University Park, Pennsylvania, USA.

William M. Henghold
Universal Technology Corporation, Dayton, Ohio, USA.

Brian B. Hundy
College of Manufacturing, Cranfield Institute of Technology, Cranfield, Bedford, UK.

Ibrahim N. Imam
Department of Engineering Mathematics and Computer Science, University of Louisville, Louisville, Kentucky, USA.

Raja K. Iyer
Center for Research for Information Systems, University of Texas at Arlington, Arlington, Texas, USA.

H. Thomas Johnson
Retzlaff Professor of Quality Management, School of Business Administration, Portland State University, Portland, USA.

Ali Khosravi-Kamrani
Department of Engineering Management, University of Missouri at Rolla, Rolla, USA.

Janardan Kulkarni
Laura Kersey Library, University of Louisville, Louisville, Kentucky, USA.

Steve R. LeClair
Air Force Materials Laboratory, Wright-Patterson Air Force Base, Dayton, Ohio, USA.

Shih-Ming Lee
Department of Industrial Engineering, Florida International University, Miami, Florida, USA.

Donald H. Liles
Automation and Robotics Research Institute, South Fort Worth, Texas, USA.

Jessica O. Matson
Department of Industrial Engineering, University of Alabama, Tuscaloosa, Alabama, USA.

Anil Mital
Ergonomics and Engineering Controls Research Laboratory, Department of Mechanical and Industrial Engineering, University of Cincinnati, Cincinnati, Ohio, USA.

Lucy C. Morse
Department of Industrial Engineering, University of Central Florida, Orlando, Florida, USA.

Suebsak Nanthavanij
Department of Mechanical and Industrial Engineering, New Jersey Institute of Technology, Newark, New Jersey, USA.

Carl-Henric Nilsson
Industrial Management, Lund Institute of Technology, Lund, Sweden.

Hakan Nordahl
Industrial Management, Lund Institute of Technology, Lund, Sweden.

Hamid R. Parsaei
Center for Computer-aided Engineering, Department of Industrial Engineering, University of Louisville, Louisville, Kentucky, USA.

Ingvar Persson
Industrial Management, Lund Institute of Technology, Lund, Sweden.

Richard W. Sapp
School of Business Administration, Portland State University, Portland, USA.

Mohamed Adel Shalabi
Department of Mechanical Design and Production, University of Cairo, Giza, Egypt.

Ramesh G. Soni
Department of Management, Indiana University of Pennsylvania, Indiana, Pennsylvania, USA.

William W. Swart
Department of Industrial Engineering, University of Central Florida, Orlando, Florida, USA.

Fariborz Tayyari
Department of Industrial Engineering, Bradley University, Peoria, Illinois, USA.

Julius P. Wong
Computer-aided Design Laboratory, Department of Mechanical Engineering, University of Louisville, Louisville, Kentucky, USA.

Ismail N. Zahran
Department of Engineering Mathematics and Computer Science, University of Louisville, Louisville, Kentucky, USA.

Acknowledgements

During the course of preparation of this book, a number of individuals made invaluable contributions. Among them are all the contributors who submitted their manuscripts and complied with the changes reviewers required and did so cheerfully and promptly. Furthermore, we realize that the quality of a book, such as this, is largely the result of professional integrity and dedication to upholding the highest standards of referees who review the contents. We have been fortunate in this regard as we have had the cooperation, encouragement, and support of the following individuals who made significant contributions to this book by either writing a chapter or reviewing submissions or both, and to them we owe a special thanks:

Layek Abdel-Malek, New Jersey Institute of Technology
Suraj M. Alexander, University of Louisville
Giovanni Azzone, Politecnico di Milano
Adedeji B. Badiru, University of Oklahoma
Robert G. Batson, University of Alabama
Umberto Bertele, Politecnico de Milano
William E. Biles, University of Louisville
James A. Bontadelli, University of Tennessee-Knoxville
Hector R. Carrasco, Florida International University
M. Jeya Chandra, Pennsylvania State University
Subrato Chandra, Florida Solar Energy Center
Adel S. Elmaghraby, University of Louisville
Ahmed K. Elshennawy, University of Central Florida
Gerald W. Evans, University of Louisville
Brian Goldiez, Institute for Simulation and Training
David J. Hamblin, Putteridge Bury Management Centre
Catherine M. Harmonosky, Pennsylvania State University
William M. Henghold, Universal Technology
Brian B. Hundy, Cranfield Institute of Technology
Ibrahim N. Imam, University of Louisville
Raja K. Iyer, University of Texas-Arlington
H. Thomas Johnson, Portland State University
Ali Khosravi-Kamrani, University of Louisville
Janardan Kulkarni, University of Louisville
Steve R. LeClair, Wright-Patterson Air Force Base
Shih-Ming Lee, Florida International University

Herman R. Leep, University of Louisville
Donald H. Liles, Automation and Robotics Research Institute
Jessica O. Matson, University of Alabama
Nancy L. Mills, University of Southern Colorado
Raul Miranda, University of Louisville
Lucy C. Morse, University of Central Florida
Suebsak Nanthavanij, New Jersey Institute of Technology
Carl-Henric Nilsson, Lund Institute of Technology
Hakan Nordahl, Lund Institute of Technology
Ingvar Persson, Lund Institute of Technology
Sabah U. Randhawa, Oregon State University
James Reeves, University of Tennessee-Knoxville
Richard W. Sapp, Portland State University
Hamid Seiffoddini, University of Wisconsin-Milwaukee
Mohamed A. Shalabi, The American University of Cairo
Ramesh G. Soni, Indiana University of Pennsylvania
G.T. Stevens, Jr., University of Texas-Arlington
William W. Swart, University of Central Florida
Fariborz Tayyari, Bradley University
Thomas M. West, Oregon State University
Bob E. White, Western Michigan University
Mickey R. Wilhelm, University of Louisville
Julius P. Wong, University of Louisville
Robert M. Wygant, Western Michigan University
Ismail N. Zahran, Department of Engineering Mathematics and Computer Science

Preface

The 1980s have witnessed a tremendous growth in the field of computer-integrated manufacturing systems. The other major areas of development have been computer-aided design, computer-aided manufacturing, industrial robotics, automated assembly, cellular and modular material handling, computer networking and office automation to name just a few. These new technologies are generally capital intensive and do not conform to traditional cost structures. The net result is a tremendous change in the way costs should be estimated and economic analyses performed. The majority of existing engineering economy texts still profess application of traditional analysis methods. But, as was mentioned above, it is clear that the basic trend in manufacturing industries is itself changing. So it is quite obvious that the practice of traditional economic analysis methods should change too.

This book is an attempt to address the various issues associated with non-traditional methods for evaluation of advanced computer-integrated technologies.

This volume consists of twenty refereed articles which are grouped into five parts. Part one, Economic Justification Methods, consists of six articles. In the first paper, Soni *et al.* present a new classification for economic justification methods for advanced automated manufacturing systems. In the second, Henghold and LeClair look at strengths and weaknesses of expert systems in general and more specifically, an application aimed at investment justification in advanced technology. The third paper, by Carrasco and Lee, proposes an enhanced economic methodology to improve the needs analysis, conceptual design and detailed design activities associated with technology modernization. The fourth paper, by Mital, examines and compares manual and flexible automated methods of assembly of several DFA products. The paper also includes the effects of changes in various production and market factors on the choice of assembly methods. In the fifth paper, Zahran *et al.* use simulation as a tool to evaluate the feasibility and the benefits of cellular manufacturing systems. The last paper in this section, by Bontadelli, focuses on the application of engineering economy, including life cycle concepts to analyze computer-controlled manufacturing systems.

Part 2 focuses on articles on models and techniques for justification of advanced production systems. In the first paper, Chandra and his co-author, Harmonosky, present some analytical techniques which may be used to obtain various types of flexibility measures and material handling costs. The second paper, outlined by Azzone and Bertele, introduces a new approach which uses the option theory and allows some of the problems which may be associated with the use of simulation based models to be overcome. The third paper, by Badiru, presents a framework for developing a multivariate learning curve model for manufacturing economic analysis. In the fourth paper, Batson and Matson explore the issue of how best to control and reduce the variance introduced by sequential fabrication steps on an individual part.

Five papers addressing various issues in costing and investment models are grouped in Part 3. Johnson and Sapp, in the first article, focus on the importance of activity-based information in identifying and managing waste in today's manufacturing environment. The second paper, by Tayyari and Parsaei, presents joint costs allocation to multiple products. In the third paper, Wong *et al.* outline the design for a software-driven totally integrated manufacturing cost estimating system. The aspect of risk evaluation of investment in advanced manufacturing technology is presented by Hundy and Hamblin in the fourth paper. The last paper in Part 3, by Nilsson *et al.*, deals with the capital-back method. This is used as a measurement of flexibility and a tool for management decision making with respect to capital budgeting.

Articles grouped in Part 4 concentrate on peripheral issues in advanced production and manufacturing systems including CIM related strategic issues for small business, application of automation, cost effectiveness of automated devices, and the safety issues in FMS environment. The first article, by Iyer and Liles, provides a synthesis of the existing computer-integrated manufacturing and computer-integrated enterprise (CIE) framework, and develops an integrating CIE framework to address small business firms' concerns in particular. Application of automated technologies to industrialized housing manufacture is the subject of a comprehensive discussion in the second paper by Elshennawy *et al.* Morse and Goldiez examine the problem concerned with the use of economic analysis to measure the effectiveness of manufacturing training devices. The last paper included in this section, by Nanthavanij, emphasizes the significance and necessity of protective devices in providing safe work conditions for workers in the flexible manufacturing environments.

The last part of this volume consists of one article by Kulkarni *et al.* This article examines the problem involved with the identification of important literature in a specific engineering field viz. advanced production and manufacturing systems.

We are indebted to the following individuals: Patricia Ostaszewski, A. Soundararajan, Ali Khosravi-Kamrani, and S. Bangalore, for their assistance and support to make this endeavor possible.

Finally, we would like to express our sincere appreciation to our reviewers who took time off from their schedule to assist us in preparation of this volume. We hope that this fine collection of the refereed articles will be of value to researchers, practitioners and those who constantly pursue ways to improve mankind.

Hamid R. Parsaei
Louisville, Kentucky
and
Anil Mital
Cincinnati, Ohio
USA

Part One

Economic Justification Methods

1 Economic and Financial Justification Methods for Advanced Automated Manufacturing: An Overview

Ramesh G. Soni, Hamid R. Parsaei and Donald H. Liles

1.1 Introduction

The economic justification process has long been identified as the biggest hurdle to the adoption of advanced automated manufacturing technology (Kaplan, 1986). In recent years, the literature has been inundated with a large number of methodologies and evaluation techniques that look promising for the economic justification process for advanced automation technology (e.g. Bennett and Hendricks, 1987; Canada, 1986; Curtin, 1984a,b; Evans and Moskowitz, 1989; Meredith and Suresh, 1986; Michael and Mille, 1985; Moerman, 1988; Parsaei *et al.*, 1988, 1989; Parsaei and Wilhelm, 1989; Wilhelm and Parsaei, 1989). Some of the techniques are, basically, sets of policies that evaluate qualitative issues related to advanced manufacturing technology. These techniques lack quantitative thoroughness but they may prove extremely effective in evaluating an investment in advanced automation. On the other hand, several other techniques have been proposed that are complex and exhaustive in nature, and require hard-core quantitative data that may be difficult to retrieve or formulate. Thus, a decision maker, who is evaluating a possible investment in an advanced automation project, may have difficulty in choosing an appropriate evaluation technique from the wide range of methodologies available for his or her situation.

The basic purpose of this paper is to develop a classification scheme to categorize the existing techniques and methodologies for evaluating investments in advanced manufacturing. By doing so, we will provide a framework for the practitioner who, than, can look at the various categories of evaluation techniques and choose one or more techniques that suit his or her situation best.

1.2 Background material

Economists agree on the fact that a growth in productivity level is essential for a society to prosper and raise its standard of living. Except for the last couple of years, the US has been experiencing a decline in the productivity growth rate for almost the last two decades. Additionally, the US is facing strong competition from Japan, Germany, and many newly industrialized nations. Now, US industries are reacting by seeking to adopt advanced automated technology (Bennett and Hendricks, 1987). However, the single biggest challenge facing US managers is the economic evaluation and justification of investment in advanced automated manufacturing technology (AAMT).

Many authors have outlined the reasons for the existence of 'the justification challenge' (Bennett and Hendricks, 1987; Canada, 1986; Kaplan, 1986; Soni *et al.*, 1988). Traditionally, investments in manufacturing equipments have been justified by doing the analyses of investment versus the resulting cost savings; however, benefits of advanced manufacturing technology lie in the strategic areas. These strategic benefits, such as, shorter lead times, consistent quality, timely delivery schedule and improved capability to react to changing demand are difficult to quantify (Curtin, 1984a; Kaplan, 1986). Also, US managers desire a short payback period, and use a high hurdle rate when analyzing an investment (Curtin, 1984a; Hayes and Garvin, 1982; Kaplan, 1986). The focus on short-term payoffs and the lack of long-term vision can hamper an investment in an advanced automation project (Kaplan, 1986). Interestingly, this short-sightedness of US managers is attributed to the management techniques that evolved in the late 1950s and have resulted in the creation of profit centres within an organization, where the promotions and rewards of a manager are based on a short-term bottom line. For a thorough discussion, see Buffa (1985), Hayes and Garvin (1982), Hayes and Wheelwright (1984).

Being plagued by the competition in the global market and being aware of the lapses in the recent management approaches, US managers are willing to take a wide array of corrective actions. These actions range in scope from operations management to human resource management, from financial management to technology management, from accounting to marketing, etc. However, the most important issue that is occupying the minds of US managers is the adoption of advanced automated manufacturing technology to gain a competitive edge in the global market. In the rest of the paper we are going to discuss the various techniques/methodologies that may be used for justifying investments in advanced automation, with the main focus on the proposed classification scheme that categorizes these justification approaches into various groups.

1.3 Classification scheme

Many loose and systematic attempts have been made in the past to categorize various economic justification approaches for advanced automation in several classes. The classification approach developed by Meredith and Suresh in 1986 has drawn particular attention of researchers and practitioners, and is often cited in the literature. They proposed three categories for economic justification approaches:

1. Economic approaches that are suitable for stand-alone automation systems;
2. Analytical approaches that are suitable for linked systems where benefits of synergies are also incorporated in the analyses; and
3. Strategic approaches that are appropriate for integrated systems.

Since the time when this classification was proposed, several new techniques for economic and financial justification for advanced automation have been recommended. Therefore, we present a new classification scheme which may possibly be more extensive in scope and may appropriately classify various techniques in different categories. In addition, the classification clarifies the data requirements for application as well as the underlying objectives of the methodologies belonging to each category.

We propose to categorize the existing justification methods into four major classes. These classes are:

1. Single-objective deterministic methods;
2. Multi-objective deterministic methods;
3. Probabilistic methods; and
4. Fuzzy set methods.

Figure 1.1 illustrates these four major classes of economic justification methods.

The first category, entitled single-objective deterministic methods, includes the **Modified Minimum Annual Revenue Requirement (MMARR)**, the **Net Present Value (NPV)** method, the **Internal Rate of Return (IRR)** method, the **cost/benefit ratio**, **payback period**, and the **integer programming** method.

The second category, multi-objective deterministic methods, includes

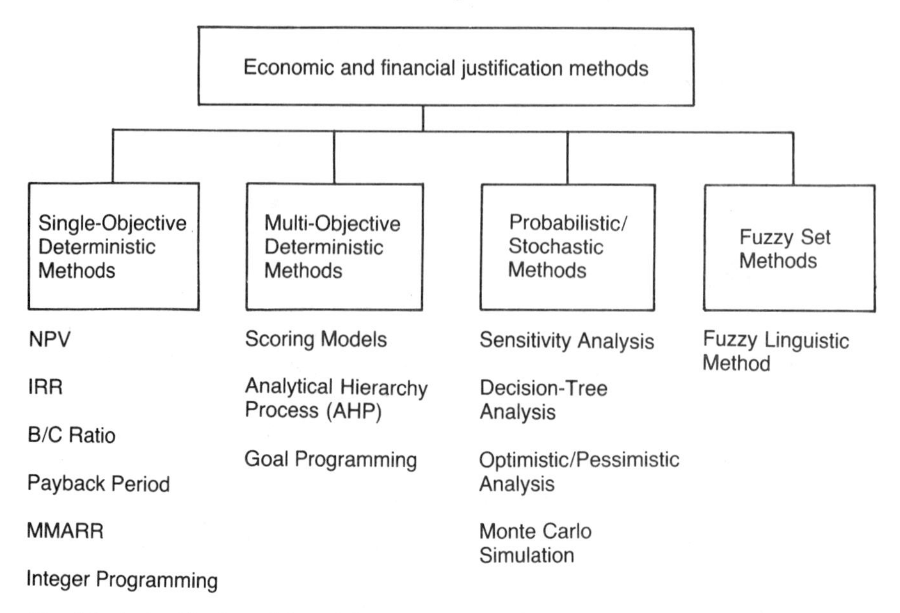

Fig. 1.1 *A classification of the existing economic justification methods.*

the **scoring methods**, the **analytical hierarchy process** and the **goal programming** method.

Under the third class, namely probabilistic/stochastic approaches, the methods included are: the **sensitivity analysis**, the **decision tree analysis**, the **optimistic-pessimistic analysis** and the **Monte Carlo simulation** method.

Finally, the fuzzy set methods are classified by themselves as a separate category. Fuzzy set models attempt to translate verbal expressions into numerical ones, thereby dealing quantitatively with imprecision in the expression of the importance of each strategic goal.

A brief discussion on each available method is given in the ensuing sections of this paper and, where suitable, illustrative examples are presented. The discussions on various methodologies and the illustrative applications are intended to clarify the suitability of the methodologies for economic justification situations. Salient features of the methodologies, data requirements, etc., are also discussed.

1.3.1 Single-objective deterministic methods

The justification methodologies classified under this category attempt to evaluate a single economic objective associated with the justification

TABLE 1.1 *The net present values for the two proposed projects (MARR = 25% and MARR = 15%)*

Project 1 (Low-cost replacement)		Project 2 (AAMT investment)	
EOY	Cashflows	EOY	Cashflows
0	−$500 000	0	−$2 000 000
1	$300 000	1	$300 000
2	$250 000	2	$400 000
3	$200 000	3	$500 000
4	$100 000	4	$600 000
5	$50 000	5	$700 000
		6	$800 000
		7	$900 000
		8	$1 000 000
NPV (25%)	$59 740	NPV (25%)	−$206 630
NPV (15%)	$163 440	NPV (15%)	$594 270

of investment in AAMT. These techniques are easy to implement and provide deterministic outcome regarding the acceptability of an AAMT project; and because these techniques are based on numbers, an analyst may have more faith on the results. The major drawbacks of this category of justification are: potential for misapplication resulting from inappropriate opportunity cost or hurdle rate; focus on short-term returns; failure to incorporate the benefits that may accrue from strategic benefits; and improper assumption of constant market share in the absence of an investment in AAMT. However, the economic justification approach developed by Soni and Stevens (1988, 1989) (MMARR method), attempts to overcome these shortcomings.

Net present value (NPV)

The net present value is the discounted cash flow to time zero. The NPV can be interpreted as the expected value of an unrealized gain from the investment, over and above a certain return – the discounting rate. Therefore, a non-negative value of NPV indicates the desirability of a project. However, estimating the future cash flows and choosing an appropriate discount rate are of utmost importance. Most books on engineering economy (Stevens, 1984) describe the procedure to determine the net present value. As pointed out by Hayes and Garvin (1982), US managers use an unrealistically high discounting or hurdle rate which may discourage an investment in an AAMT project; investments in AAMT projects tend to pay off over a longer period of time. This point is illustrated using a simple example in Table 1.1. In this example, Project

1 is low-cost equipment replacement and Project 2 is a capital-intensive investment in advanced automation. It is also assumed in this example that the advanced automation (Project 2) has a longer useful life of eight years and the benefits are not realized immediately. By using a discount rate of 25%, the NPV for Project 1 is positive; whereas the NPV for Project 2 is negative, implying the undesirability of the project. However, by using a lower discount rate of 15%, the investment in advanced automation (Project 2) appears to be more desirable. This example supports the contention that high hurdle rate may 'disproportionately distort' the future payoffs.

Internal rate of return (IRR)

The internal rate of return is that discount rate which makes the NPV of the cashflows for a given project zero. If the IRR value for a given project is higher than a selected hurdle rate, the project is acceptable. This method of evaluating an investment is technically the same as the NPV method, and therefore has all the benefits and shortcomings of the NPV approach.

Benefit-cost ratio (B/C ratio)

The B/C ratio is yet another Discounted Cash Flow (DCF) approach which is popular in the public sector. This approach does not significantly differ from other DCF approaches. The B/C ratio depends on the definition of the benefit and the cost, and therefore, a project may have more than one value of the B/C ratio (Stevens, 1984). An acceptable project has a B/C ratio of one or higher.

Payback period

The payback period is the minimum length of time required to recover the initial investment without considering the time value of money. Thus, this is also a non-DCF method. This is one of the popular methods employed in industries; the reason for its popularity can be attributed to its simplicity. However, this method, like others discussed so far, also fails to incorporate the benefits resulting from strategic issues.

Modified minimum annual revenue requirement (MMARR)

This approach to justify investments in advanced manufacturing has been developed by Soni *et al.* (1988, 1989). Although it is a DCF approach, this method can possibly overcome several shortcomings of the earlier approaches. This approach forces the analyst to consider the possibility of declining market share, and therefore declining gross incomes, in the event when investment in advanced manufacturing is not undertaken. Traditionally, most economic justification approaches assume a 'do nothing' scenario of constant market share.

This methodology, basically, compares a series of minimum annual incremental revenue requirements to a corresponding series of annual incremental gross incomes. The minimum annual incremental revenue requirement, as defined by Stevens (1984), is the minimum revenue (gross income) required to provide for all incremental costs, including capital recovery on the incremental investment, return on the incremental investment, incremental costs of goods sold, and the incremental taxes. Soni *et al.* (1988) define incremental gross income as a function of market share. They propose

> If a company invests in advance manufacturing, it can achieve cost advantage over its competitors, and therefore it can reduce its selling price. In a price-sensitive market, this would translate into a higher market share and thus, into a higher gross income. On the other hand, if the company does not invest in advanced manufacturing, it may lose its market share, and eventually may be forced out of business. The incremental annual gross income is defined as the difference between gross incomes for these two situations.

A project is considered desirable if the discounted value of the difference between the incremental gross incomes and the incremental revenue requirements is zero or more. This methodology relies on a lot of numerical data, and thus it is not very simple to implement. However, to support the complex decision of investment in AAMT projects one must be willing to carry out a detailed analysis. For a more detailed discussion on this methodology, readers can see Soni *et al.* (1980, 1990) and Soni and Stevens (1989).

Integer programming

The integer programming method attempts to choose a bundle of interrelated projects from a given set of projects; the objective is to maximize the NPV value under the given constraints of interdependence, mutually exclusiveness, multi-period budget, labour and material restrictions, etc. The integer programming formulation can simplify the computational complexity (since the computer does the 'dirty' work). However, Reeve and Sullivan (1988) point out two problems with the integer programming approach:

1. It ignores any potential cross-correlation of cash flows between the projects.

2. The discount rate must be estimated by the analyst who must, in turn, make an assumption of perfect and complete markets.

Also, the analyst must provide all the information regarding future cash flows, interdependency, etc. Therefore, the result of an analysis will be only as good as the input data provided.

1.3.2 Multi-objective deterministic methods

Just like many real-world problems, the decision to invest in advanced automated manufacturing technology involves multiple, and often, conflicting objectives. The methodologies categorized under this category attempt to help the decision maker in evaluating investments in AAMT projects. The three popularly employed techniques under this category are scoring model, analytical hierarchy process (AHP) and goal programming. These techniques are discussed next.

Scoring model

The scoring model features the ability to accommodate the considerations of intangible, or economically non-quantifiable elements in an analytical fashion. It is also capable of considering multiple criteria; it is simple to use; it can be made consistent with management policies; it can incorporate strategic considerations in the analysis; and flexible.

Scoring model differs from other economically oriented models in that the input data are usually subjective 'guesses' rather than more costly and sophisticated forms of the same guesses required by other models (Parsaei and Wilhelm, 1989).

Among proposed scoring models, the linear additive model is widely used by researchers for scoring competing alternatives. Parsaei and Wilhelm propose a justification methodology which is based on a linear additive model developed by Klee (1971). This methodology aids decision makers in evaluating the desirability of a firm's long-term and short-term advanced automated manufacturing technologies. The methodology is implemented in two phases. Phase 1 examines the desirability of various strategic (long-term) automation proposals available, while phase 2 evaluates each tactical (short-term) alternative for implementing the strategic proposal selected as most desirable in phase 1. Figure 1.2 illustrates the two phases of this decision methodology.

The two-phased approach was chosen to emphasize the bifurcated nature of the actual decision process. To implement the methodology, two sets of ordinal scale weights must be developed by decision makers (Tables 1.2 and 1.3); they are then employed in the linear additive

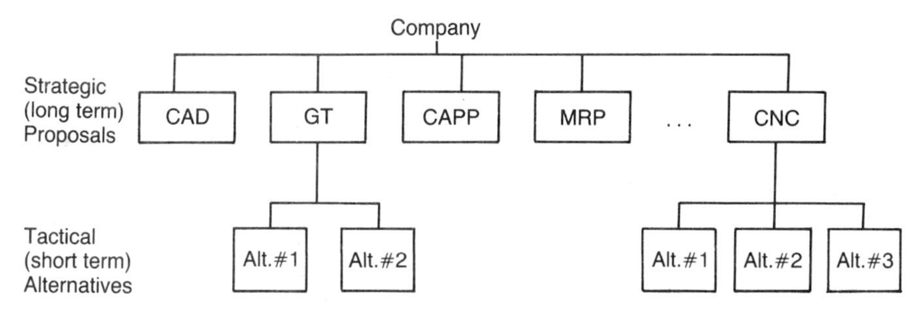

Fig. 1.2 *The graphical representation of strategic proposals and tactical alternatives.*

TABLE 1.2 *Ordinal scale weights to rank the importance of strategic or tactical decisions*

Very important	1.00
Important	0.75
Necessary	0.50
Unimportant	0.25

TABLE 1.3 *Ordinal scale weights utilized to measure the performance of each strategic and tactical decisions with respect to each attribute*

Superior	1.00
Good	0.80
Above average	0.60
Average	0.40
Below average	0.20
Poor	0.00

models in each of the two phases of the methodology (Parsaei *et al.*, 1988).

The linear additive model utilized in the two phases of this methodology can be presented by the following relationship:

$$\text{Max } D_j = \sum W_i X_{ij} \qquad i = 1, 2, \ldots, m$$

where D_j is the score earned by the jth decision alternative, W_i is the weight assigned to the ith attribute which reflect the relative importance of that attribute in the decision process, and X_{ij} is the expected performance of the jth decision alternative with respect to the ith attribute.

An illustrative example of this methodology is presented by Tables 1.4 and 1.5 (Parsaei and Wilhelm, 1989).

TABLE 1.4 *The selection process for the most desirable strategic proposal (phase 1)*

i	A_i	Attribute	Strategic Proposal #1		Strategic Proposal #2		Strategic Proposal #3		Strategic Proposal #4	
			X_{i1}	A_iX_{i1}	X_{i2}	A_iX_{i2}	X_{i3}	A_iX_{i3}	X_{i4}	A_iX_{i4}
1	0.176	Flexibility in production mix	0.80	0.140	0.80	0.140	0.60	0.105	1.00	0.176
2	0.176	Flexibility in scheduling	0.60	0.105	0.80	0.140	0.80	0.140	0.80	0.140
3	0.136	Cost reduction and savings	0.60	0.141	0.60	0.141	0.60	0.141	0.80	0.188
4	0.235	Product quality improvement	0.60	0.141	0.80	0.188	0.80	0.188	1.00	0.235
5	0.176	Reduction in WIP	0.80	0.140	0.80	0.140	0.20	0.035	0.20	0.035
		K_j		0.667		0.749		0.609		0.774
		Normalized K_j		0.861		0.967		0.786		1.000*

* The most desirable strategic proposal $\left(\text{Max. } K_j = \sum_{i=1}^{m} A_iX_{ij}\right)$.

TABLE 1.5 *The selection process for the best available alternative (phase 2)*

i	B_i	Attribute	Tactical Alternative #1		Tactical Alternative #2		Tactical Alternative #3	
			Y_{i1}	B_iY_{i1}	Y_{i2}	B_iY_{i2}	B_{i3}	B_iY_{i3}
1	0.266	Capital recovery cost	0.60	0.159	0.60	0.159	1.00	0.266
2	0.066	Floor space required	1.00	0.066	0.80	0.052	0.60	0.039
3	0.200	Expandability of system	0.80	0.160	0.80	0.160	1.00	0.200
4	0.266	Compatibility	0.60	0.159	0.60	0.159	0.80	0.212
5	0.200	Ease of operation	0.80	0.160	0.60	0.120	0.60	0.120
		L_j		0.704		0.650		0.837
		Normalized L_j		0.841		0.776		1.000*

* The best available alternative $\left(\text{Max } L_j = \sum_{i=1}^{m} B_i Y_{ij}\right)$

Analytical hierarchy process (AHP)

This approach has been developed by Saaty (1980). AHP has been applied to solve problems in many fields, and recently it has been applied to the field of justification of investments in advanced automated manufacturing technology. AHP attempts to model unstructured problems by decomposing a problem into a hierarchy of elements – objectives, attributes, sub-attributes, etc. A pair-wise comparison of elements in each hierarchy is made to determine the weights for each element; and, ultimately, relative score for each alternative is determined. For a detailed discussion of AHP, see Saaty (1980); for the application of AHP to justification of investments in advanced manufacturing, see Canada and Sullivan (1989) and Wabalickis (1988). Among the advantages of using AHP are: versatility in application; ease of implementation; and it does not require the decision maker to be completely consistent in making pair-wise comparison, but, at the same time it quantifies the judgmental consistency. However, the disadvantage of AHP is that the approach often forces the analyst to make pair-wise comparisons based on very vague or general criteria.

Goal programming

This is one of the most robust and popularly used multi-objective decision-making techniques. When making an investment decision in advanced manufacturing technology, the analyst may be faced with numerous conflicting goals, such as low investment, high flexibility, short payback period, capability to produce high-quality products, and so on. The analyst must develop ordinal priorities for these goals. The most important goal must be pursued first. After attaining the first objective, the next most important goal must be sought, and so on. The decision-maker should stop only after the most promising solution has been obtained. As suggested by Lee (1988), 'the goal programming solution process yields valuable information about goal conflicts, soundness of the priority structure for goals, and trade-offs among the conflicting objectives.' A detailed discussion on goal programming can be obtained in any standard textbook on management science (for example Turban and Meredith, 1988).

1.3.3 Probabilistic/stochastic methods

Most economic decisions are based upon the estimates of the future. These estimates are subject to inaccuracy because of uncertainty or insufficient information about the future. The uncertainty can arise due to many factors such as business risk, technology changes, global economic health, changes in the political environment and natural or manmade calamities. Various methods developed for incorporating risks and uncertainties include: decision-tree analysis, sensitivity analysis, optimistic-pessimistic analysis, and Monte Carlo simulation.

Methodologies belonging to this category describe the outcome of investments in AAMT statistically, and therefore, they incorporate a wide range of possible outcomes in an investment analysis. The popular method – sensitivity analysis, decision-tree analysis, and optimistic/ pessimistics are discussed next.

Sensitivity analysis

The analyst evaluating an investment in an AAMT project needs to take into account the uncertainty in the business environment as well as the associated economic risk. One of the simpler but indirect ways to handle uncertainty is to perform sensitivity analysis. Sensitivity analysis may indicate the factors which must be estimated with excessive care because the economic criterion (such as NPV or IRR) may be extremely sensitive to them.

Application of sensitivity analysis is illustrated with a simple example. Suppose, a company is investing $2 million in an AAMT project, with an estimated cost of capital of 15%. The estimated useful service life is five years, and the expected annual payoff is $600 000 per year during the service life. The estimated salvage value at the end of service life is $200 000. The net present value of the project is

$$\begin{aligned} \text{NPV} &= -2\,000\,000 + 600\,000(\text{P/A } 15{,}5) + 200\,000(\text{P/F } 15{,}5) \\ &= \$110\,728 \end{aligned}$$

However, the estimates of initial investment, useful service life, estimated salvage value, and cost of capital are subject to error. A sensitivity analysis is performed by changing these estimates within a 20% range. The results of the sensitivity analysis are presented in Fig. 1.3. By inspecting the graph, it is apparent that the NPV is extremely sensitive to the initial investment and the annual payoffs, and it is least sensitive to the salvage value estimation.

Decision-tree analysis

A decision tree can be used to depict acts, states-of-nature and respective payoffs graphically. The decision tree can be used to analyse the decision under risk since the probabilities of the states-of-nature must be stated clearly. A decision tree is extremely useful when a sequence of interrelated decisions must be made. Typically, a decision tree consists of

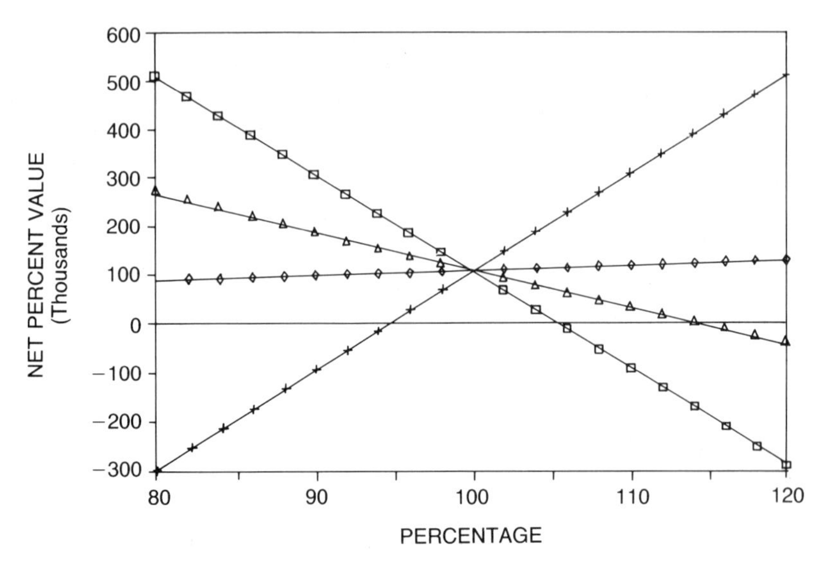

Fig. 1.3 *A graphical summary of sensitivity analysis.*

square nodes (representing a point at which a decision must be made) and, circular nodes (representing situations where outcome is uncertain).

The reader is advised that the decision-tree approach is more appropriate for situations where the analyst is making several similar decisions over a period of time; the expected minatory value is a long-run average of payoffs. For relatively infrequent, or perhaps one-shot decisions, such as investment in advanced manufacturing, the analyst must understand the implications of the expected minatory value. The expected monetary value, for one-time decisions, can be treated as a device to combine the risks and payoffs into one consolidated number.

Optimistic-pessimistic analysis

This approach is suitable for situations where the analyst does not have any knowledge of the chances of occurrence of the states-of-nature. This approach is best known as decisions under uncertainty. This is probably the most used approach in industries where the top managers are interested in knowing the best and the worst outcome of a decision. We feel that this is a quick and dirty way of analysing a problem. Other approaches such as maximin pay-off and minimax regret are also suitable for decisions under uncertainty. A discussion of these approaches can be found in any standard decision sciences book (for example, Lee 1988).

1.3.4 Fuzzy set methods

The use of fuzzy linguistic variables to evaluate advanced automation technology was proposed by Wilhelm and Parsaei (1988, 1989). This method involves the use of linguistic variables from the theory of fuzzy sets to support the phased implementation of a computer-integrated manufacturing (CIM) strategy. Fuzzy linguistic models permit the translation of verbal expressions into numerical ones, thereby dealing quantitatively with imprecision in the expression of the importance of each strategic goal (i.e. enhanced competitiveness, high product quality, low product cost, etc.) and the enabling technology involved in the implementing CIM systems (i.e. CAD, CAM, group technology, computer-aided process planning, etc.). Wilhelm and Parsaei (1989) in their proposed methodology for the evaluation of CIM systems defined two fuzzy linguistic variables: importance and capability. The use of these two linguistic variables allows the analyst to specify the importance associated with each of a set of strategic goals which impact the CIM strategy, and the tactical capability of each available CIM technology to meet the strategic CIM goals of the organization. Their study develops a heuristic algorithm based on the use of fuzzy linguistic variables to characterize the capability of available enabling technologies and to select the best technology from those available at each phase in the implementation of the CIM plan (Wilhelm and Parsaei, 1988, 1989).

1.4 Conclusion

In this paper, a classification scheme to categorize economic justification approaches for advanced automation has been presented. The four categories developed in this paper are:

1. Single-objective deterministic methods;
2. Multi-objective deterministic methods;
3. Probabilistic/stochastic methods; and
4. Fuzzy Set methods.

Several non-numeric/subjective procedures are also utilized by practitioners. However, in this classification scheme we have omitted non-numeric/subjective evaluation procedures purposely, because subjective procedures are most often used in conjunction with some form of technical approach. Hence, in this paper, our classification scheme is more or less restricted to technically oriented methods.

This classification can provide a framework for the practitioner who can select one or more evaluation method for justifying investment in advanced automated manufacturing technology. The classification scheme is also intended for clarifying the underlying objective(s) as well as the data requirements for the methodologies belonging to various categories.

1.5 References

Bennett, R.E. and Hendricks, J.A. (1987) Justifying the acquisition of automated equipment. *Management Accounting*, July, 39–46.

Buffa, E.S. (1985) Meeting the competitive challenge with manufacturing strategies. *National Productivity Review*, 155–69.

Canada, J.R. (1986) Annotated bibliography on justification of computer-integrated manufacturing systems. *The Engineering Economists*, **31** (2), 137–50.

Canada, J.R. and Sullivan, W.G. (1989) *Economic and Multi-attribute Evaluation of Advanced Manufacturing systems*, Prentice-Hall, Englewood Cliffs.

Curtin, F.T. (1984a) New costing methods needed for manufacturing technology. *AMA Forum*, **73** (4), 29–34.

Curtin, F.T. (1984b) Planning and justifying factory automation systems. *Production Engineering*, **31** (5), 46–51.

Goldhar, J.D. and Jelinek, M. (1985) Computer integrated flexible manufacturing: organizational, economical, and strategic implications. *Interfaces*, **15** (3), 94–105.

Hayes, R. and Garvin, D. (1982) Managing as if tomorrow mattered. *Harvard Business Review*, **60** (3), 71–9.

Hayes, R. and Wheelwright, S. (1984) *Restoring Our Competitive Edge*, Wiley, New York.

Kaplan, R.S. (1986) Must CIM be justified by faith alone? *Harvard Business Review*, 87–93.

Klee, A.J. (1971) The role of decision models in the evaluation of competing environmental health alternatives. *Management Science*, **18** (2), 52–67.

Lee, S.M. (1988) *Introduction to Management Sciences*, 2nd edn, Dryden Press, New York.

Meredith, J.R. and Suresh, N.C. (1986) Justification techniques for advanced manufacturing technologies. *International Journal of Production Research*, **24** (5), 1043–57.

Michael, G.J. and Millen, R.A. (1985) Economic justification of modern computer-based factory automation equipment: a Status report. *Annals of Operations Research*, **3** (3), 25–34.

Moerman, P.A. (1988) Economic evaluation of investments in new production technologies. *Engineering Costs and Production Economics*, **13**, 241–62.

Noble, J.L. (1989) Techniques for cost justifying CIM. *The journal of Business Strategy*, January/February, 44–9.

Parsaei, H.R., Korwowski, W. and Wilhelm M.R. (1988) A methodology for economic justification of flexible manufacturing systems. *Computers and Industrial Engineering*, **15**, 1–4, 117–22.

Parsaei, H.R. and Wilhelm, M.R. (1989) A justification methodology for automated manufacturing technologies. *Journal of Computers and Industrial Engineering*, **16** (3), 341–6.

Reeve, J.M. and Sullivan, W.G. (1988) Strategic evaluation of interrelated investment projects in manufacturing companies. *Cost Management Systems Research Report Submitted to CAM-I, International*, February.

Saaty, T.L. (1980) *The Analytical Hierarchy Process*, McGraw-Hill, New York.

Soni, R. and Stevens, G.T., Jr (1989) Justification of flexible manufacturing systems using price, quality, and flexibility gains. *ASEE Annual Conference Proceedings*, June 25–9, Lincoln, Nebraska.

Soni, R., Liles, D.H. and Stevens, G.T., Jr (1988) The economic justification of investments in advanced manufacturing. *Proceedings of the Institute of Industrial Engineering Conference*, October 30–November 2, St Louis.

Soni, R., Liles, D.H. and Stevens, G.T., Jr (1990) Computer aided decision model for economic justification of investments in advanced manufacturing. *Justification Methods for Computer Integrated Manufacturing Systems* (ed.) H.R. Parsaei, T.L. Ward and W. Karwowski, Elsevier Science, Amsterdam. 160–75.

Stevens, G.T., Jr (1984) *Engineering Economy*, Reston, Virginia, USA.

Suresh, N. and Meredith, J.R. (1985) Justifying multimachine systems: an integrated approach. *Journal of Manufacturing Systems*, **4** (3), 212–29.

Turban, E. and Meredith, J.R. (1988) *Fundamentals of Management Science*, 4th edn, Business Publications. Plano, Texas, USA.

Wabalickis, R.D. (1988) Justification of FMS with the analytical justification process. *Journal of Manufacturing Systems*, **71**, 175–82.

Wilhelm, M.R. and Parsaei, H.R. (1988) A model for selecting and evaluating flexible automation systems. *The Proceeding of the 1988 IIE Integrated Systems Conference*, October 30–November 2, St Louis.

Wilhelm, M.R. and Parsaei, H.R. (1989) A fuzzy linguistic approach to implementing a strategy for computer integrated manufacturing. *International Journal for Fuzzy Sets and Systems* (In Press).

2 Expert Systems Applied to the Economic Justification of Advanced Manufacturing Systems

William M. Henghold and Steve R. LeClair

2.1 *Introduction*

This chapter relates aspects of applied artificial intelligence (AI), in the form of expert or knowledge based systems, to economic justification of advanced manufacturing systems. In keeping with the theme of the book, we focus on the general task of venture justification. However, the approach taken extends beyond the scope of venture justification applications and addresses some fundamentals and capabilities pertinent to all expert system applications. This breadth of discussion is necessary in demonstrating clearly what expert systems technology can do versus what it is reported to have done.

It is assumed that readers are not AI experts but rather are interested in the potential of applying a new technology to their environment. For this reason a background section is presented first. This section provides a rather cursory look at some fundamentals which form a baseline for the remainder of the chapter. For readers desiring more in-depth treatment a few references are provided at the end of the chapter (Hayes-Roth *et al.*, 1983 and Liebowitz 1989).

With some understanding of the technology, the next section examines the problem of assessing when expert systems are well suited to the problem of interest. Here we present some general guidelines on the selection of expert systems technology, as opposed to more conventional techniques such as algorithms and simulation, in addressing the needs and considerations of economic justification activities.

If expert systems technology is determined to be well suited to the activity, then the issue of maintenance and system improvement must be addressed. Therein, we present a top-level look at an expert system architecture which couples symbolic and numeric problem-solving techniques in the design of a **multi-expert** system that performs maintenance of the knowledge base and enables experience-based learning as a means to improve the application heuristics continually.

Finally, we review a specific application of the above self-improving system which uses three independent, potentially conflicting experts (lines of reasoning). Each expert line of reasoning is specifically applied to the venture justification of investments in advanced manufacturing.

2.2 Background

AI is, and will continue to be, many things to many people. In general, it can be thought of as the study of mental faculties through the use of computational models. In the broadest sense AI, as a field of computer science, is applied to varied applications such as computer vision, natural language understanding and robotics, process control. For the purposes of this chapter, AI will be limited to knowledge-based computer systems used as decision aids. A knowledge-based system or, more popularly, an expert system (ES), attempts to duplicate learned or experiental expertise. These computer based systems seek to represent and apply experiental knowledge relative to some specified area of expertise to aid decision makers in solving problems in that same specified area.

The expert systems market is the largest and most pervasive subset of all AI applications. Systems have received considerable media attention due to their ability to engage in judgmental reasoning similar to that of some domain expert and to exhibit comparable or superior levels of performance. The basic structure of an ES has been described differently by various authors. In its simplest form, an ES utilizes a computer-interpretable representation of domain-specific knowledge and a control structure referred to as an inference engine, to extract problem-specific information via a user (decision maker), to arrive at problem specific conclusions. A simple model in this form is shown in Fig. 2.1.

The predominant method of building expert systems is first to choose a tool or shell to be employed in the construction of these systems. Such tools provide a variety of means of representing knowledge, e.g. objects, frames, or IF-THEN type constructs called production rules. In addition, each tool offers a variety of reasoning techniques, i.e. reasoning about the represented knowledge and inferring conclusions (e.g. goal-driven or backward chaining). The representation and reasoning tools constrain the way the designer (knowledge engineer) attempts to build a system.

The Fig. 2.1 also highlights the fact that the database, the control structure (rule interpreter) and knowledge base are separate. What is

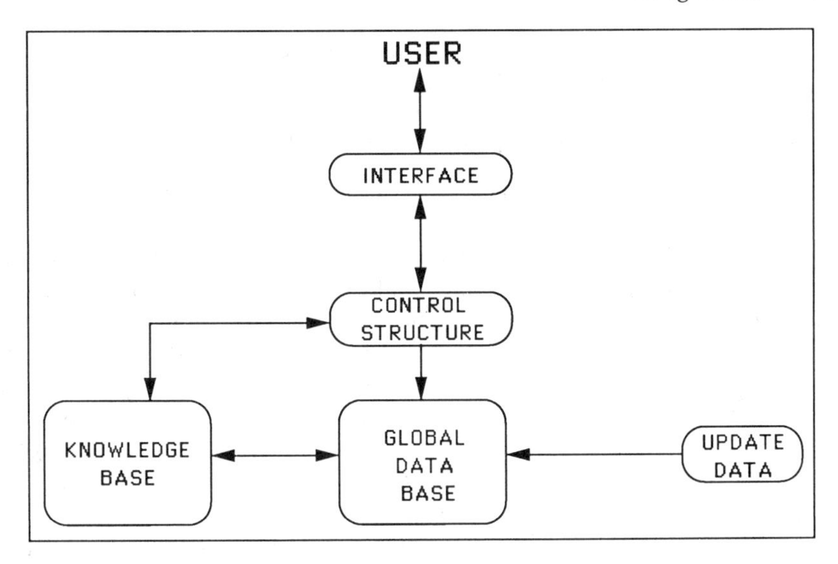

Fig. 2.1 *A model of a knowledge-based system.*

not so obvious is the fact that the knowledge base is static and that it often contains embedded metarules (or rules about rules). These inherent characteristics of an ES have a profound effect on the need for and ease of system maintenance as they apply to sustained performance over time.

If one looks at the history of AI, a picture that is best described as 'spotty' emerges. Early on, there were some rather strange claims of computers doing all the work in the near future. Practitioners held the belief that they could arrive at a few immutable laws of reasoning, add some computers and out would pop superhuman performance. Once they began to realize how difficult the problems associated with 'intelligence' were, the practitioners backed off on their claims, but not without a certain amount of damage to the field.

A second attempt at industrializing AI is currently being pursued by industry, government and academia alike. This attempt still fuels unrealistic expectations with media hype and sales jargon. But there are some valuable learned lessons from the first attempt which may improve the chances for success this time around. If one tries to capture the essence of these learned lessons as they apply to present-day knowledge-based systems, there are three points to remember above all others. These are:

1. Expert systems are very brittle, i.e. they reason about knowledge in a limited domain and at the boundaries or outside of that domain they are often useless.

2. The decision-making power of the ES is dependent upon the scope and depth of knowledge that it possesses.
3. The correctness of the solution is a function of how closely the particular problem matches the system knowledge.

These three points re-emphasize the fact that knowledge is central to intelligence.

Even with these learned lessons firmly in mind, present-day systems do not model expertise as well as one would hope. There is a general lack of flexibility in problem solving (i.e. the external world and therein knowledge about the world changes over time but systems do not). Expert systems cannot yet converse in an intelligent way, explain and justify their conclusions in an acceptable manner, or learn on their own. This is of fundamental importance in venture-justification problems as managers will be extremely interested in a natural language dialogue, why a conclusion was reached, and a self-improving system.

It is well to close this discussion with a note of caution regarding the users of ESs. Consider the following quote:

> . . . is easy to use because it has a spreadsheet like interface. You don't need to learn a computer language or to program. Enter examples, as you would data into a spreadsheet, and let the program create the expert system.
>
> *Product Advertising Brochure*

Such advertising speaks directly to an ongoing problem of *caveat emptor*. If it is taken as gospel that ES development is as easy as spreadsheet building, then management would be remiss if it were to provide internal funding aimed at ES development problems. Therein, if management does not realize that there is a lot more to be gained by interaction with computational tools than just entering examples on a spreadsheet then ES applications may fail for lack of a suitable human-machine environment. Thus, regardless of the significant progress made in ES modelling of expertise in arriving at first-rate decision-making aids, one must continually guard against the threat of oversell.

2.3 The application problem

The potential success of a knowledge-based system depends on appropriate selection of a specific application domain. Failure can almost be assured by approaching the development from a perspective characterized by: 'Here is my problem. Go solve it using AI techniques.' In reality, the following process of elimination must exercised for each potential application:

1. Many problems simply are not amenable to the application of AI.
2. Of those that are, some could be better solved by algorithmic techniques.
3. And, finally, of those for which AI is most appropriate, only some are suited to solution by present knowledge-based system technology.

A much more reasonable approach to application selection is characterized by: 'AI technology appears to offer a chance to solve some problems that are important to me. Why don't we investigate which application(s) can do us the most good and has the best chance of success.'

At the highest level, successful application potential is maximized by passing three tests. These are that the application is technically feasible, venture justifiable and user acceptable. That is, it can be done, is worth the effort, and will be used as opposed to languishing on some disk file. The latter concept means that the barriers to application of AI are manageable.

The management domain, of which economic justification is a subpart, is characterized by decisions. The types of decisions being made are often concluded from qualitative rather than quantitative measures, where the complexity of the problems are characterized by their nonlinear behaviour, lack of solution formalisms, and uncertain and/or incomplete information. Such problems are too complex for mathematical models or simulation and are characteristic of those problems for which AI techniques such as heuristic search and pattern matching are well suited.

While there is some expectation of the suitability of AI to management-type problems, there are as well some barriers. Dhar (1987) captures these barriers quite succinctly – successful expert systems tend toward individual expertise, normative models and a one-shot problem-solving orientation.

But managerial problem solving involves distributed expertise (e.g. diffusion of responsibility across multiple organizations) and evolutionary problem solving in a historical context. Thus while knowledge-based systems map facts to conclusions given a stable model of some domain, large portions of the management world simply don't have the required stable model. This is a real implementation barrier. AI systems must come to grips with decision-making environments where highly respected experts often arrive at different conclusions and the fact that the process by which the decision-maker decides whose advice to follow is at best highly subjective.

Another way to surface potential problems associated with the use of ESs is to look at decision types and see what ES techniques have to offer. Baldwin and Kasper (1986) have come up with a useful categorization:

1. Operating decisions characterized by routine guidelines;
2. Coordination decisions which are recurring and routine but which require interpersonal skills, large numbers of factors, etc.;
3. Exception decisions which are *ad hoc*; and
4. Strategic decisions which are *ad hoc* and have significant organizational impact.

Clearly the state of ES technology is most amenable to the operating environment. Procedural tasks, while not mainstream AI, are the most likely to be barrier free from a capability sense. Venture justifiability is another matter as it is often a cross between coordination decisions and strategic concerns.

Venture justification in many organizations often highlights an important issue encountered during practically all decision-making situations – how to deal with conflicting advice. Whether in a large corporation, a small job shop or in the more formalized structure of accounting and financial businesses, conflict or disagreement among individuals and/or groups is more the norm than the exception. Instead of a single, omniscient expert, most venture justification problems utilize many knowledgeable people, often with differing perspectives, regarding a solution. For example, in evaluating advanced manufacturing technology, one expert may consider issues focused on near-term quality improvements, another on long-term quality improvements, another on long-term survival of the firm, and still another on cash-flow stability over the next five years. In addition, priorities associated with these views may change because various investments affect the priorities differently. Difficulties are compounded if innovations cannot be shown quantitatively to increase a company's value.

Most, if not all, man-made enterprises or institutions are designed around 'organized' conflict, were individuals constantly strive to achieve various subsets of goals that may conflict with one another. Organized conflict can be employed as an oppressive system of checks and balances or as a means to stimulate or motivate learning in the form of new ideas and solutions.

2.4 A multiple knowledge-source learning system

As indicated previously, application domains do not remain static. Thus new knowledge must be learned over time. Since the birth of AI as a discipline in the mid 1950s, learning has been a central research area. Two reasons can be given for this interest in learning: first the ability to learn is a measure of intelligent behaviour; and, second, learning is essential for building systems which improve their performance. This last reason is often glossed over with 'ease of modification' words. System maintenance in a changing world is a complex problem which is a prime impediment to expert systems reaching their full potential. What is not immediately apparent is that the conflict mentioned in the previous section can be used to facilitate system maintenance and improvement, i.e. learning.

The conventional way to develop a multiple-expert knowledge system is to seek consensus among the experts during knowledge acquisition (i.e. before building the knowledge base). The system then results in a single line of reasoning which represents a blending of the expertise of multiple experts. But by blending expertise, the knowledge sometimes becomes illogical and error-prone because it is difficult to achieve consensus without awkward compromise of each expert's reasoning.

The approach described here is based on the work of one of the authors in the development of multi-expert knowledge systems (LeClair, 1985). A multi-expert (multiple-expert, multiple-line-of-reasoning) knowledge system enables the accommodation of potentially conflicting expertise in the same knowledge base. Each expert is accommodated as a complete, self-contained, single-expert, single-line-of-reasoning system. At the same time the sharing of knowledge between experts is possible. In addition, the system is able to recognize conflict between experts and provides the decision maker with a means of evaluating the conflict such that a preferred line of reasoning can be selected.

The engineering tools used to develop multi-expert knowledge systems are: a rule-based goal-driven inference engine, a rule tester based on sequential analysis, and a rule inducer based on an information-entropy algorithm.

Figure 2.2 represents a system architecture for the multi-expert knowledge system (MKS) paradigm. Use can be thought of as a three-stage process consisting of rule acquisition, test and organization. During acquisition the independent lines of reasoning are exercised using the inference system. These are coded as IF-THEN rules using an inference system called EXPERT-4 developed by Mr Jack Park. In the case of conflicting advice, a mechanism is provided to resolve the conflict. The result is a single pattern leading to a particular solution which has the effect of combining the knowledge from the individual experts.

Learning systems require a critic to handle the important credit-assignment task. To determine acceptance or rejection or new lines of reasoning, a test based on sequential analysis is used in the system. This test enables both learning and unlearning of evolving domain knowledge. The method chosen to assign credit is the sequential probability ratio test. First the user, acting as a teacher, must provide confirmation or disconfirmation of the system-level advice. That is, whether or not through the passage of time a particular recommendation proved to be good or bad

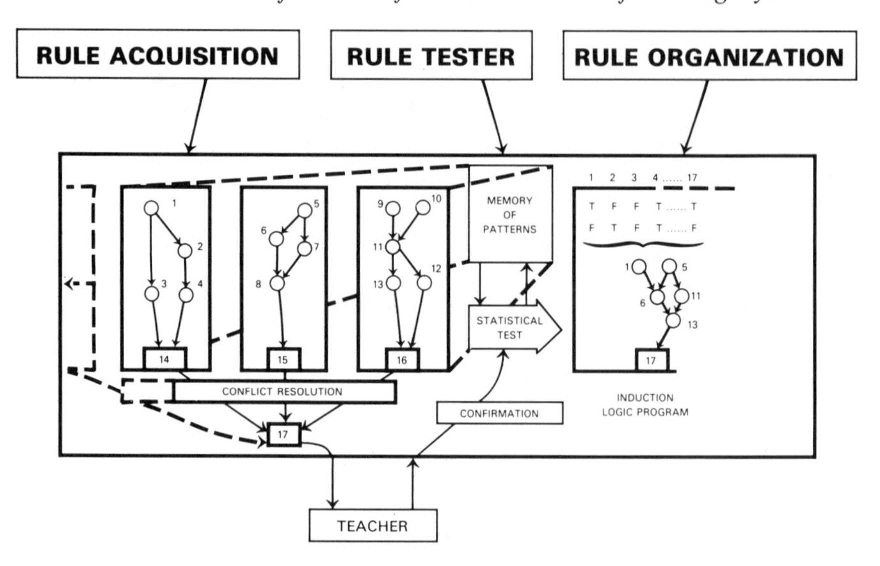

Fig. 2.2 *The concept of a multi-expert system.*

advice. Rule testing is then performed on the basis of the sequential probability ratio test. Here, depending on the number of occurrences of a particular rule as well as on the number of disconfirmations, the rule is either accepted, rejected or placed in a 'continue-testing' mode.

Rule organization, in the form of a decision tree of a learned line of reasoning, is provided by induction. The induction is performed on a training set formed by the total population of accepted rules (i.e. those that pass the sequential probability ratio test.) The tool used is based upon Iterative Dichotomiser (ID3) algorithm (Quinlan, 1985). The rule-refining process induces general principles from the specifics contained in the example set.

The MKS paradigm may be used in many different application domains as determined by the knowledge base. The limiting factor is that the system is designed for classification types of problems. That is, a pattern of attributes (IF portions of rules) which lead to a particular class (conclusion.) This specification is by no means as limiting as one might think. Classification is a significant subset of human problem-solving. Examples include:

1. The collection of instrument readings that indicate that a reactor is safe rather than unsafe.
2. The symptoms that indicate that a disease is of a particular type.
3. The set of indicators that show that the fault in some inoperative robot is in the gripper circuit card.

4. The economic conditions that dictate a buy decision for some stock rather than one to sell or hold.

The combination of acquisition, test and organization provides an integrated process which allows for the generation of new knowledge. Although planned for, the integration (to close the loop) to allow learned knowledge to become a virtual line of reasoning, which competes with provided lines of reasoning, is not implemented at this time.

2.5 The case of advanced manufacturing system justification

As indicated previously, the MKS can be applied in various domains. One of the initial applications was to provide consultant service for the area of venture justification of advanced manufacturing technology. This is a classification problem dealing with a set of problem attributes (criteria) that lead to a recommended state of GO, NOGO or DEFER, for the investment being considered. These states are considered to be mutually exclusive and, as such, the categorizations are representative of problem-solutions in the management domain. In addition, the problems exceed the bounds of well-established routine, and decision aids (i.e. expert systems) must be supportive rather than a replacement alternative.

It is important to remember that the result of a session is a categorization and not a rank ordering. Thus, if two investment opportunity cases are run and both lead to GO recommendations, nothing is inferred about which case is better.

The implemented venture-justification system uses three independent lines of reasoning. As a result of this fact and the domain choice, the expert system is called XVENTURE3 where the X denotes the system's experimental nature. Domain expertise was provided by Professor Thomas Boucher of Rutgers University, Ms Callie Berliner and Mr James Brimson formerly of the Computer Aided Manufacturing-International Organization, and Professor William Sullivan of the University of Tennessee.

The expert system contains on the order of 300 production rules. It runs on an IBM PC or AT type personal computer with a minimum memory of 320K.

Figure 2.3 shows a high-level categorization of some of the factors considered by the three experts (labelled A, B, and C) as they try to put the attributes of the project under consideration into a form leading to one of the three mutually exclusive goal states of GO, NOGO or DEFER. It should be emphasized that the examples which are shown are not exhaustive but rather only illustrative.

As would be expected with the justification of advanced manufacturing

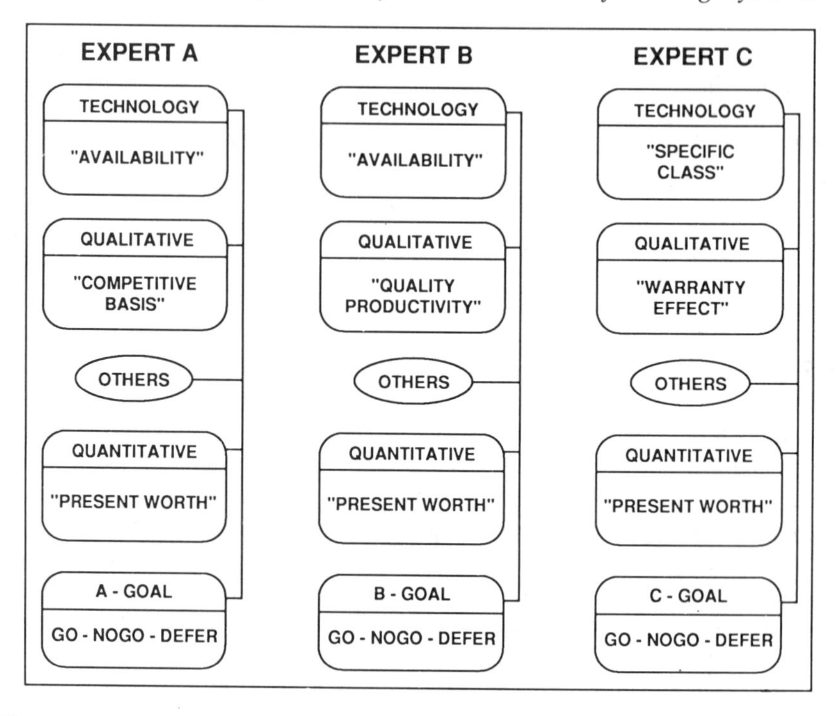

Fig. 2.3 *Factors in an expert line of reasoning.*

technology, there are a combination of potential technical considerations, qualitative considerations and quantitative considerations. Further, the considerations differ on a per expert basis.

One line of reasoning depicted stresses the importance of company objectives. Another expert's perspective looks at strategic acceptability, organization for production, machine technology base, and decision-making structure. The third seeks to combine strategic and tactical considerations and includes such items as technology and business-plan match and cost-accounting system changes.

Two of the experts are interested in the technology availability. There are differences, however, as one asks about whether the technology is freely available while the other is interested in the project's proprietary or non-proprietary nature as well as the developmental status. The third line of reasoning does not inquire directly about availability but rather asks for a specific technology designation. The user is first questioned about broad categories (e.g. touch or non-touch) and, depending on the first designation, chooses from specific examples within those categories (e.g. robotics or computer-aided inspection).

As would be expected, the expert lines of reasoning deal heavily with qualitative issues. These cover a wide spectrum. One expert explicitly makes a differentiation based upon the industry's competitive basis (e.g. price, service or quality). Another expert asks for a description of the project's efforts on flexibility, capacity, quality and productivity. The third expert allows for an explanation of the project's impact on the costs of correcting quality problems after delivery to the customer (warranty costs, for example).

It is important to note that goal states of all three types can be arrived at without resorting to any quantitative question. However, if quantitative factors are required, each expert considers quantitative issues in terms of the present value or present worth. The present-worth algorithm includes a broad spectrum of costs such as the lost sales from not making the investment. This forces the user away from the faulty logic that doing nothing is without risk in a competitive market.

It is worthwhile to look at a simplified version of one of the experts to gain further insight into some of the considerations. Figure 2.4 provides such a view into one of the lines of reasoning. It is a paraphrased depiction of the rule base and can be thought of as a kind of pseudocode. In reference to Fig. 2.4, the boxes represent facts or IF-type assertions and the double boxes represent the goal states or hypotheses. These goals are tested using a backward-chaining inference system.

While the figure appears complicated, it is really quite simple to read. For instance, the following rule can be traced:

IF your technology is ⟨specific choice⟩
AND your improvement areas are ⟨specific choice⟩, ⟨specific choice⟩, and ⟨specific choice⟩
AND technology benefits do NOT MATCH improvement desires
THEN the recommendation is NOGO

where the match is provided as part of the expert's logic via truth table.

Another way to state this type of rule in more specific terms is IF your primary concern is information timeliness and accuracy, THEN don't waste time working a robotics project. Or, more generally speaking, make sure there is technology-benefit to improvement-need congruence for your investment.

Another example rule which can be traced from Fig. 2.4 is:

IF the match conditions are met
AND the technology is NOT for new products
AND the present value of the investment is NOT greater than zero
AND one of your improvement areas is quality
AND the effect on rework is 'large'
THEN the recommendation is GO

The above rule provides specific focus on some of the intangible measures of investment worth. Further, it shows that a present value less than zero

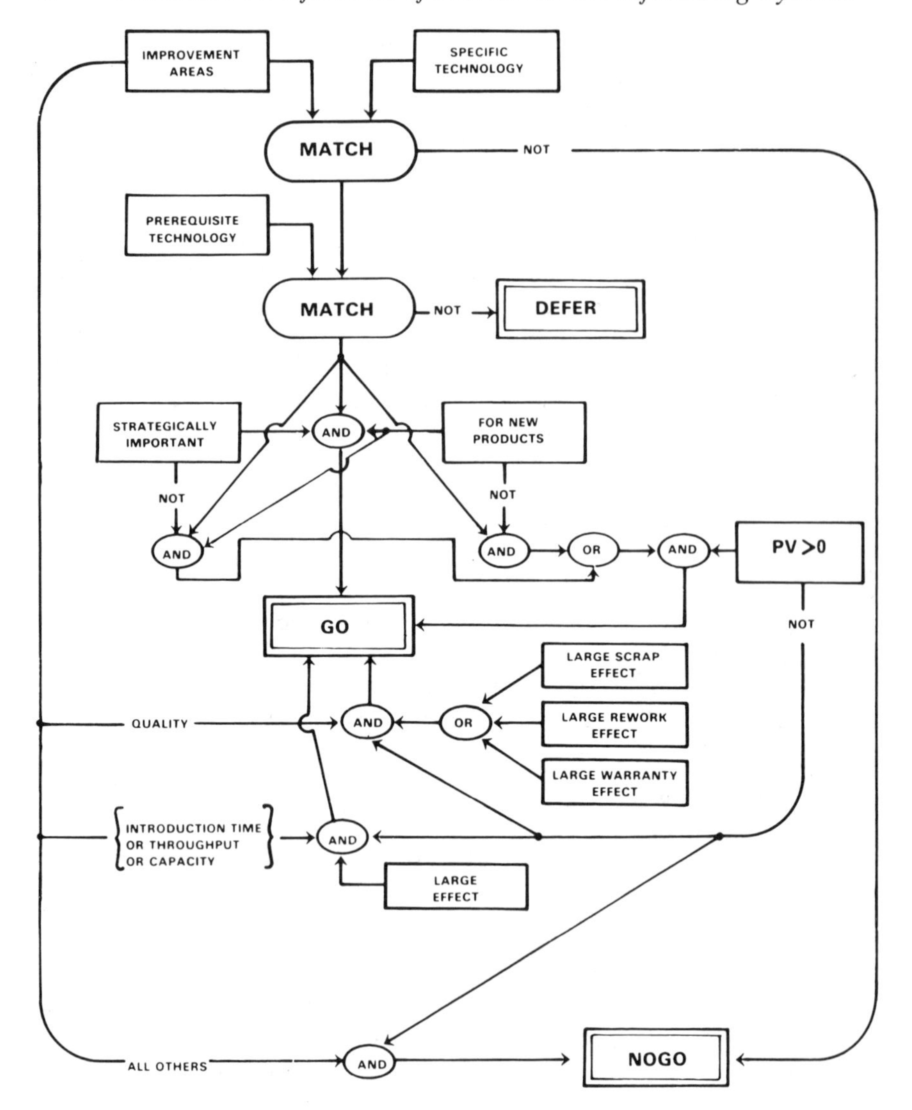

Fig. 2.4 *A pseudo-line of reasoning.*

does not preclude a GO recommendation in areas where benefits which are hard to quantify are thought to be significant.

As indicated previously, the induction of new knowledge requires a training set of examples. To this end an experiment was conducted to gather a large number of test cases of the system performance, versus real project justification scenarios, from various companies' experience bases.

A number of participants were selected based upon their experience and willingness to participate. Users were asked to exercise the system on their own experience-driven examples. Run-time data was then collected and fused into a single training set.

Test project selection is a critical factor. The system must be tested against past projects for which the goodness or badness of the resulting decision is known. These may include implementations that proved beneficial or detrimental as well as decisions not to implement that turned out to be wise or unwise. The user's ability to confirm or disconfirm at run-time is essential to building a large training set in a short period of time. The temporal effects (time between decision and implementation of advanced technology) could become quite pronounced and therefore inhibit the timely learning of new knowledge related to manufacturing venture justification decisions.

At the time of this writing the response from test-case participants has been quite fragmentary. Not enough cases have been returned to perform meaningful induction across a fused training set. We continue to gather data and solicit your participation in the future because the more knowledge (cases and outcomes) that we can provide for the system the more profound is the learned knowledge.

Test-case participants were also provided with a critique form. The form provided a means to gather additional feedback on the system as a whole. Participants could therefore include comments on:

1. Any one of the experts' criteria for evaluating ventures in advanced manufacturing technology versus the user's criteria which may be more or less problem specific;
2. Problems they may have had with the system's *rather rudimentary* explanation facility; and
3. The user interface, etc.

Some comments from the critique sheets show up consistently and therefore are worth mentioning. As indicated the experimental system uses production rules and runs on personal computer class machines. The former are most applicable to shallow reasoning and the latter limit certain utilities. The participants want much better explanation facilities and, in particular, a facility to respond to 'why' type questions. These desires are in spite of the fact that participants were told of the system's experimental nature and that learning was the focus of the experiment. This search for explanation is to be expected for management domain problems. Managers want to 'understand'. Even in more powerful development environments, explanation is basically a rule-unwinding scenario. Further, the management function will always limit systems to consultative decision support as opposed to replacement.

2.6 Conclusion

In concluding this chapter it is both appropriate and important to consider one final thought. Artificial Intelligence, expert systems and more specifically multi-expert knowledge systems involve many new problem-solving approaches and technologies. And, although these new approaches and technologies are of significant potential value, it is still hard to know exactly the perspective from which they should be viewed. This dilemma was characterized by Lady Lovelace over 100 years ago, as follows:

> In considering any new subject, there is frequently a tendency, first, to overrate what we find to be already interesting or remarkable; and, secondly, by a sort of natural reaction, to under value the true state of the case, when we do discover that our notions have surpassed those that were really tenable.

Lady Lovelace was talking about Babbage's Analytical Engine. But she could just as easily have been referencing Artificial Intelligence, Expert Systems, or Multi-expert Knowledge Systems applied to Venture Justification of Advanced Manufacturing Technology.

2.7 References

Baldwin, D. and Kasper, G.M. (1986) Toward representing management-domain knowledge. *Decision Support Systems*, **2** (2).

Dhar, Vasant (1987) On the plausibility and scope of expert systems in management. *Journal of Management Information Systems*, **4** (1), 25–41.

Hayes-Roth, F., Waterman, D.A. and Lenat, D.B. (eds.) (1983) *Building Expert Systems*, Addison-Wesley, Reading, Massachusetts.

LeClair, S.R. (1985) *A Multiexpert Knowledge System Architecture for Manufacturing decision Analysis*, Ph.D. Thesis, Arizona State University, Tempe, Arizona.

Liebowitz, J. (1989) *Structuring Expert Systems: Domain Design and Development*, Prentice-Hall, Englewood Cliffs, New Jersey.

Quinlan, J.R. (1985) *Induction of Decision Trees*, Technical Report 85.6, The New South Wales Institute of Technology Report Series.

2.8 For interested readers

The expert system described herein is available for use by readers. There is a nominal charge to cover postage, handling and disk costs only. Interested readers should write or call the authors at:

Universal Technology Corporation
4031 Colonel Glenn Highway
Dayton, OH 45431-1600
Phone (513) 426-8530
FAX (513) 426-7753

2.9 Acknowledgement

This work is sponsored by the Manufacturing Technology Division of the Air Force Materials Laboratory, Wright-Patterson AFB, Ohio 45433.

3 Economics of Advanced Manufacturing Systems During System Design and Development

Hector Carrasco and Shih-Ming Lee

3.1 Introduction

Although today's top executives recognize that the move towards automated, high-technology manufacturing systems is necessary to improve productivity and remain competitive, analysts are finding such modernization projects increasingly difficult to justify with traditional techniques. Previous studies (Parsaei *et al.*, 1988; Michaels *et al.*, 1984a) have demonstrated that the automated factory cannot be effectively planned, managed or justified with the same techniques used in the existing labour-intensive operations. Computer integrated manufacturing systems are capital intensive. Traditional economic analysis methodologies, when used with the inflated discount rates that are applied today as a hedge against inflation and risk, fail to justify high-technology equipment unless it is accompanied by unrealistically high benefit estimates.

Informal surveys of analysis methodologies used in industry to justify capital investments indicate that few companies apply unique methods in justifying high-technology equipment. In fact, many companies continue to rely on an undiscounted payback period as their primary justification tool, requiring payback periods of as little as two years. More technically advanced companies appear to favour more sophisticated methods such as the internal rate of return; however, even the most technically advanced do not apply stochastic methodologies.

To recognize potentially excellent applications for new technology, the measures of cost and performances evaluations must be improved. Little or no tracking of actual costs is carried out in a form that is useful to justify or even manage an automated factory. Also, the verification and validation of projected economic factors are not widely practised since the general perception is that, once the investment is made, it is a sunk cost. Cost/benefit tracking is necessary to control high-technology manufacturing systems. In addition, the confirmation of expected returns from an investment enhances management support of new projects and their justification methods. Traditional project analysis and management techniques have not proved satisfactory for:

1. Needs analysis and requirements definition in the development and evaluation of alternatives which are cost effective;
2. Cost/benefit and other economic analysis techniques, which help evaluate critical manufacturing elements; and
3. Tracking the resulting costs and benefits of the implemented system.

An effective cost/benefit analysis is a critical prerequisite for the successful implementation of the automated factory (Michaels *et al.*, 1984b). However, due to the cumbersome and tedious nature of a detailed economic analysis, designers seldom do a complete project analysis until the design nears completion. The proposed methodology would enable the designer to build a computerized economic model early in the design process. The model would then evolve with the system design, enabling its use throughout the development life cycle of the manufacturing system.

3.2 The life cycle and economic analysis

When a problem is encountered or a new or enhanced system is deemed necessary, there are accepted procedures for arriving at a solution. There are many models of this development process. While the number of tasks and their names may vary, the basic structure remains the same. As shown in Table 3.1, there has first to be an understanding of the problem, then solutions are generated. After an optimal/satisfactory solution has been generated the system must be constructed or implemented. Finally the system is utilized and maintained. These steps are shown as sequential; however, in actual practice some tasks may overlap and even recycle. In addition, after the system is in place and operating new needs arise. Needs for enhancement, expansion, or complete replacement eventually develop, requiring a new development process, hence the term **project life cycle** is used.

3.2.1 Current role of economic analysis

It is accepted by most analysts and managers that cost estimates become more reliable as the design details become more specific. As a result,

TABLE 3.1 *Typical system development phases and life cycle model steps*

System development phases	Life cycle steps
Conceptual	1. Needs analysis
	2. requirements definition
Design	3. Preliminary design
	4. detailed solution
Construct and integrate	5. Construction and verification
	6. integration and validation
Implement and maintain	7. Implementation and user acceptance
	8. maintenance and support

during the early development stages of major manufacturing modernization projects, little reliance is made on economic decision analysis techniques. This is true because few of the costs are known with sufficient precision to perform a significant, deterministic analysis. Design decisions during the early development stages are usually based on 'educated' guesses or at best simplified models having limited scope. A complete, detailed economic analysis is delayed until the design nears completion. At this point all major design decisions have been made and the purpose of the economic analysis is to accept or reject the completed design.

3.2.2 Consequences of current approach

This traditional approach of economic justification has proven to be inadequate for justifying comprehensive systems because in many cases they fail to substantiate the positive decision that management had made prior to the detailed cost/benefit justification attempt. Some analysts point out that modernization is a strategic decision, therefore traditional economic justification is not suitable. Many blame the problem on the inability to quantify and formally consider the so-called intangible benefits. Whatever the cause, the end result is that managers view the economist's models and their predictions as suspect.

This is not an attempt to question the concept of time value of money, simply a statement that, as the economic factors failed to justify the decisions already made, economics has not been the driving force in strategic decisions (White, 1989). This does not conform to the basic financial theory that the objective of the firm is to maximize its future value to the shareholders (Bussy, 1978).

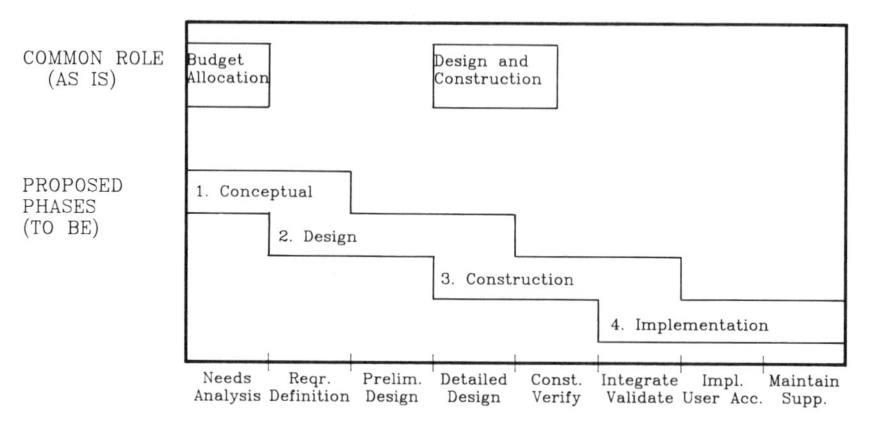

Fig. 3.1 *Economic analysis during a life cycle.*

3.3 Proposed economic analysis methodology

The underlying principle of the proposed economic analysis methodology is the application of sound economic analysis methods throughout the life cycle of a manufacturing system. As mentioned earlier, few economic considerations are addressed and specifically used for design development during the early life cycle steps of industrial projects. Figure 3.1 describes the perceived classical application of economic decision analysis in a system development effort (Blank and Carrasco, 1985). Most of its use occurs in the detailed design and system construction steps, with more of a budgetary approach used (as apposed to an actual economic decision analysis approach) in the conceptual phase of a project. In the proposed methodology, each of the phases is shown on a different level indicating that it requires a unique approach to economic evaluation. However, there is an overlap and joining of the phases to represent the needed evolution or flow of the cost estimation process and economic decision analysis throughout the life cycle of the system. The role of economic analysis in these development phases is discussed in detail in the following sections.

3.3.1 Conceptual phase

During the conceptual phase, a manufacturing system can be modeled as a single block with the desired inputs and outputs as shown in Fig. 3.2. Typically, at this point, upper management decisions are made based on past experiences that are not easily quantified. As mentioned earlier, many budget decisions are based on an educated guess, which once made becomes a dominant factor in the design of the system.

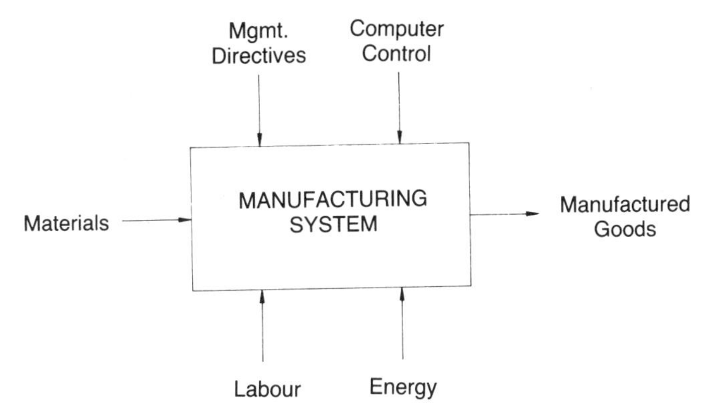

Fig. 3.2 *Model of a basic manufacturing system.*

The only economic data that can be obtained with any degree of certainty is the cost of the existing system or, for new facilities, the estimated cost of not having the system. A total system cost estimate, including risk, for the proposed system is needed. This can be addressed by examining the costs of existing systems with similar inputs and outputs, using in-house projects, current literature, and other available sources. The most likely cost and the worst and best case scenarios alone can be of significant value with more sophisticated analysis possible if sufficient information is available to provide cost distributions.

If the economic analysis of previous designs along with ongoing projects is maintained in a manufacturing system database, and if costs are tracked and updated during the construction and implementation stages, data would be accumulated for use in the evaluation of future projects. When the next project is being considered, the analyst can scan the database to select projects that are similar. When several projects have been identified, pertinent economic factors such as development cost, initial cost and annual cash flows can be used to estimate the expected costs and benefits of the new project early in the conceptual phase.

3.3.2 Design phase

In the design phase, the requirements and limitations established during the requirements definition step must be satisfied with a physical system. This physical system can be said to consist of men, machines and materials; however, they can be broken down into smaller components. For example a machine may consist of several sub-assemblies held together by an assortment of nuts and bolts. In general, the design of a manufacturing system does not require this fine detail. The term **configuration unit** will be defined as the smallest component that is of interest in the system design of a manufacturing facility.

During the preliminary design stage, after the designer or analyst has determined the type of configuration units that will be needed, a database containing generic configuration units could be accessed. For example, if a preliminary design requires several machining centres, as shown in Fig. 3.3, the generic representation of the CNC machines, with average tooling, could be selected from the database, providing the expected costs for typical machining centres. In this way, economic analysis can be performed during early design although sufficient details may not be available for the selection of specific equipment models. Although the scenarios depicted may vary widely, many components can be common to several different designs. The use of data files, storing cost and characteristic data for major components, would enable the analyst to assemble and analyse various scenarios very quickly. Benefit/cost analysis with risk addressed could then be used to select promising alternatives while sensitivity analysis could be added to determine sources of variance, thus identifying critical design parameters.

As the detailed design progresses, the generic equipment can be replaced with specific configuration units with the relevant cost factors such as the actual purchase price, tooling cost, installation cost, operating cost, overhead and equipment life being substituted for the average costs used in the earlier analysis. In this way, whenever a configuration unit is specifically identified, changed, added or removed, the pertinent data is updated and a new economic analysis generated. This would allow the economic model to evolve with the system design providing the designer with feedback regarding the economic consequences of each decision.

3.3.3 Construction and implementation phases

Finally, during the system construction and implementation phases, all of the cost and revenue factors that are relevant and measurable should be recorded and used to update the economic analysis. Since the capital budgeting process is sequential, new information should be evaluated and incorporated into the valuation formula. As the cash flows are realized, they provide new information which can be used to update the economic model, revising prior estimates.

Care should be taken to include all planning and design costs from the initiation of the project. The actual cost of all equipment and its installation should be included. Contracted and in-house engineering, design modifications, programming, personnel training, shut-down costs, testing and integration costs, and any other costs that can be attributed directly to that account should be entered. After the installation is complete,

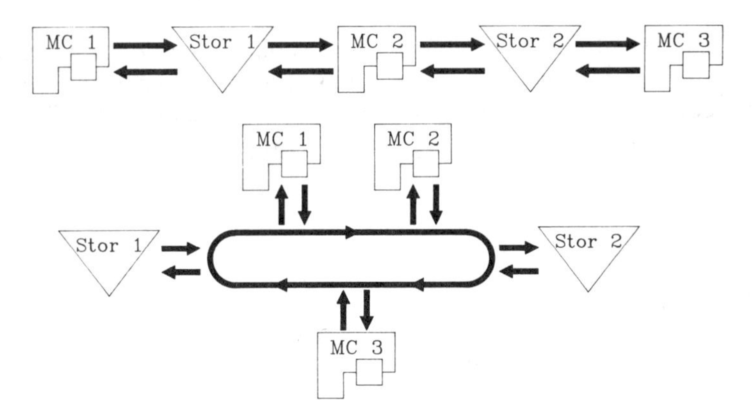

Fig. 3.3 *Model of a generic manufacturing system.*

operating costs should also be tracked. Examples of the operating costs that should be included are direct labour, direct material, equipment, engineering, energy use, in-process inventory, scrap, administrative support, and indirect labour and material costs. Tracking these observed costs provides a basis for verification and/or updating of cost estimates made for this and future projects. Actual costs can also be compared with expected distributions and the costs that differ significantly can be identified and the causes determined.

Confirmation of expected savings will promote the use of economic methodologies by management in the evaluation of new projects. The availability of accurate manufacturing cost information would also aid in the evaluation of costs typically treated as overhead costs. This allows the assignment of costs directly to machining centres, thus improving the estimate of value added by machine/process times.

3.4 Conclusion

As a direct result of the perceived shortcomings of traditional economic analysis methods, an extended economic analysis methodology has been proposed. The primary contribution is the extension of the justification process to all phases of the system development life cycle.

Although traditional methods are perceived as inadequate for today's strategic decisions, the use of discounted cash flows techniques should not be abandoned. The inability to justify high-risk, high-cost projects is largely due to the discount rates required by most organizations. Risk and uncertainty are typically addressed by the application of risk-adjusted discount rates (unrealistically high hurdle rates). The extension of economic analysis to aid in optimizing the early design decisions, along with the availability of more cost information, should allow for an improved analysis of risk, freeing the discount rate to serve its intended purpose of simply equating present and future cash flows. These high hurdle rates have severely handicapped strategic, long-term projects.

3.5 References

Blank, L. and Carrasco, H. (1985) The economics of new technology: system design. *1985 Annual Industrial Engineering Conference Proceedings*, 161–8.

Bussy, L.E. (1978) *The Economic Analysis of Industrial Projects*, Prentice-Hall, Englewood Cliffs, New Jersey.

Michaels, L.T., Muir, W.T. and Eiler, R.G. (1984a) The relationship between technology and cost management. *Materials Handling Engineering*, **39** (1), 57–64.

Michaels, L.T., Muir, W.T. and Eiler, R.G. (1984b) Improving technology cost-benefit analysis. *Materials Handling Engineering*, **39** (2), 49–54.

Parsaei, H.R., Karwowski, W., Wilhelm, M.R. and Walsh, A.J. (1988) A methodology for economic justification of flexible manufacturing systems. *Computers & Industrial Engineering*, **15** (1–4), 117–22.

Sullivan, W.G. (1986) Models IEs can use to include strategic, non-monetary factors in automation decisions. *Industrial Engineering*, **18** (3), 42–50.

White, B.E. (1989) Justification of CIM systems: is special treatment necessary? *Proceedings of the IIE Integrated Systems Conference & Society for Integrated Manufacturing Conference*, 588–92.

4 Economics of Flexible Assembly Automation: Influence of Production and Market Factors

Anil Mital

4.1 Introduction

Global competition is forcing manufacturers continually to look for ways to reduce costs and improve product quality. Productivity enhancement has now become a dire necessity instead of just a passing fancy. Factory automation, considered a way out of present woes, is revolutionizing manufacturing worldwide. More and more manufacturers are considering automating their operations so as to become more productive and competitive on a worldwide basis. This drive to automate production operations is, however, not confined to manufacturing operations (metal cutting and metal processing). Increasingly, we come across industries that have automated their materials handling and product assembly or subassembly operations. While considerable progress has been made towards automating materials handling and manufacturing operations, automating assembly operations has been slow and difficult. One of the major difficulties confronting engineers is the basic design of the product: products are either not designed for automated assembly or are poorly designed.

Since assembly automation is considered fairly important to the economic health of a manufacturing outfit, more and more designers are adopting design-for-assembly (DFA) procedures. As a result, certain guidelines have been developed to simplify the design and assembly of products. For instance, snap-in retainers are a common feature in most products designed for automated assembly. Besides the design simplicity and assembly convenience, DFA products tend to lower material and inventory costs, by virtue of requiring fewer components, and materials handling costs. It is also believed that DFA products have lower assembly

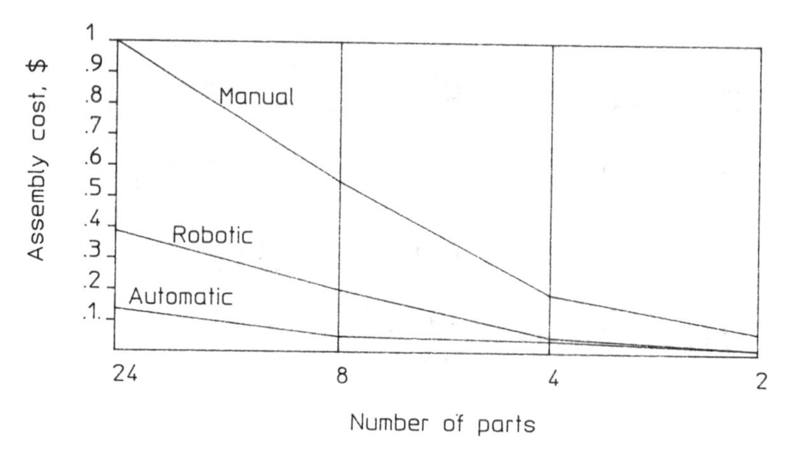

Fig. 4.1 *Effects of product design on the cost of assembly (Boothroyd, 1988).*

costs. This belief, however, has no proven basis. On the contrary, one recent report suggests that when DFA products are assembled manually, the cost of assembly is no more than the cost when such products are assembled by robots (flexible assembly automation) or automatically by hard-core dedicated equipment (automatic assembly method) (Boothroyd, 1988; Fig. 4.1). Other works suggest that for DFA products manual mode of assembly may be cheaper and more efficient than the flexible automated assembly mode (Mital *et al.*, 1988; Mital and Genaidy, 1988; Genaidy *et al.*, 1990). If so, the flexible automated assembly method may be redundant.

The basis for suggesting that assembling DFA products manually may be more economical than assembling them by flexible automated method lies in human and robot performances. Even though very limited information comparing human and robot performances is available, the general consensus is that for assembly tasks robots are generally slower than humans, particularly when the products have been designed for assembly.

Paul and Nof (1979) compared performances of humans and robots for a pump assembly task. The times for human performance were based on Bolles' and Paul's (1973) study. It was found that robots required eight times the time taken by humans. Wygant (1986) compared the actual total time taken by robots and humans for drilling a 1.27 cm deep hole in a casting. The time, estimated by ROBOT MOST, was 0.22 minute for humans and 0.50 minute for robots. The robots, thus, took 127% more time than humans to complete the task. Mital (1989), Mital and Mahajan (1989), and Mital *et al.* (1988) also compared assembly time taken by robots and humans for a variety of products using MTM (Maynard *et al.*, 1948) and RTM (Paul and Nof, 1979) predetermined time systems. In every case, humans outperformed robots. Genaidy *et al.* (1990) conducted a direct time study and observed that for a simple assembly task robots took approximately 342% more time than humans. For palletizing and stacking tasks, Mital (1990), using direct time study, observed that robots took almost 645% more time than the humans. The major reason for this large difference in performance times appears to be the learning on the job: while humans do learn on the job and become more efficient with time, robots do not. A robot's performance does not vary much, but is expected to deteriorate as it ages.

Besides performance, there are numerous other factors, such as wage and interest rates and quantity to be produced, that may influence the choice of a assembly method. The purpose of this work is to demonstrate the effects of various production and market factors on economic decision-making as it relates to product assembly methods. In particular; the influence of the following factors on the cost of manual and flexible automated assembly methods is considered:

1. Performance time;
2. Production quantity;
3. Wage rate;
4. Interest rate; and
5. Robot reliability.

It is also shown that for the same product, local values of some of these factors can lead to entirely different decisions in different countries.

4.2 Performance time of alternative assembly methods and its impact on cost

In order to evaluate the impact of performance time on assembly cost, two DFA products (a standard 13-amp plug, Fig. 4.2, and a coffeemaker, Fig. 4.3) were considered and their assembly cycle time and annual assembly costs were determined. For the flexible automated assembly method, the assembly cycle times were estimated by the robot time measurement (RTM) system (Paul and Nof, 1979). The methods time measurement (MTM) system (Maynard *et al.* 1948) was used to estimate assembly cycle times for the manual assembly method.

4.2.1 Standard 13-Amp Plug

The assembly sequence for the standard 13-amp plug was as follows:

1. Insert the cover screw in the fixture.
2. Insert the base over the cover screw.

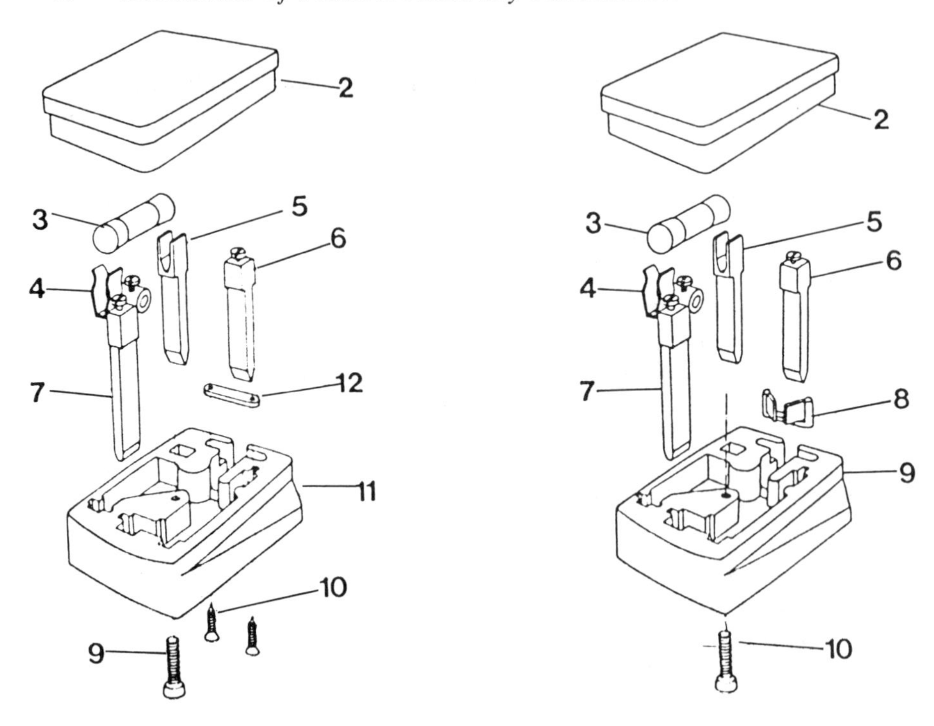

Fig. 4.2 *The standard 13-amp plug redesigned for assembly (Boothroyd and Dewhurst, 1982; Redford and Lo, 1986).*

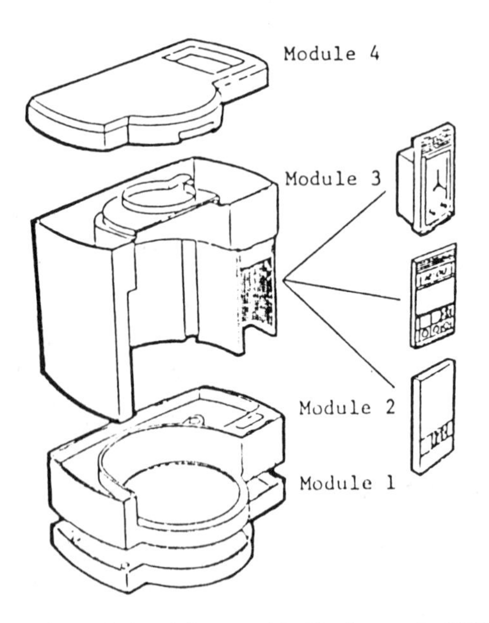

Fig. 4.3 *Norelco coffeemaker redesigned for assembly (Darlow et al., 1987).*

3. Insert, in any sequence, the ground pin, the live pin, the neutral pin, and the redesigned cord grip.
4. Insert the fuse.
5. Place the cover.
6. Drive the cover screw.

A special fixture was needed for the last step for flexible automated assembly. It was estimated that 1 000 000 units of the electrical plug could easily be sold each year. For the assembly, such as shown in Fig. 4.2, a Seiko robot could be used. The workstation layout for assembly is shown in Fig. 4.4.

Given these, the cycle time and assembly cost per year were determined for both flexible automated and manual methods. Table 4.1 shows the assembly sequence, RTM and MTM elements, and associated times (in TMU-time measurement unit).

From these cycle times and production volume, annual costs for the two assembly methods were estimated as follows.

Flexible automated assembly

Production volume = 1 000 000 units/year
Assembly cycle time = 0.4263 minute
Capacity/workstation = 2.3458 pieces/minute
= 281 496 pieces/year
(assuming 2000 hours/year and 1 shift)

If only one shift is operated throughout the year, the number of workstations required to meet the demand would be 3.5524 or 4. However, the least investment alternative to meet the demand would be to have two work stations and operate two shifts each day.

Fixed cost/workstation:

Robot cost	=	$35 000	
Feeders (10)	=	$170 000	($17 000/feeder – feeders are needed to feed components)
Gravity chutes	=	$500	($50/chute)
Bins	=	$500	($50/bin)
Work bench	=	$500	
Fixtures	=	$2000	
Total	=	$208 500	

Variable cost/workstation:

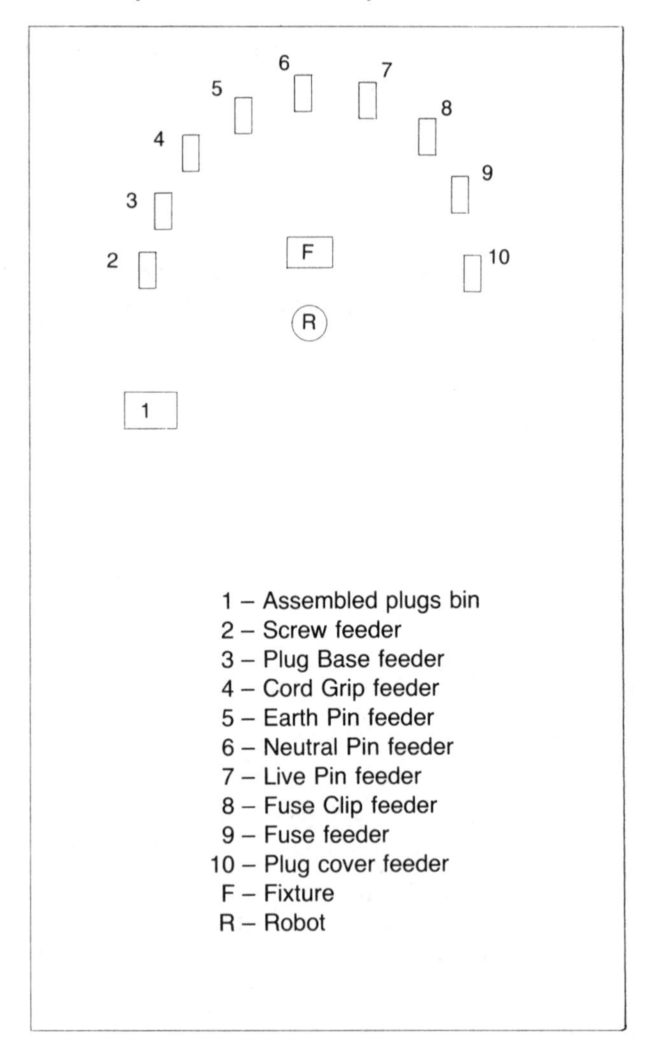

Fig. 4.4 *Workstation schematic for assembling the standard 13-amp plug.*

Maintenance = \$1000
Energy = \$3500 (@ \$0.07/kWh)
Total = \$4500

Total investment = \$417 000

Given annual interest rate of 12%, 10-year life of the investment, and robot salvage value of \$5000, the annual cost of flexible automated assembly comes out to be:

TABLE 4.1 *Estimated cycle times for robotic and manual assembly of the 13-amp plug*

Assembly sequence	RTM element	Time (TMU)	MTM element	Time (TMU)
1. Place cover screw in fixture (right hand)	R2 76	26	R 12 C	14.2
	ST 0.5	3	G1A	2
	GR 1.5	7	M 12 C	15.2
	M2 30	23	P2 SS	25.6
	SF 0.5	3		
	RE 3.0	10		
2. Place plug base (right hand)	R2 30	23	R 12 C	14.2
	ST 0.5	3	G1A	2
	GR 3.0	10	M 12 C	15.2
	M2 30	23	P2 NS	21
	SF 0.5	3		
	RE 0.5	4		
3. Place cord grip (right hand) and ground pin (left hand)	R2 30	23	R 12 C (L & R)	14.2
	ST 0.5	3	G1A (L & R)	2
	GR 1.5	7	M 12 C (L & R)	15.2
	M2 30	23	P2 SS	25.6
	SF 0.5	3	M 5 A	7.3
	RE 0.5	4	P2 SS	25.6
	R2 30	23		
	ST 0.5	3		
	GR 3.0	10		
	M2 30	23		
	SF 0.5	3		
	RE 3.0	10		
4. Place neutral pin (right hand) and live pin (left hand)	R2 30	23	R 12 C (L & R)	14.2
	ST 0.5	3	G1A (L & R)	2
	GR 3.0	10	M 12 C (L & R)	15.2
	M2 30	23	P2 SS	25.6
	SF 0.5	3	M 5 A	7.3
	RE 3.0	10	P2 SS	25.6
	R2 30	23		
	ST 0.5	3		
	GR 3.0	10		
	M2 30	23		
	SF 0.5	3		
	RE 3.0	10		
5. Place fuse clip (right hand) and fuse (left hand)	R2 30	23	R 12 C (L & R)	14.2
	ST 0.5	3	G1A (L & R)	2
	GR 3.0	10	M 12 C (L & R)	15.2

TABLE 4.1 *Continued*

Assembly sequence	RTM element	Time (TMU)	MTM element	Time (TMU)
	M2 30	23	P2 SS	25.6
	SF 0.5	3	M 5 A	7.3
	RE 3.0	10	P2 SS	25.6
	R2 30	23		
	ST 0.5	3		
	GR 1.5	7		
	M2 30	23		
	SF 0.5	3		
	RE 0.5	3		
6. Place cover	R2 30	23		58.7
	ST 0.5	3		
	GR 3.0	10		
	M2 30	23		
	SF 0.5	3		
	RE 3.0	10		
7. Drive (tighten) cover screw	Index work head	10		34.7
	T (rotate assembly)	20		
	Index work head	10		
8. Drop (release) assembly	R2 1	10.5	G1A	2
	GR 0.5	4	R 30 B	24.3
	M2 76	26	RL 1	2
	RE 1.5	7		
Total cycle time (TMU) = 710.5				500.8
Total cycle time (minute) = 0.4263				0.300

Note: Descriptions and symbols in parentheses are for manual assembly. Distances are in centimetres for RTM; in inches for MTM.

$$\text{Annual cost} = 9000 + 417\,000(\text{A/P}.12,10) - 10\,000(\text{A/F}.12,10)$$
$$= \$82\,239$$

Manual assembly

$$\text{Assembly cycle time} = 0.300 \text{ minute}$$
$$\text{Capacity/workstation} = 3.3333 \text{ pieces/minute}$$
$$= 400\,000 \text{ pieces/year}$$
(assuming 2000 hours/year and 1 shift)

The least cost alternative to achieve the required production volume would be to have three workstations (1 worker/workstation) and only one shift per day. Unit workstation cost would be:

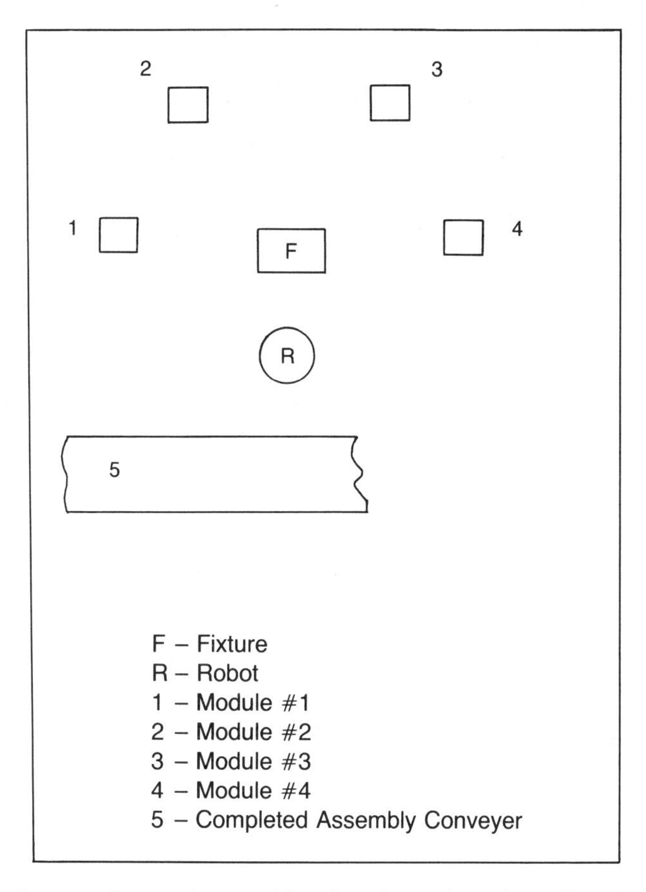

Fig. 4.5 *Workstation schematic for assembling the redesigned Norelco coffeemaker.*

Fixture = \$500
Work bench = \$500
Bins (10) = \$500

Total = \$1500

Total cost for 3 workstations = \$4500

Annual labour cost = \$60 000 (\$20 000/worker)

Given annual interest rate of 12%, 10-year life of the investment, no salvage value, and \$10/hour wage rate, the annual assembly cost comes out to be:

Annual cost = 60 000 + 4500(A/P.12,10) = \$60 796.50

TABLE 4.2 *Estimation of cycle times for robotic and manual assembly of the redesigned coffeemaker*

Assembly sequence	RTM element	Time (TMU)	MTM element	Time (TMU)
1. Place module 1	R2 100	26	R 30 B	25.8
in fixture	ST 0.5	3	G1A	2
	GR 1.5	7	M 30 C	22.7
	M2 100	26	P1 NS	10.4
	SF 0.5	3	RL 1	2
	RE 1.5	7		
2. Place module 2	R2 100	26	R 30 B	25.8
	ST 0.5	3	G1A	2
	GR 1.5	7	M 30 C	22.7
	M2 100	52	P2 NS	21
	SF 0.5	3	RL 1	2
	RE 1.5	7		
3. Place module 3	R2 100	26	R 30 B	25.8
	ST 0.5	3	G1A	2
	GR 1.5	7	M 30 C	22.7
	M2 100	26	P2 NS	21
	SF 0.5	3	RL 1	2
4. Place module 4	R2 100	26	R 30 B	25.8
	ST 0.5	3	G1A	2
	GR 1.5	7	M 30 C	22.7
	M2 100	26	P2 NS	21
	SF 0.5	3	RL 1	2
Total cycle time (TMU)	=	314		283.4
Total cycle time (minute)	=	0.1884	0.170	

Note: distances are in centimetres for RTM; in inches for MTM.

The cost advantage of the manual assembly method over the flexible automated assembly method, in this case, is equal to $21 442.5 per year.

4.2.2 Norelco Coffeemaker

The coffeemaker was redesigned for assembly by a Rensselaer Polytechnic team (Darlow *et al.*, 1987) and is shown in Fig. 4.3. The newer design has 26% fewer parts and 70% fewer fasteners. The final assembly consists of four modules as shown. The workstation layout for assembly is shown in Fig. 4.5.

Just as in the case of electrical plug, cycle times for flexible automated and manual assembly methods were estimated. Table 4.2 shows these estimations.

It was estimated that 500 000 coffeemakers could easily be sold each year. From this and estimated assembly cycle times the cost of flexible automated and manual assembly methods were calculated as follows.

Flexible automated assembly

Production volume = 500 000

Cycle time = 0.1884 minute

Capacity/workstation = 5.3079 units/minute

= 636 948 units/year
(assuming 2000 hours/year and 1 shift)

Fixed cost for the workstation:

Robot cost = $35 000

Feeders (5) = $85 000 (1 for finished assembly and 4 for modules)

Fixture = $1000

Work bench = $500

Total = $121 500

Variable cost for the workstation:

Maintenance = $1000

Utilities = $3500 (@ $0.07/kWh)

Total = $4500/year

Given annual interest rate of 12%, 10 year investment life, robot salvage value of $5000, the annual cost comes out to be:

Annual Cost = 4500 + 121 500(A/P.12,10) − 5000(A/F.12,10)

= $25 720.5

Manual assembly

Cycle time = 0.17 minute

Capacity/workstation= 5.8824 units/minute

= 705 888 units/year
(assuming 2000 hours/year and 1 shift)

Fixed cost:

Fixture = $500

Work bench = $500

Bins (5) = $250

Total = $1250

Variable cost:

$$\begin{aligned} \underline{\text{Labour cost} = \$20\,000} \quad & (\$10/\text{hour}) \\ \text{Total} = \$20\,000/\text{year} & \end{aligned}$$

Given annual interest rate of 12%, workstation life of 10 years, and no salvage value, the annual cost comes out to be:

$$\text{Annual cost} = 20\,000 + 1250(\text{A/P}.12,10) = \$20\,221.25$$

The manual assembly, in this case, would have a cost advantage of $5499.25 per year. It should be realized that while the cycle time for the manual assembly method is expected to decrease with worker learning it does not change for the flexible assembly method (Genaidy *et al.*, 1990). In reality, while the cycle time for a robot would not change, its availability certainly would. The actual production capacity, therefore, would be less than 1/cycle time. Reduced robot capacity would tend to favour manual assembly.

Based on the estimates of cycle times and fixed and variable costs, it was observed here that the manual assembly of DFA products is more economical than their assembly by flexible automated (robotic) methods. However, before such a general conclusion can be drawn, the effect of factors such as production volume and labour costs, which heavily influence fixed and variable costs, on assembly costs must be determined. For instance, if in the case of the electrical plug, the production volume is increased to 1 688 976 units (total capacity of robotic workstations for three shifts), the economic advantage would favour flexible automated assembly (assembly cost would be approximately $87 000 for the flexible assembly method and $101 000 for manual assembly). Reduced labour costs, on the other hand, would turn the economic advantage in favour of manual assembly. If the production volume increases to a point that new investment in robots and workstations becomes necessary, the economic advantage would again change in favour of manual assembly.

Similar observations can also be made in the case of the coffeemaker. For instance, if the production volume doubles, the economic advantage of manual assembly will be lost and flexible automated assembly would become economically desirable (robotic assembly cost would be approximately $30 000 per year; manual assembly cost would be approximately $40 000 per year). Clearly, it is not possible to make a sweeping statement that flexible automated assembly is always more economical or manual assembly is always more economical. Before deciding on the assembly method, careful economic analysis must be carried out and the effect of changing market conditions considered. The impact of production and market factors are considered in subsequent sections.

4.3 Impact of variations in production volume on costs of alternative assembly method

Market demand has the most profound effect on production volume and, hence, on investment needs and the choice of assembly method. As the production volume changes, workstation needs and, therefore, equipment and manpower needs, also change. For increased production more resources will be needed. If the production needs go down, resource requirement will also diminish. In order to determine the impact of production volume on assembly costs, variations in the production volume of a DFA product (compressor cylinder head valve assembly – Romeo and Camera, 1980), while other factors such as the interest and wage rates remained fixed, were considered. The DFA product chosen is shown in Fig. 4.6 and has 12 elements (exhaust valve leaf, leaf lock, two rivets, valve plate, valve body, gasket, suction valve plate, and four screws) and 300 assembled units are needed each hour (Romeo and Camera, 1980). The assembly sequence is as follows: pick valve plate and place on rivets already fed and placed in the rivetting machine; pick bent exhaust valve leaf and place it on valve plate; pick leaf lock and place it on valve plate and leaf; perform rivetting; pick up all rivetted components; rotate the subassembly 90°; and place it upon valve body and gasket located at another station (all pick up and placement operations are done by one robot arm); pick and place gasket on valve body; pick suction valve plate and place it upon the rivetted subassembly; pick up the entire group of parts and place it upon four preplaced screws in a pneumatic screwing machine; complete screwing operation; pick up and place the assembled group on a belt (all pick up and placement operations are done by the second robot arm). Special feeders and chutes are required for feeding parts.

Given the size of the various components and the equipment, work layout was configured (Fig. 4.7) and distances moved by the robot and the human worker were determined. These distances and the type of movement were used to estimate assembly cycle times. For flexible automated assembly, use of a PUMA assembly robot was considered. The estimates of assembly times were based on MTM for manual assembly method and RTM for flexible automated assembly method and are given in Table 4.3. From these cycle times, the costs of flexible automated assembly and manual assembly of the compressor cylinder head valve assembly were calculated. Since only variations in production volume were of interest, while it was changed, other factors were kept constant. The interest rate was 12% and the wage rate was $10 per hour (these values were the market values in the United States in February, 1989). The cost calculations for current production volume (300 units per hour) are shown below.

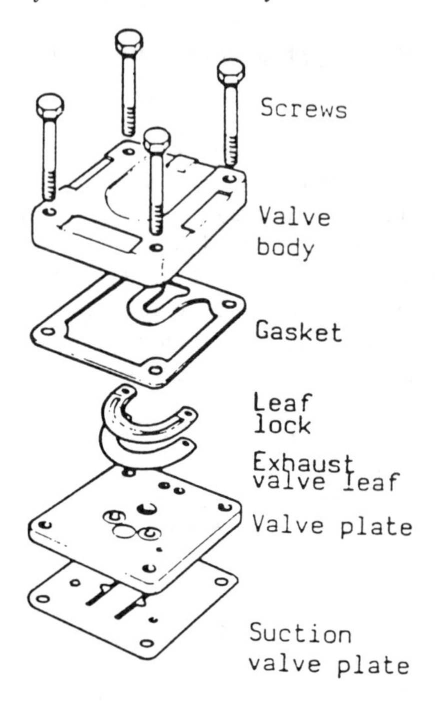

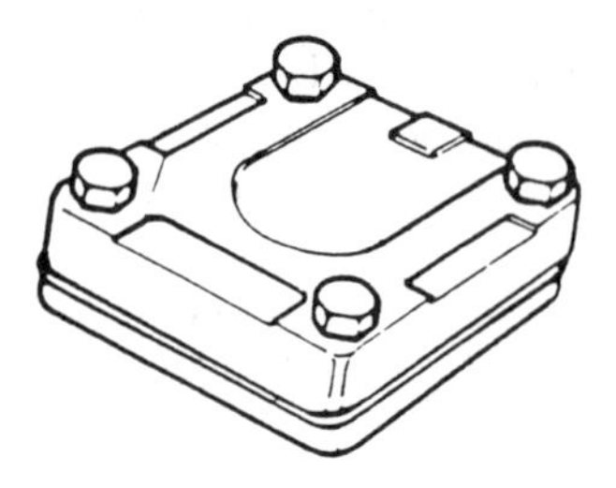

Fig. 4.6 *Compressor cylinder head valve assembly and components (Romeo and Camera, 1980).*

Flexible automated assembly

$$\begin{aligned}
\text{Assembly cycle time} &= 0.2589 \text{ minute} \\
\text{Production capacity/workstation} &= 1/\text{cycle time} \\
&= 3.8625 \text{ units/minute} \\
&= 231.75 \text{ units/hour} \\
\text{Production volume} &= 300 \text{ units/hour}
\end{aligned}$$

Fig. 4.7 *Workstation configuration for assembling the compressor cylinder head valve assembly.*

This means that if only one work shift is operated per day, two workstations will be needed. The least investment alternative, however, would be to have two shifts per day (approximately 11 hours/day).

Fixed cost for the workstation:

Robot (PUMA assembly robot) =	$37 000	(Unimation, Pittsburgh, USA)
Feeders (5) =	$85 000	(@ $17 000/feeder)
Rivetting and screwing fixtures =	$2000	
Bins and work bench =	$1000	
Miscellaneous (foundation, etc.) =	$10 000	
Total investment =	$135 000	

Variable cost for the work station:

Maintenance =	$2000
Energy =	$4000
Total =	$6000

TABLE 4.3 *Assembly sequence and RTM and MTM elements for assembling compressor cylinder head valve assembly*

Assembly sequence	RTM element	Time*	MTM element	Time*
1. Pick valve plate and place on rivets (robot arm 1)	R2 100	26	R 30 C	26.7
	ST 0.5	3	G1A	2
	GR 0.5	4	M 30 C	30.7
	M2 75	27	P1 S	5.6
	ST 0.5	3	RL1	2
	RE 0.5	4		
2. Pick bent exhaust valve leaf and place it on valve plate (robot arm 1)	R2 75	26	R 30 C	26.7
	ST 0.5	3	G1A	2
	GR 0.5	4	M 30 C	30.7
	M2 75	27	P1 S	5.6
	ST 0.5	3	RL1	2
	RE 0.5	4		
3. Perform rivetting		30		30
4. Pick up rivetted Components, rotate 90 degrees and place it upon valve body and gasket located elsewhere (robot arm 1)	R2 30	23	R 10 A	8.7
	ST 0.5	3	G1A	2
	GR 0.5	4	T 90	5.4
	Rotate assy.	20	M 30 C	30.7
	M2 100	29	P1 S	5.6
	ST 0.5	3	RL1	2
	RE 0.5	4		
5. Place gasket on valve body (robot arm 2)	R2 100	26	R 30 C	26.7
	ST 0.5	3	G1A	2
	GR 0.5	4	M 30 C	30.7
	M2 75	26	P1 S	5.6
	ST 0.5	3	RL1	2
	RE 0.5	4		
6. Pick up suction valve plate and place it upon rivetted subassembly (robot arm 2)	R2 100	26	R 30 C	26.7
	ST 0.5	3	G1A	2
	GR 0.5	4	M 30 C	30.7
	M2 75	27	P1 S	5.6
	ST 0.5	3		
	RE 0.5	4		
7. Pick up all grouped parts and place them upon 4 screws (robot arm 2)	R1 1	10.5	G1A	2
	ST 0.5	3	M 30 C	30.7
	GR 0.5	4	P1 S	5.6
	M2 100	29	RL1	2
	ST 0.5	3		
	RE 0.5	4		
8. Complete screwing operation		40		40
9. Pick up and place subassembly on the belt (robot arm 2)	R2 100	26	R 10 C	12.9
	ST 0.5	3	G1A	2
	GR 0.5	4	M 30 C	30.7
	M2 100	29	RL1	2
	ST 0.5	3		
	RE 0.5	4		
Total cycle time (TMU) =		431.50		408.33
Total cycle time (minute) =		0.2589		0.2450

Note: Elements 1 and 4 are done concurrently by the robot arms. Manually, element 5 is performed while rivetting (element 3) is being performed. Distances are in centimetres for RTM and inches for MTM.

* In TMU.

Annual cost = 6000 + 135 000(A/P.12,10)
(assuming investment life to be 10 years)
= $29 895

Manual assembly method

Assembly cycle time = 0.245 minute
Production capacity/workstation = 1/cycle time
= 4.0816 units/minute
= 244.896 units/hour
Production volume = 300 units/hour

Therefore, two workstations will be needed if only one shift is run per day. With one workstation, operators would need to work 10 hours per day or 1.25 shift. The least investment alternative would be to have one workstation and run the operation 10 hours each day.

Fixed cost for the workstation would be:

Bins (9) =	$450
Work bench =	$500
Rivetting and screwing fixtures =	$2000
Total =	$2950

Variable cost would be:

Labour cost =	$25 000/year
Utilities =	$1000/year
Total =	$26 000/year

Annual cost = 26 000 + 2950(A/P.12,10) = $26 522

In the above manner, the costs of manual and flexible automated assembly methods were computed for production volumes ranging from 100 units per hour to 800 units per hour. These costs are summarized in Table 4.4.

As shown in Table 4.4, production volume has a cyclic effect. It alternatively favours manual and flexible automated assembly methods. Thus, fluctuations in market demand would have a profound effect on the economic desirability of a particular assembly method. However, once a method is implemented, it is not possible to change it without incurring additional costs. If a method is chosen on the basis of a given production volume, at least the following questions should be asked: 'Can that many be sold?' and 'Would this mean not meeting the market demand?'

TABLE 4.4 *Variations in assembly cost with production volume*

Volume (#/h)	FAA ($)	MA ($)	Choice
100	27 150	21 522	MA*
200	28 415	21 522	MA*
300	29 895	26 522	MA*
400	32 986	42 522	FAA
500	34 441	42 522	FAA
600	36 531	52 522	FAA
700	60 956	55 522	MA*
800	62 837	71 544	FAA

Note: FAA is flexible automated assembly method; MA is manual assembly method.
* Shows that manual assembly method is more economical.

4.4 Impact of variations in the interest rate on costs of alternative assembly methods

Investments are invariably made by borrowing money. The cost of money, therefore, is a major consideration in choosing an alternative. In general, lower interest rates would favour large investment in equipment while high interest rates would discourage it. When interest rates are high, methods that have low capital cost tend to become more attractive. The choice of product assembly method represents a similar situation. If, in the previous example (compressor cylinder head valve assembly), the production volume and wage rate are kept fixed and interest rates are varied, the influence of interest rate on economic decision-making can be determined.

In general, variations in interest rate from 4% to 18% would cover a very broad range. The costs of flexible automated assembly and manual assembly methods will be determined as follows.

Flexible automated assembly

For the compressor cylinder head valve assembly:

Assembly cycle time = 0.2589 minute

Production capacity/workstation = 1/cycle time = 231.75/hour

This means more than one shift would be necessary. The least investment alternative would be to have one workstation and to operate it for 11 hours per day.

Fixed cost/workstation = $135 000 (this includes 5 feeders, foundation, rivetting and screwing fixtures, the robot, and bins and work bench)

TABLE 4.5 *Variations in assembly cost with interest rate (i)*

i (%)	FAA ($)	MA ($)	Choice
4	22 645	26 364	FAA
6	24 346	26 401	FAA
8	26 115	26 440	FAA
10	27 978	26 480	MA
12	29 895	26 522	MA
14	31 879	26 566	MA
16	33 931	26 610	MA
18	36 037	26 656	MA

Note: FAA is flexible automated assembly method; MA is manual assembly method.

Variable cost/workstation = \$6000 (this includes maintenance and energy costs)

Total annual cost = 6000 + 135 000(A/P_i,10) (assuming 10-year life of the investment)

By substituting different values for i, the cost of assembly can be determined. The costs for different values of i are shown in Table 4.5.

Manual assembly

For the manual assembly method, the production capacity of the workstation would be 1/0.245 = 244.896 units/hour. The least cost alternative would be to run one 10-hour shift. The fixed cost (rivetting and screwing fixtures, work bench, and bins) would total \$2950 and the variable cost (operator wages and utilities) would be \$26 000 per year.

Total annual cost = 26 000 + 2950(A/P_i, 10)

Again, by substituting different values for i, the cost of assembly can be determined. Table 4.5 shows the assembly cost for different values of i.

Table 4.5 clearly shows that as the interest rate goes up, investment becomes undesirable. Since the flexible automated assembly method requires substantial investment (in robots, feeders, etc.), it continues to become economically less desirable as interest rates go up. The manual assembly method, on the other hand, needs little investment and so is economically more attractive when the cost of money is high.

4.5 Impact of variations in the wage rate on costs of alternative assembly methods

In both previous analyses, the wage rate was kept fixed at \$10 per hour. This is the typical wage rate of a manufacturing worker in the United

TABLE 4.6 *Variations in manual assembly cost with wage rate ($/h)*

Wage rate	FAA ($)	MA ($)	Choice
6	29 895	16 522	MA
8	29 895	21 522	MA
10	29 895	26 522	MA
12	29 895	31 522	FAA
14	29 895	36 522	FAA
16	29 895	41 522	FAA

Note: FAA is flexible automated assembly method; MA is manual assembly method.

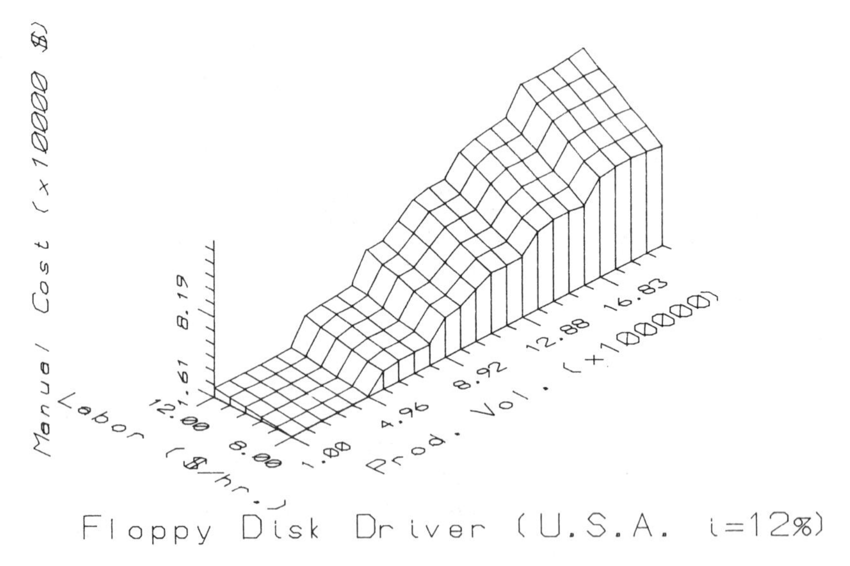

Fig. 4.8 *A 3-D plot of assembly cost (manual method), production volume and labour cost for the floppy disk drive designed for assembly.*

States at the present time (International Labour Organization's Year Book on Labour Statistics). Since wages fluctuate as a result of market conditions (supply of skilled workers, union–management agreements, etc.), it is important that we understand how this factor influences the cost of manual assembly method (cost of the flexible automated assembly method will not be affected by changes in the wage rate).

In order to determine the influence of wages, the wage rate was varied from $6 per hour to $16 per hour and the costs of assembling the compressor cylinder head valve manually were determined. Table 4.6 summarizes these costs.

As shown in Table 4.6, higher labour costs encourage automation while

Fig. 4.9 *A floppy disk drive designed for assembly (Albo, 1983).*

low labour costs favour manual assembly. Also, the flexible automated assembly method is not sensitive to wage rate fluctuations.

4.6 Interactive effects of production and market factors

In our analyses, we have investigated variations in one factor while other factors have remained at fixed levels. In reality, all these factors change at the same time and the effect of simultaneous variation (interactive effect) of all production and market factors on assembly costs must be considered. A 3–D plot, such as the one shown in Fig. 4.8, may be helpful in analysing the interactive effects. Figure 4.8 shows variations in the cost of assembling a floppy disk drive, shown in Fig. 4.9, manually, with production volume and wage rate for a given interest rate. Similar graphs may be prepared for other interest rates and other factors and compared with similar cost graphs for the flexible automated assembly method.

4.7 Designing for assembly pays

Regardless of the assembly method, it pays to design products for assembly. Table 4.7 compares the assembly costs of two products (a standard

TABLE 4.7 *Estimated time (minute) and annual cost ($) of assembling the original and redesigned 13-amp electrical plug and Norelco coffeemaker*

Product	Design	Assembly method	Assembly time	Cost
13-Amp plug	Original	Flexible automated	0.6315	102 700.00
		Manual	0.4868	80 547.50
	DFA	Flexible automated	0.4263	82 239.00
		Manual	0.3000	60 796.50
Coffeemaker	Original	Flexible automated	0.7848	95 685.00
		Manual	0.5901	60 684.37
	DFA	Flexible automated	0.1884	25 720.50
		Manual	0.1700	20 221.25

Production volume: plug – 1 000 000 units/year; coffeemaker – 500 000 units/year.

13-amp plug and the Norelco coffeemaker), for both flexible automated and manual methods, before and after redesigning them for assembly. The redesigned products are shown in Figs. 4.2 and 4.3, respectively (for original designs see Redford and Lo, 1986 and Darlow *et al.*, 1987).

It is clear from Table 4.7 that the assembly of redesigned products is more economical than the assembly of original designs. The economic advantage of redesigning the two products also extends to their assembly manually.

In the case of the electrical plug, the economic advantage favoured redesign by $20 461, in case of flexible automated assembly, and by $19 751, in case of manual assembly. In case of the coffeemaker, the economic advantage favoured redesign by a much larger amount ($69 364.50 for flexible automated assembly and $40 463.12 for manual assembly). These savings clearly demonstrate that it is highly desirable to redesign products for assembly.

4.8 Equipment reliability and economic advantage

In all cost calculations, it has been assumed that the production capacity of a robotic workstation remains unchanged. In reality, this is not so. The robot reliability is not 100%; as a matter of fact, if we consider dropped products, drifting, etc., in addition to downtime, it may be as low as 50% (Soni, 1988). This means, for flexible automated assembly we must have larger built-in production capacity than indicated by the reciprocal of assembly cycle times. Additional investment in this capacity would have a very profound effect on the assembly cost of the flexible automated assembly method and, therefore, must be considered.

For instance, in the case of 13-amp plug, if the robot reliability is indeed 50%, the workstation capacity would be reduced to 140 748 pieces per year per shift. In order to meet the demand, eight workstations would be needed. The minimum investment alternative in this case would be to have three workstations operating three shifts per day. The annual cost would be:

$$\text{Annual cost} = 13\,500 + 625\,500(\text{A/P}.12,10) - 15\,000(\text{A/F}.12,10) = \$114\,389$$

The advantage of manual assembly over flexible automated assembly will increase to $53 592.9 per year, a substantial improvement. In case of compressor cylinder head valve assembly, the cost of flexible automated assembly method will go up to $32 895 (Table 4.6) and the manual assembly will remain attractive up to a wage rate of $12 per hour, instead of up to $10 per hour. The flexible automated assembly method will also become unattractive at interest rates as low as 6% (Table 4.5). Clearly, the reliability issue is a critical one and must not be overlooked.

4.9 Impact of local values of production and market factors on the choice of assembly method

Previous discussion clearly demonstrates that the choice of assembly method for DFA products is a function of production and market factors. Since these factors vary from place to place and country to country, they are expected to influence assembly costs significantly. Thus, an assembly method, while economically desirable in one location may not be economically desirable in other locations. To demonstrate this point, assembly costs of two DFA products (the compressor cylinder head valve, shown in Fig. 4.6, and the floppy disk drive, shown in Fig. 4.9) in four different countries (Japan, United Kingdom, United States, and the then Federal Republic of Germany) were considered. Both the manual method and flexible automated assembly method were considered.

RTM was used to estimate the assembly cycle time for the flexible automated assembly method. For manual assembly method, MTM was used for estimating cycle times in the UK and USA (in both these countries MTM is very popular – Knott, 1988). Work Factor (Konz, 1983) was used to estimate performance times of German workers (Overly, 1985; Rameirez, 1985). Performance times obtained by work factor were reduced by 10% to obtain performance times of Japanese workers (Rameirez, 1985). Table 4.8 shows the manual and robotic assembly cycle times for both DFA products in all four countries.

A PUMA assembly robot was used for assembling both products. Given the size of the various components and the equipment, workstation

TABLE 4.8 *Estimated cycle times for flexible automated and manual assembly of compressor cylinder head valve assembly and floppy disk drive assembly*

DFA product	Performance time (minute)	
	Robot	Human
1. Compressor head valve subassembly		
FRG	0.2589	0.1960**
Japan	0.2589	0.1715***
UK	0.2589	0.2450*
USA	0.2589	0.2450*
2. Floppy disk drive assembly		
FRG	0.4515	0.3290**
Japan	0.4515	0.2891***
UK	0.4515	0.4113*
USA	0.4515	0.4113*

* Based on MTM.
** Based on work-factor.
*** Work-factor time estimate was reduced by 10%.

TABLE 4.9 *Summary of wage and interest rates in Japan, UK, USA, and FRG*

Country	Interest rate (%)	Hourly wage rate (US Dollars)#
FRG	6.500	9.13
Japan	3.375	14.27*
UK	13.000	6.78
USA	12.000	10.00

Note: 1 US dollar = 128.6 Japanese Yen = 1.86 West German mark = 0.5726 UK pound.
* Computed from monthly wages ($2379.14) and assuming 2000 hours per year.
\# Since the number of work hours per week and the number of work days per year varied from country to country, but not very widely, to simplify comparison it was assumed that total number of work hours per year was 2000.

layout for assembly was configured (for example, see Fig. 4.7). From this, distances moved by the robot and human arms were determined. These distances and the type of movements were used to estimate cycle times.

From the assembly cycle times and production volumes, annual costs for the two assembly methods were estimated. The hourly earnings of workers were estimated from ILO's *Year Book of Labour Statistics*. Table

TABLE 4.10 *Summary of annual assembly costs (US dollars) for compressor cylinder head valve assembly and floppy disk drive assembly*

DFA product	Robotic assembly	Manual assembly
1. Compressor Head Valve Assembly		
FRG	24 507	19 671
Japan	21 891	29 892
UK	31 802	18 494
USA	29 895	26 522
2. Floppy disk drive assembly		
FRG	33 047	21 821
Japan	29 506	29 779
UK	43 201	20 014
USA	40 568	28 854

4.9 gives the summary of worker hourly earnings and interest rates in each of the four countries (interest rates were taken from the *Wall Street Journal*). The number of work hours per week was also determined from ILO's *Year Book of Labour Statistics*. It should be noted that even though the actual number of hours worked per week differed from country to country (on an average 40.4 in the then FRG to 41.6 in the UK), for convenience's sake work weeks in all four countries were considered to have 40 hours. For the same reason, the number of work days per year was also considered to be the same in all countries (in reality the number of actual work days in each country differs slightly due to different national holidays, etc.).

From these data, the annual assembly costs for the two DFA products were determined. Table 4.10 summarizes the assembly costs for both robotic and manual assembly methods in each of the four countries.

As shown in Table 4.10, the manual assembly option is more economical in the then FRG, UK and USA, while the flexible automation assembly option is cheaper in Japan. The difference in costs between the manual and flexible automated options was relatively small in the FRG as compared to such cost difference in either the UK or USA. When the initial investment increased, for example in the case of floppy disk drive assembly, the flexible automated assembly option was still cheaper in Japan. The increased investment, however, resulted in greater economic advantage for manual assembly in the FRG, UK, and USA. This was in spite of the fact that Japanese workers took less time to complete the assembly task.

Also, as shown in Table 4.9, interest rates and hourly wage rates in all four countries are substantially different. Low interest rates in Japan lead to lower investment costs. Higher hourly rates also favour automation. Relatively smaller hourly wages and higher interest rates in western countries tend to favour manual assembly. The FRG, which has the lowest interest rate among the three western countries considered here and also has relatively lower hourly wages compared to Japan and the USA, could opt for either the manual or flexible automated assembly option especially if the initial investment is not very high. The cost analysis clearly shows that if one country chooses a particular method of work and finds it economically beneficial, it does not mean the same method would also be economical in other countries. The impact of local values of production and market factors could be very profound and must be taken into consideration prior to choosing an assembly method.

4.10 Final remarks

The purpose of this chapter was to show that production and market factors have profound effect on the choice of assembly method. In particular, interest rate, wage rate, production volume, equipment reliability, and performance times are essential considerations in computing assembly costs.

While there is no doubt that designing products for assembly is cost effective, the cost analyses conducted in this work cast considerable doubt on the belief that the flexible automated assembly method will always lead to increases in productivity. As a matter of fact, in many instances, manual assembly of DFA products was found to be economically more attractive. The disadvantages for the flexible automated assembly method are primarily created by the slow performance of robots, their low reliability, high interest rates, and declining wages. In general, low interest rates and high labour costs tend to favour the flexible automated assembly method. High interest rates, low wages, low equipment reliability, and the ability to learn on the job favour the manual assembly method.

It is clear that a sound economic analysis is a prerequisite for choosing an assembly method. There is no guarantee that advanced technologies, such as that considered here, will always enhance productivity. Moreover, the choice of method is seriously influenced by the local value of production and market factors. Thus, what may be economical in Japan may not necessarily be so in the United States. Realization of this alone is a major accomplishment on the part of those considering alternative methods of work.

4.11 References

Albo, R.T. (1983) Robotic assembly of floppy disk drives. *Proceedings of the 13th International Symposium on Industrial Robots and Robots 7*, Chicago, Illinois, **1**, April, 5.1–11.

Bolles, R. and Paul, R. (1979) *The Use of Sensory Feedback in a Programmable Assembly System*, Memo AIM-220, Stanford Artificial Intelligence Laboratory, California.

Boothroyd, G. (1988) Making it simple – design for assembly. *Mechanical Engineering*, February, 28–31.

Boothroyd, G. and Dewhurst, P. (1982) *Design for Assembly Handbook*, University of Massachusetts.

Darlow, M.S., Gabriele, G.A. and Steiner, M.W. (1987) A design for assembly case study: redesign of a coffeemaker. *Proceedings of the IXth International Conference on Production Research* (ed. Anil Mital), University of Cincinnati, Ohio, 2514–21.

Genaidy, A.M., Duggal, J.S. and Mital, A. (1990) A comparison of robot and human performances for simple assembly tasks. *International Journal of Industrial Ergonomics*, **5**, 73–81.

Knott, K. (1988) Personal communication, Pennsylvania State University, State College.

Konz, S.A. (1983) *Work Design: Industrial Ergonomics*, Grid Publishing, Columbus, Ohio.

Maynard, H.B., Stegemerten, G.J. and Schwab, J.L. (1948) *Methods-Time Measurement*, McGraw-Hill, New York, 12.

Mital, A. (1989) Manual versus flexible assembly: a cross comparison of performance and costs in four different countries. Extended summaries of papers presented to the Xth International Conference on Production Research, Taylor & Francis, 358–9.

Mital, A. (1990) *A Comparison of Robot (HT-3) and Human Performances for Palletizing and Stacking Tasks*, Technical Report, Ergonomics Research Laboratory, University of Cincinnati, Ohio.

Mital, A. and Genaidy, A.M. (1988) 'Automation, robotization in particular, is always economically desirable' – fact or fiction? *Ergonomics of Hybrid Automated Systems I* (ed. W. Karwowski, H.R. Parsaei and M.R. Wilhelm), Elsevier, 743–50.

Mital, A. and Mahajan, A. (1989) Impact of production volume and wage and interest rates on economic decision-making: the case of automated assembly. *Proceedings of the Conference of Society for Integrated Manufacturing*, 558–63.

Mital, A., Mahajan, A. and Brown, M.L. (1988) A comparison of manual and automated assembly methods. *Proceedings of the IIE Integrated Systems Conference*, 206–11.

Overly, D. (1985) Personal communication, Ford Motor Company, Dearborn, Michigan.

Paul, R.P. and Nof, S.Y. (1979) Work methods measurement – a comparison between robot and human task performance. *International Journal of Production Research*, 17, 277–303.

Rameirez, W. (1985) Personal communication, Ford Motor Company, Batavia, Ohio.

Redford, A. and Lo, E. (1986) *Robots in Assembly*, Open University Press, Milton Keynes, 123 and 130.

Romeo, G. and Camera, A. (1980) The DEA Pragma assembling system – a couple of applications in compressor subassemblies. *Proceedings of the 10th*

International Symposium on Industrial Robots & 5th International Conference on Industrial Robot Technology, 491–500.

Soni, A.H. (1988) Personal communication, University of Cincinnati, March.

Wygant, R.M. (1986) Robots v. humans in performing industrial tasks: a comparison of capabilities and limitations, Unpublished PhD dissertation, University of Houston, Texas.

5 Justification of Cellular Manufacturing Systems

Ismail N. Zahran, Adel S. Elmaghraby and
Mohamed Adel Shalabi

5.1 Introduction

Cellular manufacturing is the major implementation of a broad production philosophy called group technology (GT). The concept of GT is based on grouping similar work, and utilizing the property of similarity in reducing the efforts associated with the processes of part design, planning and manufacturing (Burbidge, 1979). In part design, GT is concerned with discounting the unnecessary variety by eliminating the redundancies in part designs, and diminishing the need for new designs through the utilization of previously developed ones. This is achieved by forming families of similar designs, and assigning a code to each of these families according to a coding scheme to express its features. As a new part is proposed, its geometry is analysed and a code is assigned to it on the basis of the coding scheme. The available library of designs is searched for a design conforming to the features expressed by the new part code instead of developing a new design (Groover and Zimmers, 1984). The same procedure is performed in the area of part process planning, with the exception of exchanging part designs with part process plans (Hyer and Wemmerlov, 1985).

The layout of multi-product batch manufacturing systems is traditionally designed on a functional basis, with the machine park divided into a set of work centres, where a work centre is a set of machines functioning similarly. In a typical functional system, batches of various parts flow between different work centres for the processing of their operations. In systems producing a large variety of parts, each having a complex processing route, the expected consequences of the functional system organization are:

1. Long batch travelling time;
2. Long batch queuing time;
3. High work-in-process (WIP) levels;

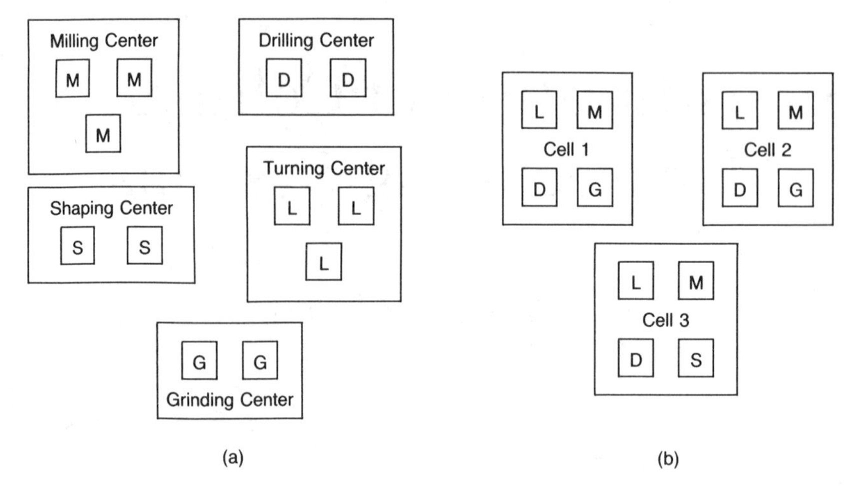

Fig. 5.1 *A sample manufacturing system.*

4. Complexity in production planning and control;
5. Unidentifiable responsibilities for scrap work;
6. Higher stocks to ensure against the overdue dates; and
7. Low labour satisfaction.

Cellular layout was introduced to replace the functional organization of multi-product batch manufacturing systems. In a typical cellular system, parts are classified into families with each family consisting of parts requiring approximately the same set of processing operations. In turn, machines are grouped in cells in such a way that each cell involves the set of machines required for the production of a family of parts. The machine park is reorganized physically on this basis. This is equivalent to partitioning the large functionally organized system into a set of sub-systems, where more production control, and effective regulation of part flows can be achieved. The difference between functional and cellular layouts is shown in Fig. 5.1, using a sample manufacturing case. This case will be used later to demonstrate the simulation methodology of predicting the benefits of cellular systems.

Two complex problems are usually encountered when it is intended to implement cellular manufacturing. The first is the identification of the machine cells and part families, and the second is the prediction of the savings or the improvements in the work environment to be achieved by the implementation. Many techniques are now available for the identification of families of similar parts and the machine cell associated with each family. These techniques depend on analysing the part processing routes for such identification. Some of the most cited techniques are as follows.

5.1.1 Flow analysis techniques

For these techniques, data is represented in the form of either a machine-component matrix, or a part route list. Either the matrix is rearranged or the list is scrutinized for the identification of part families and machine cells. Examples of these are:

1. Production flow analysis (Burbidge, 1979);
2. Component flow analysis (El-Essawy and Torrance, 1972); and
3. Systematic flow analysis (Zahran, 1985).

5.1.2 Cluster analysis techniques

In these methods, the grouping of machines into cells is based on a similarity measure computed from the part processing routes. On the basis of the values of these similarities, machines are clustered together into cells. Examples of these techniques are:

1. Single linkage cluster analysis (McAuley, 1972);
2. Average linkage cluster analysis (Seifoddini and Wolfe, 1987; Seifoddini, 1988);
3. Syntactic pattern recognition (Wu *et al.*, 1986);
4. Graph theoretic method (Rajagopalan and Batra, 1975); and
5. Weighed similarity measure method (Mosier and Taube, 1985).

5.1.3 Machine-component analysis methods

This is a set of the diverse heuristic solutions to the problem of identifying part families and machine cells. Examples of these heuristics are:

1. Cluster identification method (Kusiak and Chow, 1987; Kusiak, 1988);
2. Rank order clustering algorithm (King, 1980; King and Nakornchi, 1982);
3. Modified rank order clustering algorithm (Chandrasekharan and Rajagopalan, 1986);
4. Hospitality-flexibility method (Purcheck, 1985); and
5. Load-based technique (Shalaby and Zahran, 1990).

Few attempts were made to address the benefits to be achieved from the cellular manufacturing implementation. Willey and Dale (1978) and Dale (1980) introduced a method for measuring the applicability of cellular manufacturing for a given system. Their approach is to collect data from a number of companies before and after the implementation of cellular manufacturing. The data is associated with a large number of performance measures (42 different measures) such as the throughput time, proportion of orders delivered on time, and setting time per batch. They presented these data in graphs to illustrate the relationship between the values of these measures before the implementation and the percentage changes in the same measures after the implementation. Using these graphs, the amount of expected improvement is estimated. Boucher and Muckstadt (1985) built a cost estimation model to evaluate the conversion from a functional layout into a cellular one. They calculated savings based mainly on the assumption of reduced setup time, and negligible queuing and moving times in the cellular layout.

5.2 Benefits of cellular systems

Many benefits are sought by implementing a cellular system. These benefits may be operational, such as a reduction in time, or economical, such as a reduction in cost, or social, such as improvement in the work environment. The benefits of cellular systems could be attributed to the set of unique features of such systems. These features are as follows:

1. The dedication of a machine cell for each part family;
2. The similarity of parts contained in each part family;
3. The physical proximity of the machines in a cell;
4. The relatively smaller size of each one of the machine cells, compared to the size of the machine park in the equivalent functional system; and
5. The relatively smaller size of each one of the part families, compared to the size of the part mix in the equivalent functional system.

The expected benefits of cellular systems compared with the functional systems are mainly as follows:

1. Reduced part travelling time;
2. Reduced part setting time;
3. Reduced part queuing time;
4. Reduced batch total throughput time;
5. Reduced work in process;
6. Reduced inventory levels;
7. Simplified production control;
8. Reduced handling costs of tools;
9. Increased possibilities of job enrichment and job enlargement;
10. Increased labour satisfaction; and
11. Improved due date achievements.

In the following sections, each of these benefits will be discussed separately for the sake of illustrating the influence of implementing cellular systems.

5.2.1 The effect of cellular layout on total travelling time

Due to the physical proximity of machines in a cell, the total distances to be travelled by the parts assigned to this cell will be very short if compared with the distances travelled by the same parts in the functional system. Two pieces of information are required in order to compute the distance travelled by a part in either the original functional system, or the proposed cellular system. These are 1. the sequence of machines required for part processing, and 2. the factory layout.

As a direct result of reducing the total distance travelled by a batch, the total batch travelling time will be reduced in the cellular system.

5.2.2 The effect of cellular layout on the setting time

The time required to set up the machines is expected to be less in the cellular system than that in the corresponding functional one. The similarity of the parts consecutively loaded into the machines assigned to a cell will result in less changes in machine setups. The cellular layout must be accompanied by a scheduling system which orders similar batches consecutively.

Implementing the concept of composite component can also minimize the setup times. This concept implies the setting of the whole cell to the complete production of a hypothetical component whose processing route represents all the processing needs of the whole part family assigned to the cell.

5.2.3 The effect of cellular layout on batch queuing times

The queuing time of a batch of a certain part is the sum of times spent by the batch waiting in front of the machine centres to perform its processing operations, or waiting for a transporter to move it to the next processing facility. When arriving at the processing facility, a batch queuing time is simply the sum of setting and machining times required by its predecessors in the queue plus the unavoidable delays such as machine maintenance time. In cellular manufacturing systems, the setting times will be lowered, and there is an opportunity for effective scheduling of batches on machines involved in a cell due to the smaller number of machines in the cell and the relatively small number of different parts involved in the part family.

5.2.4 The effect of cellular layout on the total throughput time

A batch throughput time is the total time spent by the batch in the manufacturing system from when it is received as raw materials or semifinished items until it is delivered in its completely processed form. The throughput time of a batch consists of four elements: 1. travelling time, 2. queuing time, 3. setting time, and 4. machining time. The first three of these elements can be reduced in the cellular system. The sum of these three elements represents a considerable proportion of the whole part throughput time. Therefore, by reducing the travelling, queuing and setting times, the total throughput time will be reduced.

5.2.5 The effect of cellular layout on WIP levels and costs

WIP cost is the cost of holding the work in the shop floor. This cost for a certain part is almost a function in the following three parameters:

1. The value of the raw material;
2. The value added to the part after each operation; and
3. The total throughput time.

Work-in-progress cost is equivalent to the revenue that can be expected if the value of the parts existing in the shop were used for other purposes. Work-in-progress cost can be computed using three different approximation methods. The first is to measure WIP as the number of batches existing on the shop floor. The second method could be called the time-weighted incremental value method. It considers the value added to the part after the performance of each one of its processing operations. Figure 5.2(a) explains how the value of WIP for a certain part 1 is incremented after each operation according to this method. From this figure, the average WIP cost for a unit of part 1 can be calculated as:

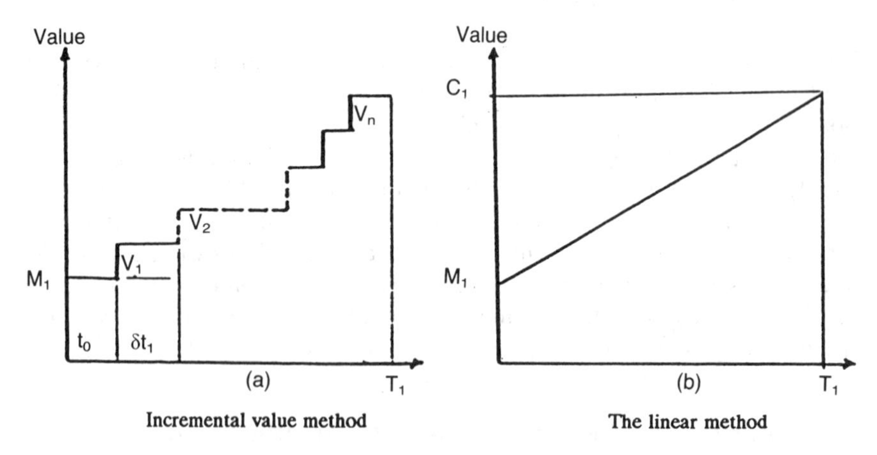

Fig. 5.2 *Representation of change in WIP: (a) incremental value method, and (b) the linear method.*

$$W_1 = \left(i \left\{ (M_1 t_0) + \sum_{j=1}^{n-1} \left[\left(M_1 + \sum_{x=1}^{j} V_j \right) \cdot \delta t_j \right] \right\} \right) \Big/ T_1$$

where: i = percentage opportunity cost; n = number of operations; M_1 = raw material cost; t_0 = time spent by the raw material in the system before starting the first operation; V_j = value added to the part after completing operation j; δt_j = the time span between completing process j and completing process $j + 1$; T_1 = total throughput time; and W_1 = WIP cost due to a unit.

In this method, it is assumed that the value of the part increases with the completion of the operation. The progress in the part value during the performance of the operation is ignored because the unit processing time is negligible if compared with the total throughput time.

The third method for calculating the average WIP is to consider that the increase in part value from its raw material value to its value as a finished item is of a linear form. This is illustrated in Fig. 5.2(b). Therefore, the average WIP cost for part 1 is given by:

$$W_1 = i[(M_1 + C_1)/2]$$

where C_1 is the finished part value.

Regardless of the method used for computing WIP, cellular manufacturing reduces the length of part stay in the shop by reducing the travelling, setting and queuing times. This helps in rushing the parts out of the shop, and keeping the level of WIP low.

5.2.6 The effect of cellular layout on inventory costs and levels

The majority of analysts restrict the computation of the optimum lot size into two cost items. These are the setting cost and the carrying cost. The famous EOQ formula is usually computed on the basis of a compromise between these two cost elements. In addition to these two elements, other elements must be considered. For instance, the cost of moving the lot inside the shop must be taken into account. Parts are moved usually from one operation to the next in batches, not as individual units. So, the cost of handling, as well as the setting cost, must be considered as one of the batch costs. The WIP cost must be also considered in the lot size computation since the WIP cost is a function of the lot size, the unit value and the total throughput time. The average annual cost related to a cycle stock of a part is:

$$\text{Total cost} = \text{Setup cost} + \text{Holding cost} + \text{Moving cost} + \text{WIP cost.}$$

Thus, for part 1:

$$C_1 = A_1 \frac{D_1}{Q_1} + iC_1 \frac{Q_1}{2} + H_1 \frac{D_1}{Q_1} + W_1 Q_1$$

where: A_1 = machine setting cost; D_1 = annual demand; Q_1 = batch size; i = percentage carrying cost; C_1 = unit value; H_1 = batch total handling cost; and W_1 = WIP cost per unit.

Differentiating with respect to the lot sizes Q and equating with zero to get the optimum value of Q_1 that minimizes the total cost of a batch, we obtain:

$$Q_1 = \sqrt{[2(A_1 + H_1)D_1/(iC_1 + 2W_1)]}$$

Three of the inputs to this equation could be modified in the cellular system. These are the setting cost, handling cost and WIP cost. This offers a great opportunity to produce low volume batches at reasonable costs to satisfy the growing customer needs for item variety. At the same time, the reorder point for a part is computed as the demand of its throughput time. Considering that there are 250 working days per year and the demand rate is constant, the reorder point could be computed for a part 1 as follows:

$$O_1 = (T_1 D_1)/250$$

where T_1 is the the throughput time of part 1; and D_1 is the the annual demand of part 1.

Therefore, a definite change in the reorder point will occur as a result of reducing the throughput time. The reduction in reorder point will reduce the carrying cost of units held as inventory, and the cost of space dedicated for inventory.

5.2.7 The effect of cellular layout on job due date satisfaction

It is expected that the satisfaction of due dates will be better in a cellular system since the batch throughput time is expected to be shorter. The improvement can take the form of either the delivery of batches being ahead of or at the due dates, or being beyond the due dates but within an overdue period shorter than that normally achieved in the functional system. The improvement in job delivery will result in less penalties for late delivery, and better reputation for the company in the market.

5.2.8 The effect of cellular layout on the scope of production control

In cellular systems the scope of production control is to be reduced to a set of relatively small machine cells with each cell assigned the responsibility of producing a limited number of parts. The reduction in the scope of production control is expected effectively to assist in simplifying job scheduling and sequencing processes. Also, with the limited entry of parts

TABLE 5.1 *Reported benefits from cellular manufacturing*

Type of benefit	Percentage improvement		
	Mean	Min.	Max.
Reduction in throughput time	45.6	5.0	90.0
Reduction in WIP inventory	41.4	8.0	80.0
Reduction in materials handling	39.3	10.0	83.0
Improvement of labour satisfaction	34.4	15.0	50.0
Reduction in number of fixtures	33.1	10.0	85.0
Reduction in set-up time	32.0	2.0	95.0
Reduction in space needs	31.0	1.0	85.0
Improvement of part quality	29.6	5.0	90.0
Reduction in finished inventory	29.2	10.0	75.0
Reduction in labour cost	26.2	5.0	75.0
Increase in equipment utilization	23.3	10.0	40.0
Reduction in pieces of equipment required to manufacture cell parts	19.5	1.0	50.0

into the machine cells, the patterns of job flow are expected to be simplified.

5.2.9 The effect of cellular layout on the job satisfaction

The most significant social benefit of cellular manufacturing is the higher job satisfaction due to the employee involvement in the production of complete items instead of the performance of separate operations on incoming batches as in the functional system.

5.3 Reported implementations of cellular systems

Some of the benefits obtained from several implementations of cellular manufacturing have been discussed in the literature. A recent study by Wemmerlov and Hyer (1989) surveyed the benefits achieved from cellular manufacturing in 32 US firms. The results were very impressive. The average percentage improvements are shown in Table 5.1, based on Wemmerlov and Hyer (1989).

Dale (1984) studied the change in setup and throughput times in 35 companies implementing cellular manufacturing principles. He concluded that the average reductions were 17% in setup time and 55% in throughput time.

In another cellular manufacturing implementation studied by Nagarkar (1979), the throughput time declined by 75%.

5.4 Classification of justification methods

According to Meredith and Suresh (1986), the changes in manufacturing systems that might need justification, could be classified into the following three categories:

1. *Stand-alone*. This is concerned with the addition of a new machine or a robot to replace a worn-out or an obsolete piece of equipment.
2. *Linking*. This addresses the combination of several stand-alone pieces of equipment to form a cell.
3. *Integration*. This includes linking and computerizing the processes of design, planning, handling, machining, and the support systems.

Corresponding to this classification scheme, Meredith and Suresh also suggested that the problem of justifying a technological or organizational change can be approached in one of the following two methods:

1. *Economic*. Where the sole concern is the cost savings to be achieved from the change. Purely economic techniques are suitable for the stand-alone changes. This category includes the evaluation of one or more of the following:

   (a) Payback period,
   (b) Internal rate of return,
   (c) Rate of return, and
   (d) Net present worth.

2. *Analytic*. These techniques consider multiple factors that are not necessarily monetary when justifying a technological change. Typical factors that could be considered are: savings in time, simplification in planning and control procedures, social considerations, and modernization of the production processes. This category includes the following methods:

   (a) Value analysis,
   (b) Portfolio analysis (non-numeric methods, scoring models, mathematical programming methods), and
   (c) Risk analysis.

5.5 Justification of cellular systems

According to our discussion of the benefits of cellular systems, the majority of these benefits are not necessarily monetary benefits. These benefits are related to: 1. savings in the setting, handling, and queuing times; 2. simplifying the processes of production control; and 3. changing the work environment in a way that satisfies the labour force and makes it more involved in the process of production. Managers are usually interested in the benefits expressed as savings in unit costs. As indicated by Parsaei and Wilhelm (1989), the use of tactical techniques, purely economic, evaluation methods to justify the proposed technological changes, will result in the rejection of strategically vital proposals. Methods that are concerned with multiple measures that are not necessarily monetary must be used when justifying new technologies and new system organization methods, in particular if the benefits sought are not directly monetary ones.

The performance of the system after implementing the improvement to be justified must be predicted prior to evaluating the benefits versus the costs. The only ways available for this prediction are the actual implementation and computer simulation. The results to be achieved from experimenting with an actual implementation may be more precise, but it usually needs lots of money and long time to give results. Also, it may be impossible to implement all of the recommended improvements together. In contrast, computer simulation can give results with a certain confidence level without actual change in the system configuration. With computer simulation, the performance of the system can also be inspected over a long period of time, perhaps years, and the system inputs could be also changed by the user to check the influence of each factor on the system's performance.

Computer simulation is an ideal solution, and perhaps the only one, for the problem of predicting the benefits of cellular systems. The other solution, i.e. the actual implementation, is practically impossible to apply due to the cost requirements. On the other hand, the benefits of cellular systems need a long time to become effective.

The results of simulating a proposed cellular system may be compared with measurements from the existing functional one, or with results of simulating it, too. It may be more economical to simulate the existing functional system instead of spending months in the collection of performance measurements.

The results of few simulation experiments were published in the recent years for cellular systems. As classified by Banerjee and Flynn (1987), these experiments either deal with comparing cellular systems versus functional systems (Ang and Willey, 1984; Flynn and Jacobs, 1986), or the development and evaluation of scheduling and maintenance policies to be applied in the new cellular system (Banerjee and Flynn, 1987; Mosier *et al.*, 1984).

5.6 Guidelines for simulating cellular systems

The most challenging stage in simulating any real-life system is the model-building stage, in particular when the system has many details. Production systems usually have many elements that must be described by the model.

These include machines, handling systems, labour power, and the dynamic entities or parts to be processed. Each of these elements is to be described by a set of attributes. For example, a machine must be described by its type of process, its location inside the shop, and probably its failure rate. The simulation model must also be provided with the rules applied in managing the real-life system. In a production system, these rules are almost always called job-sequencing or prioritizing rules. The data required to simulate the performance of the cellular system include the following:

1. The composition of each cell by machines;
2. The layout of proposed cellular system;
3. The processing routes of the parts;
4. Batch sizes and rates of batch arrivals to the system for various parts;
5. The setting and machining times on various machine types; and
6. The rules of job sequencing.

It is recommended to simulate the whole cellular system by one model, instead of simulating each cell separately. This will simplify the following:

1. Collecting overall statistics for the performance measures. Obtaining overall statistics is usually more preferred than obtaining statistics for each one of the individual cells.
2. Overcoming the problem of the exceptional parts. Exceptional parts are those which cannot be fitted into any one of the proposed cells. Two cells or more must cooperate in the production of each of these parts.

The layout could be described by the coordinates of the location of each machine in the shop floor relative to a chosen origin. These coordinates could later be used to compute the distance to be travelled by a part from one machine to another. Processing routes can be represented by the real routes with their frequency of occurrence as derived from the real-life system, or by assigning a probability to each machine as being the one to perform the next operation. Setting and machining times on each machine type could be represented by specifying a probability distribution for each of them.

The cells may be either flexible, in the sense that a cell can manufacture any part arriving at the system if it is able to offer all of its processing requirements, or inflexible in the sense that it is dedicated to a prespecified family of parts. Even in the flexible systems, some rules will exist for allocating the parts into cells, such as allocating the part to the cell with the least load. So, in simulating any of the flexible or dedicated cellular systems, the model must describe the rules of allocating parts to cells. Also, the rules of sequencing the jobs assigned to each cell must be described by the model.

TABLE 5.2 *The Machine data*

Machine symbol	Machine type	Number available	Mean setting time (h)	Mean machining time (h)
L	Lathe	3	3.5	0.45
M	Mill	3	3.2	0.55
D	Drill	2	2.0	0.22
S	Shaper	2	2.8	0.40
G	Grinder	2	0.8	0.15

5.7 A sample case

To demonstrate the use and validity of all of these concepts, a hypothetical case was designed and simulated. In the following sections, the case and the simulation model will be described, and the results of running the model will be presented and discussed.

5.7.1 Case description

This case involves five machine types whose layout is assumed originally to be designed on a functional basis, and a cellular layout is suggested for them. The functional layout and the corresponding cellular layout for the sample case are shown in Fig. 5.1. In the functional layout, these machines are grouped in five machine centres on the basis of the machine function. In the cellular layout, these machines are grouped into three cells. Machine data are given in Table 5.2. The processing routes were recognized by the simulation model as a set of probability distributions. The first one describes the probability associated with each machine to be the first machine in the part processing route. Then a distribution was built for each machine type to show its possible subsequent machines in part processing routes, and the probability associated with each one. This method assists in checking also the flexibility of the production systems, more than does the conventional method of restricting the analysis only to the set of used routes. These distributions are shown in Table 5.3.

Parts arrive at the system in batches with interarrival times uniformly distributed between four and eight hours. Batch sizes for the arriving parts are Poisson distributed with the mean of 30 units. When a batch arrives at the shop, a random number of processing operations is generated from a uniform discrete distribution with an upper limit of 10 operations and a lower limit of one operation. The first operation is generated based on the probability of it being on the first machine. Then, if this is part of multiple operations, the second operation is generated from the distribu-

TABLE 5.3 *The probability of part routing*

Machine	Probability of being the first machine	Probability of being next machine L	M	D	S	G
L	0.30	0.27	0.00	0.43	0.00	0.30
M	0.20	0.18	0.20	0.22	0.00	0.40
D	0.18	0.20	0.30	0.20	0.30	0.00
S	0.15	0.00	0.00	0.30	0.30	0.40
G	0.17	0.10	0.20	0.28	0.22	0.20

tion of the followers associated with the first machine, and so on until the whole part processing route is generated for the arriving part.

The batch requests each machine involved in its route. If the machine is not available, the batch joins the queue in front of the machine until it becomes available. As soon as the batch seizes the machine, a machine-setting time and unit machining time are generated. The machine is first set up, and the parts are then processed unit by unit until the batch is completed. If more operations are necessary for the batch, it is transferred to the required machine with a travelling speed of 1 km per hour, and the cycle is repeated until the batch completes its whole set of operations. When a machine completes a batch it looks for another batch, requesting it. If more than one batch are requesting the same machine, the one which arrived first seizes it.

In the cellular system, the batch has to look for a cell that can satisfy all of its processing needs. If more than one cell can offer the processing needs of the batch, the batch is assigned to one of them randomly. If the batch cannot be assigned to a cell for a complete processing, it travels from one cell to another to complete its operations. If more than one cell can satisfy the imminent operation of such an exceptional batch, the one with the shortest queue in front of the required machine is selected.

No job preemption is allowed. As a batch seizes a machine, it does not relinquish it until all units are processed. Also, batch splitting and operation overlapping are not permitted. Machine failures are not considered to make the results completely attributed to the differences in the layouts.

5.7.2 Results and discussion

Two simulation models were built, one for the functional system and the second for the cellular one. The SIMSCRIPT II.5 simulation programming language was used in building both of the two simulation models. SIMSCRIPT II.5 was selected because of its modularity, and its self illustrative programming commands (Russell, 1983).

TABLE 5.4 *Simulation results for alternative layouts*

Measure	Functional layout	Cellular layout	Percentage improvement
Average travelling time	0.1274	0.0891	30.06
Average setting time	12.9114	4.3197	66.54
Average waiting time	105.8983	9.2336	91.28
Average throughput time	173.3687	68.5594	60.45
Average number of batches waiting	413.0000	36.6420	91.12
Machine utilization	0.7906	0.8122	2.73
Average WIP	423.8230	46.3020	89.08
Average due date satisfaction	0.7090	0.9980	40.76

The same stream of batches was generated for both of the two layouts. The model was run for 2000 hours. This is equivalent to a working year with 250 working days and eight working hours per day. Average statistics were collected for the following performance measures:

1. Batch travelling time;
2. Setting time;
3. Waiting time;
4. Throughput time;
5. Number of batches waiting;
6. Machine utilization;
7. Work in process; and
8. Due date satisfaction.

The results are given in Table 5.4. Batch travelling time was measured as the time consumed by the batch in travelling between machine centres in the functional system, and the time consumed in travelling between cells in the cellular system. The moves within a cell were ignored because of the proximity of machines in each cell. Setting time is the sum of times consumed in preparing the whole series of machines involved in the processing route of the batch. Waiting time is the sum of times spent by a batch waiting for the machines to become idle. Throughput time for a batch is the sum of travelling time, waiting time, machine setting time and processing time. The number of batches waiting is the number of batches in the queue in front of the machine centre in the functional layout, and the number of batches in the queue in front of the individual machine in the cellular layout. Machine utilization is the percentage time the machine spent in setting or batch processing. Work in progress was measured as the number of different batches existing in the system, either processed by a machine, or waiting in a queue. A due date was assigned to each batch as twice the total time required for setting the machines and processing the batch on the various machines involved in its route. So, the due date satisfaction was measured for both of the functional and cellular systems as the percentage of batches whose total throughput times are less than or equal to their due date.

The average travelling time was low in the cellular system because most of the batch movements were inside the cells. Only 34.4% of the parts were required to move from one cell to another for completing their processing, which is an indication of a good cellular design. Average setup time was less in the cellular system because of the similarity of parts assigned to each cell. Average waiting time was also less in cellular systems because of the availability of machines after reducing the time required for setting the machines. Average throughput time was less in the cellular system as a result of reducing the travelling, waiting and setting times.

With reducing the average batch throughput time, the number of batches waiting was reduced on average into approximately one tenth of the average number waiting in the functional system. The average WIP was also reduced approximately by the same ratio, and the due date satisfaction was improved as a direct result of reducing the average throughput time.

Machine utilization was found to be higher in the cellular system. This was mainly due to reducing the time span required to complete the batches generated in 2000 hours. While the stream of batches generated in the cellular system required a total time of 2026 hours to be completed, the same stream of batches required 2354 hours in the functional system. Machine utilization was computed as the percentage of the total run length that the machines were busy either in setup or in batch processing.

5.8 Conclusions

Cellular layout provides a viable alternative to the design of systems working in discrete batch manufacturing which are functionally designed. Appropriate justification is needed prior to the conversion of a functional system into a cellular one. The justification involves two tasks. The first is the prediction of the savings or improvements, while the second is the evaluation of these savings and improvements versus the cost.

Computer simulation provides a good tool to predict some estimates of improvement sought by a cellular design. In this paper, a sample case was presented for analysing and simulating a system. The results obtained from the simulation favour cellular design. These results are dependent on the specific case which was selected to represent a system suitable for cellular design. However, the methodology is general and will apply equally to any batch manufacturing system.

Simulation results were designed to provide needed measures for a multi-attribute decision-making process since the problem of selecting a manufacturing system configuration is one of multiple criteria.

Comparison of a cellular system and a functional system is not an absolute measure since a poorly designed system regardless of its layout cannot be tolerated. In conclusion, design techniques cannot be overlooked and selection of appropriate design techniques is as important as the selection of layout philosophy.

5.9 References

Ang, C.L. and Willey, P.C.T. (1984) A comparative study of the performance of pure and hybrid group technology systems using computer simulation techniques. *International Journal of Production Research*, **22** (2), 193–233.

Banerjee, A. and Flynn, B.B. (1987) A simulation study of some maintenance policies in group technology shop. *International Journal of Production Research*, **25** (11), 1595–1609.

Boucher, T.O. and Muckstadt, J.A. (1985) Cost estimating methods for evaluating the conversion from a functional manufacturing layout to group technology. *IIE Transactions*, **17** (3), 268–75.

Burbidge, J.L. (1979) *Group Technology In Engineering Industry*, Mechanical Engineering Publications, London.

Chandrasekharan, M.P. and Rajagopalan, R. (1986) MODROC: an extension of rank order clustering for group technology. *International Journal of Production Research*, **24** (5), 1221–33.

Dale, B.G. (1980) How to predict the benefits of group technology, *Production Engineering*, February, 51–4.

Dale, B.G. (1984) Work flow in manufacturing systems. *Engineering Management International*, **3**, 3–13.

El-Essawy, I.F.K. and Torrance, J. (1972) Component flow analysis – an effective approach to production systems design. *Production Engineering*, **19**, 165–70.

Flynn, B.B. and Jacobs, F.R. (1986) A simulation comparison of group technology with traditional job shop manufacturing. *International Journal of Production Research*, **24** (5), 1171–92.

Groover, M.P. and Zimmers, E.W. (1984) *CAD/CAM Computer Aided Design and Manufacturing*, Prentice-Hall, Englewood Cliffs, New Jersey.

Hyer, N.L. and Wemmerlov, U. (1985) Group technology oriented coding systems: structures, applications, and implementation. *Production and inventory management*, second quarter, 55–78.

King, J.R. (1980) Machine-component grouping in production flow analysis: an approach using a rank order clustering algorithm. *International Journal of Production Research*, **18** (2), 213–32.

King, J.R. and Nakornchi, V. (1982) Machine-component group formation in group technology: review and extension. *International Journal of Production Research*, **20** (2), 117–33.

Kusiak, A (1988) EXGT-S: a knowledge based system for group technology. *International Journal of Production Research*, **26** (5), 887–904.

Kusiak, A. and Chow, W.S. (1987) Efficient solving of the group technology problem. *Journal of Manufacturing Systems*, **6** (2), 117–24.

McAuley, M.E. (1972) Machine grouping for efficient production. *Production Engineering*, **19**, 53–7.

Meredith, J.R. and Suresh, N.C. (1986) Justification techniques for advanced manufacturing technologies. *Justifying a new technology* (ed. J. Meredith), Industrial Engineering and Management Press, Georgia, pp. 82–97.

Mosier, C. and Taube, L. (1985) Weighted similarity measure heuristics for the group technology clustering problem. *OMEGA*, **13** (6), 577–83.

Mosier, C.T., Elvers, D.A. and Kelly, D. (1984) Analysis of group technology scheduling heuristics. *International Journal of Production Research*, **22** (5), 857–75.

Nagarkar, C.V. (1979) Application of group technology to manufacture of sheet-metal components. *Annals of CIRP*, **28**, 407–11.

Parsaei, H.R. and Wilhelm, M.R. (1989) A justification methodology for automated manufacturing technologies. *Computers and Industrial Engineering*, **16** (3), 363–73.

Purcheck, G. (1985) Machine-component group formation: an heuristic method for flexible production cells and flexible manufacturing systems. *International Journal of Production Research*, **23** (5), 911–43.

Rajagopalan, R. and Batra, J.L. (1975) Design of cellular production systems. A graph theoretic approach. *International Journal of Production Research*, **13** (6), 567–79.

Russell, E.C. (1983) *Building Simulation Models with SIMSCRIPT II.5*, CACI, Los Angeles, California.

Seifoddini, H. (1988) Comparison between single linkage and average linkage clustering techniques in forming machine cells. *International Journal of Production Research*, **15** (1–4), 210–16.

Seifoddini, H. and Wolfe, P.M. (1987) Selection of a threshold value based on material handling cost in machine-component grouping. *International Journal of Production Research*, **19** (3), 266–70.

Shalaby, M.A. and Zahran, I.M. (1990) A load based technique for cellular manufacturing systems design. *Proceeding of the Second International Conference on Management of Technology*, Miami, Florida.

Wemmerlov, U. and Hyer, N.L. (1978) Reseach issues in cellular manufacturing. *International Journal of Production Research*, **25** (3), 413–31.

Willey, P.C.T. and Dale, B.G. (1978) A management tool for predicting the benefits to a company of group technology. *Proceeding of 19th MTDR Conference*, UMIST, September.

Wu, H.L., Venugopal, R. and Barash, M.M. (1986) Design of a cellular manufacturing system: a syntactic pattern recognition approach. *Journal of Manufacturing Systems*, **15** (2), 81–8.

Zahran, I.M. (1985) Systematic flow analysis – an engineering approach to the design of cellular manufacturing systems. *Current Advances in Mechanical Design and Production, Proceeding of the Third Cairo University Conference*, Cairo, 579–86.

6 Application of the Engineering Economic Analysis Process and Life Cycle Concepts to Advanced Production and Manufacturing Systems

James A. Bontadelli

6.1 Introduction

When a management decision is being contemplated regarding investment in an advanced production and manufacturing system project, the potential impact on the firm's competitiveness may be significant. But what if the economic analysis of the project has not been done well? Not only is the decision maker handicapped (probably without realizing it), but the longer term survival of the firm itself may be endangered.

The costs and benefits associated with any project requiring significant capital investment are important, and carefully considered by management. Thus, relevance of the economic dimension in project decision making is obvious. However, the analysis process (as well as the credibility of results) is often challenged. This has been true for advanced production and manufacturing systems. For example, the applicability of discounted cash flow (DCF) techniques to these systems has been questioned, and remains a topic of discussion (Hodder, 1986; Kaplan, 1986). Surely, such a fundamental approach and logic is not the problem. Rather, how the analysis process, including these techniques, is applied in a situation, and the available information, will determine the credibility of results. Therefore, the application of engineering economy principles and methodology including life cycle concepts, while avoiding certain problems inherent in the analysis of these systems, is essential to sound decision making in this area.

The focus of this article is application of the engineering economic analysis process, including life-cycle considerations, to the analysis of advanced systems in production and manufacturing. The basic steps in the process are discussed, and some areas of methodology where problems may occur in these applications are examined in more detail. Also, the importance of prior product and manufacturing systems planning is briefly discussed as an essential condition for accomplishing a sound economic analysis. The discussion is integrated for the reader through the technical flow of the analysis process itself, and an example application.

6.2 Prior planning

Inadequate product and manufacturing systems planning within a company will jeopardize the subsequent evaluation of projects using advanced technology. Prior planning and the information made available from it are so important that we will briefly discuss this topic first. Without it, the economic analysis of advanced systems is based on a foundation of sand.

The primary responsibility of management is to ensure that a company remains competitive in its markets at a profit. Technological innovations in both products and services and in production and manufacturing systems have become paramount to market competitiveness. Marketing and creative financing are not sufficient. Thus, planning for continuous improvement is critical to meeting customer expectations and ensuring company survival.

TABLE 6.1 *Life cycle phases and steps*

Phase	Step	Activity
I Acquisition	1	Needs assessment
		Definition of requirements
	2	Conceptual design
		Advanced development
		Prototype testing
	3	Detailed design
		Production or construction planning
		Facility and resource acquisition
II Operation	4	Production or construction
	5	Operation or customer use
		Maintenance and support
	6	Retirement and disposal

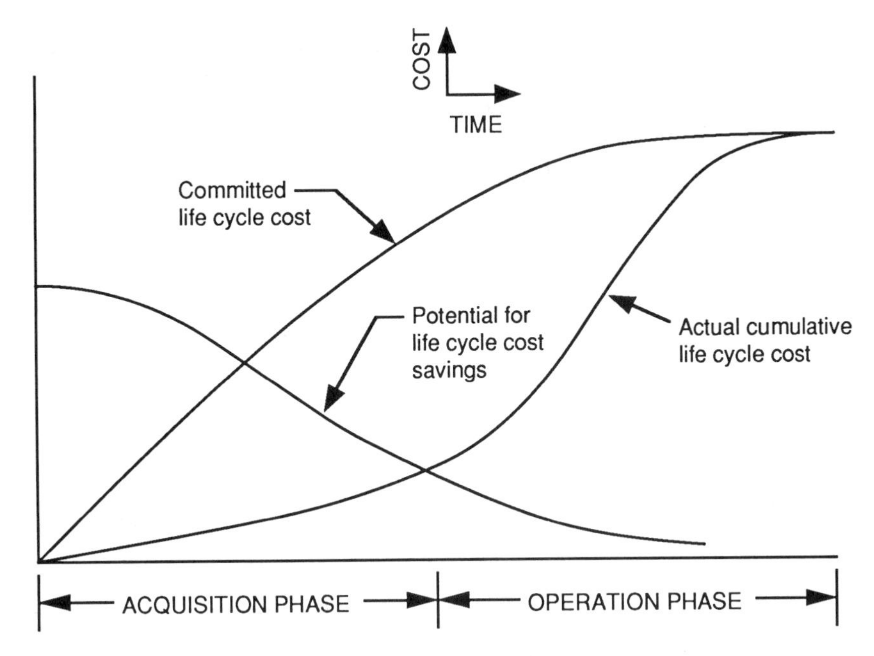

Fig. 6.1 *Relative life cycle cost profiles.*

Effective prior planning within a company is also a major factor in total production cost control. This basic point can be illustrated by looking at the phases and steps in the life cycle of a product or system, and at the related cost profiles. The general phases and steps of the life cycle are listed in Table 6.1. The relative profiles for committed life-cycle costs, actual cumulative life-cycle costs, and the *potential for achieving life-cycle cost savings* are shown in Fig. 6.1.

It is through a product and manufacturing systems planning process that company needs and requirements in these areas are defined. But, as indicated in Fig. 6.1, the first phase of the life cycle is where the greatest potential for achieving cost savings occurs. This potential for savings relates to both the improvement projects and their subsequent impact on product costs. In addition, achieving the potential benefits of advanced systems in a manufacturing environment, including reduced costs, *requires that the individual projects be integrated within an overall structure.* A company product and manufacturing system plan fulfills this role.

We will not attempt in this brief discussion to address explicitly the contents of an adequate company plan, or the process needed to develop and then update it. The objective is to focus on its importance. Simply stated, without an adequate plan, product needs from both the market and manufacturing viewpoint are not explicit. As a result, the manufacturing system requirements for competitiveness cannot be defined. Consequently, the advanced systems projects cannot be designed and integrated conceptually. And, unless this is done with an adequate plan we cannot put a solid foundation under the economic analysis of an individual project as part of a larger whole.

6.3 The analysis process

The underlying viewpoint of the discussion in this section is that when the engineering economic analysis process is applied correctly, credible and useful economic information to assist the decision making process will result. As previously indicated, any basic problems are not in the analysis techniques but in how we apply them in a particular project situation.

The description of the steps in the engineering economic analysis process may vary by author. In this discussion we will use the following seven steps to define the process (DeGarmo *et al.*, 1989):

1. Recognition and formulation of the problem (or improvement opportunity);
2. Development of the feasible alternatives;
3. Development of the net cash flow for each feasible alternative;
4. Selection of an economic criterion;
5. Analysis and comparison of the feasible alternatives;
6. Selection of the preferred alternative from an economic viewpoint; and
7. Post-evaluation of results.

It is in the first four steps of the process where most of the significant problems occur in the economic analysis of advanced system applications for production and manufacturing.

Life-cycle concepts are included in steps 1, 2 and 3. However, control of life-cycle costs, as previously indicated, begins with the improvement requirements. Additional project results to be achieved are developed in step 1. The feasible alternatives in step 2 are developed to meet these needs. Then, the estimated net cash flow developed in step 3 for each feasible alternative includes these previous project life-cycle considerations plus those required in step 3.

Before discussing the process steps and some related problems involved with new-technology projects, the role of economic analysis results in project decision making needs to be reviewed. We often use the term **justification** in reference to the reason for an economic analysis of a project. Unfortunately, this burdens the economic results with an inappropriate role in the decision process. The purpose of an economic analysis is not project justification, but to assist decision making by

TABLE 6.2 *Some potential problems*

Engineering economic analysis step	Potential problem(s)
1	Not putting enough effort into the step; rushing to steps 2 and 3
	Sound base for project economic analysis not established
2	Extent of knowledge within the company about advanced systems
	Actual state of operating experience with automation
	Achieving progressive technical compatibility (in the manufacturing system)
	Not considering a range of automation options
	Not using a structured process to select the feasible project alternatives
3	Establishing length of the analysis period
	Inconsistent use of an estimating baseline
	Overestimating revenues from continuation of current operations
	Underestimating time and costs to implement advanced technology
	Not identifying and estimating all indirect project benefits
	Assuming that reductions in product costs fully convert to profits
	Not including inflation and price changes
4	Setting the MARR, or hurdle rate, too high
5 6	Normally, there are no particular problems from an economic viewpoint; however, multi-criteria project decision more complex
7	Not accomplishing post-evaluation of projects consistently or well

providing a reliable estimate of a project's economic impact on the company. However, the economic factor is only one of several involved in a final project decision. It is important that management maintain this perspective and have credible economic analysis results available in a timely manner to assist their decision making.

The potential problems highlighted in this article related to an economic analysis of an advanced systems project are summarized for each step of the engineering economic analysis process in Table 6.2.

6.3.1 Step 1: recognition and formulation of the problem (or improvement opportunity)

What we want to achieve in this first step of the process is getting the project well defined and understood, and explicit in written form. Even though this is desirable for any capital project, it is particularly important when new technology is involved.

In Section 6.2, we discussed briefly the product and manufacturing system plan for a company. Much of the information needed in this first step, such as the improvement requirements, should have been established in the plan. This would be based on the future product mix, quality, price, lead time, and so on, needed to meet customer expectations and to remain competitive.

The additional information developed in this step includes the project goals, objectives, and other results to be achieved. Also, constraints such as project schedule and total budget need to be identified at this time. In addition, it is often necessary to do a more detailed project feasibility study in this step. Technical and implementation risk must be acceptable at this stage; otherwise, the economic results will have too much uncertainty.

A common problem in this step is simply not putting enough effort into it; We often rush into the second and third steps of the process. Consequently, a sound base is not established for the project and the economic analysis. Awareness of this potential problem by management and technical staff personnel is the first preventative measure. Then, company procedures governing capital projects which require development of an adequate project definition are an effective tool to achieve consistency in this step.

6.3.2 Step 2: development of the feasible alternatives

A project decision involves making a choice among two or more alternatives. Developing and defining the *feasible* alternatives for implementing the project are important because of the impact on the quality of the decision. A problem that often occurs in this step, similar to step 1, is not devoting adequate effort to develop the alternatives that merit detailed analysis. We want high assurance that the mutually exclusive set of feasible alternatives developed in this step will include the best alternative for project success.

In developing the feasible alternatives for an advanced production and manufacturing system project, we need to consider a range of potential options. In advanced systems, there is a tendency by both technical and management personnel to focus on the technology instead of on the product and operating requirements. This occurs when prior planning has not been done well, and when the company does not have adequate knowledge and experience with the new technology.

A general technique for accomplishing this second step in the engineering economic analysis process uses the *systems viewpoint and approach* to select the few feasible project alternatives for detailed analysis and comparison. In this approach, judging the feasibility of a potential alternative before detailed analysis is dependent on understanding its effects in the operating environment as well as the potential interactions between this project decision and others.

The explicit consideration of the effects of each potential alternative in selecting the few for detailed analysis can be summarized as follows:

1. Create all the potential alternatives for accomplishing the project that merit further consideration. The range of these may be determined partially by the company product and manufacturing systems plan and the project definition (step 1). For example, the improvement requirements established for the project may preclude further consideration of the no-change alternative, or making only minor changes to the present operation.
2. Identify the factors external to the company which are not already included in the improvement requirements that may impact the project (e.g. environmental regulations).
3. Identify the factors internal to the company which are not already included in the improvement requirements (e.g. workforce capabilities).
4. Accomplish basic calculations that will assist in determining whether an alternative can meet essential performance criteria.
5. Screen each potential project alternative against the external and internal factors and the essential performance criteria. The knowledge, experience, and judgment of the appropriate people are used to accomplish the screening process.
6. Select the small group of alternatives judged feasible at this point for detailed analysis and comparison. These should yield a preferred alternative for accomplishing the project successfully.

There are several considerations that need scrutiny when developing the feasible alternatives for an advanced technology project. Each is a potential problem area. The first is *having adequate knowledge within or available to the company about the advanced production and manufacturing technology used in the project*. This is a prerequisite. It is not possible to accomplish step 2 of the process without outside assistance when expertise within the company is limited.

Related to the first concern is *the state of actual operating experience with automation within the company*. This is an important factor. The potential problem is the actual amount of operating, maintenance, engineering and other support capabilities and experience within the company. Success with advanced technology is very dependent on the company's collective learning curve situation.

Achieving progressive technical compatibility of the improvement projects in the company's plan is critical. In addition to addressing this need in preplanning, it must be achieved project by project and thus reflected in the feasible alternatives of each. Unless this is accomplished, progress toward a computer-integrated manufacturing environment will not occur.

The feasible project alternatives selected for detailed analysis should include a range of levels of automation. A balance of several factors needs to be achieved with the preferred alternative: maintaining competitiveness; controlling project scope within achievable goals; high assurance of implementation success; and so on. We do not want to focus on maximum automation if the preferred alternative should be changes to existing operations which result in some improvements, yet meet the planned requirements.

6.3.3 Step 3: development of the net cash flow for each alternative

When steps 1 and 2 have been done adequately, this becomes the pivotal step in the engineering economic analysis process. The accuracy of the prospective cost and revenue estimates for the project developed in step 3 will determine the ultimate credibility of the economic analysis.

Before identifying some problems and errors that can occur in this step of the process, several related topics are briefly discussed. The first is the *viewpoint or perspective* used in developing the various cash flows that determine the net cash flow for each feasible alternative. The perspective used for capital investment projects is that of the company (owners of the firm), and it needs to be applied consistently in estimating all costs and revenues.

The second and third topics are the *length of the analysis period* and the *project cost and revenue structure*, respectively. Life-cycle considerations apply to both of these items. A problem in selecting the length of the analysis period for an advanced production and manufacturing system project is the basis to be used. Should the period be based on the project equipment, software, and so on, or should the products and related manufacturing requirements served by the project be the basis?

There is not a general answer to this question. It depends upon the situation. For example, if the project application is expected to provide flexible capabilities for present product and service requirements as well as future needs not specified at this time, the analysis period should be based on the project equipment, software, and so on. It would be the estimated useful life to the company of these items in their primary planned use. However, if the project is dedicated to presently planned product or service requirements, the analysis period should be based on the period of these needs. In neither situation is the physical birth-to-death concept of the life cycle necessarily used. Thus, in advanced technology projects the length of the life cycle is normally defined differently than it would be for a highway bridge or office building. In these situations, the physical birth-to-death concept is more applicable.

The project cost and revenue structure is used for identifying and categorizing all cash flows that will occur during the defined life cycle. These will be included in the economic analysis. Detailed data are developed and organized within this structure and are used to estimate the future costs and revenues. Then, these are combined to develop the net cash flows for the feasible project alternatives.

The net cash flow for an alternative represents the best estimate of future costs and revenues from the perspective being used. Therefore, the estimated changes in costs and revenues associated with an alternative have to be relative to a baseline consistently used for all feasible alternatives.

The *estimating baseline* can be defined in two different ways. The first method is the total cost and revenue approach. As the name implies, with this method the total net cash flow for each feasible alternative is estimated; even if one is the no-change alternative. The second method that can be used is the differential approach. Using this method, the net cash flow for the no-change alternative is defined as zero whether or not it is one of the feasible alternatives. Then, the net cash flow for each alternative represents the estimated differences in costs and revenues from present operations. If the project is a new situation and not relative to a current operation, then the total cost and revenue approach would be used. Since economic comparison of the feasible alternatives for a project involves only the differences in the net cash flows, either estimating baseline method will work if properly applied. It is easy inadvertently to use a different estimating baseline for one or more of the costs and revenues. This error can cause inconsistent differences between the net cash flows of the alternatives.

Some of the additional problems and errors (summarized in Table 6.2) that can occur in step 3 when analysing advanced production and manufacturing system projects are discussed briefly below.

Overestimating revenues from continuation of current operations

Competition in the market will continue to increase. Degradation in sales revenue will undoubtedly occur unless continuous improvements in products and manufacturing systems are made by a company. If the no-change alternative is being analysed, careful review of the future market-share and selling-price assumptions is required. Other competitive firms, domestic and foreign, will be improving their products and services to increase market share, and reducing their manufacturing and operating costs to decrease selling price.

Underestimating time and costs to implement advanced technology

This problem is caused by several factors. The implementation of advanced systems with increased automation reduces direct labour. However, requirements for additional engineering, maintenance and other specialized personnel increases. Thus, new-technology projects require careful analysis of their impact in all functional areas.

Overhead and support-cost estimating is a problem area. Allocation schemes based on direct labour, material, machine operating hours, and so on, are not adequate for estimating these project cash flows, and can distort significantly the estimates made. Only direct analysis of the project impacts on a functional basis, using activity-based cost-accounting concepts, will provide credible overhead and support-cost estimates (Cooper, 1988; Johnson, 1989).

Another problem is underestimating the time and effort needed to realize cost reductions from indirect factors in the manufacturing sytem. For example: achieving work-in-process (WIP) inventory reductions may require changes to input specifications and adjustments by suppliers; changing maintenance practices to achieve high production equipment reliability and availability usually requires significant changes in the way maintenance work is managed; and acquiring new skills through upgrade training and new hires takes time.

Estimating errors in these prospective cost and revenue areas bias the economic results in favour of the new-technology alternatives.

Estimating the indirect benefits often associated with advanced systems

Some of the indirect benefits that can result with successful implementation of an advanced system project are improved quality, shorter production lead times, and more flexibility in meeting customer product and service needs. How do we include these benefits in the economic analysis? The general approach is to estimate the economic impact of these factors to the extent feasible (e.g. in terms of changed sales), and then to describe explicitly the other impacts in a manner that will assist decision making. These three factors are critical to competitiveness, and improvements in them will affect sales.

There are additional indirect benefits that may occur in the production and manufacturing operation. For example, reductions often occur in both floor space requirements and WIP inventories. These type of benefits need to be included in the economic analysis. In the case of floor space, if there is another need for it, then an appropriate value can be used to account for the savings to the company. If there is not another need for it, then the reduced costs (savings) in maintenance, utilities, and so on can be used in the analysis. Reductions in work-in-process inventory result in a one-time saving to the company, plus reducing average investment in working capital with related annual savings. Similarly, other indirect benefits resulting from the project need to be identified and included in the economic analysis.

Assuming that reductions in product costs fully convert to profits

Competitors are not going to 'sit still' if another company succeeds in gaining market share. They will usually reduce profit margins to retain market share. Thus, reductions in unit product costs may not result in proportional increases in revenues since selling prices may have to be lowered to remain competitive.

Including inflation and price changes in the economic analysis

A common error is to estimate future cash flows in present dollars (i.e. real dollars with the current year being the base year) and then use a nominal interest rate (which includes an inflation component) for the time value of money. This is an easy mistake to make since we understand costs and revenues best in terms of the purchasing power of present dollars; and the minimum attractive rate of return (MARR) set by a company for capital projects is normally a nominal rate.

Consequently, if inflation and price changes are expected (the normal situation), the estimates in real dollars are numerically less than equivalent actual dollar estimates. In addition, the nominal MARR is numerically greater than the real interest rate. As a result, the real dollar estimates do not account for inflation and price changes and understate future actual dollar costs and revenues. Then, the present worth of the project's net cash flow is further understated by the nominal MARR value. The result is a bias against capital investment and, thus, new-technology projects.

How can this problem be avoided? An economic analysis strategy which works well in practice is to include inflation and price changes in the original cost and revenue estimates (i.e. develop the estimates in actual dollars), and use a nominal MARR value for the time value of money. Then, selected cash flows can be converted to real dollar equivalents to assist interpretation of results.

6.3.4 Step 4: selection of an economic criterion

The minimum attractive rate of return (MARR), or hurdle rate, is the primary economic criterion. And, a value for this discount rate that is not appropriate for the project decision will cause problems. The higher the MARR the more bias there is against new capital investment.

The penalty from the use of a MARR criterion that is too high is disproportionate in the case of advanced systems versus most capital projects within a company (Kaplan, 1986). New-technology applications in manufacturing usually provide more flexibility to produce future generations of products. Consequently, they have longer useful lives. In addition, the initial implementation of an advanced systems project usually requires an extended operating period to realize the full potential cost savings and other benefits.

The combination of a MARR criterion which is too high, equipment and other investments with useful lives resulting in a longer economic analysis period, and the delayed realization of cost reductions and other benefits, is deadly. If not handled well in an economic analysis, the impact on new-technology projects becomes a major problem.

What can we do to lessen the potential problem? The most important action is to focus on the MARR value initially. For example, common uses of a high hurdle rate are to ration capital and to compensate for risk. These are not good applications of the MARR criterion, particularly when advanced systems are involved. An economic criterion based on either the marginal or weighted cost of capital to the company is more appropriate for these projects.

6.3.5 Step 5: comparison of the feasible alternatives

When the previous four steps in the engineering economic analysis process are done well, this step usually does not present problems. Any combination of the primary analysis techniques can be used to compare the alternatives: equivalent worth methods (present worth, future worth and annual worth); internal rate of return; external rate of return; and the simple payback and the discounted payback methods to provide supplementary information. Other calculations such as average return on investment can be made if desired by company management.

6.3.6 Step 6: selection of the preferred alternative

From an economic viewpoint, the selection of the preferred alternative will be based on the results from step 5. However, the project decision by company management is more complex since economics is only one consideration. Its task is to balance the various factors related to a project and reach the best decision for the company. If prior planning has been done adequately, the information available from it will assist the decision-making process.

6.3.7 Step 7: post-evaluation of results

This step should be accomplished after some results from implementing the preferred project alternative become available. However, there is often a problem with this step: it is seldom done consistently or well in practice. Without accomplishing evaluations of project results versus original estimates, it is not possible to identify areas where errors have been made and the apparent causes. Developing this part of the company's institutional memory is essential to the improvement of cost and revenue estimating. The information will help improve the accuracy and credibility of future economic analyses, and will be particularly helpful with advanced systems projects.

6.4 Example application

It is not feasible with a brief example application to illustrate the total engineering economic analysis process and each of the potential problem areas discussed in section 6.3. However, a representative advanced manufacturing system project in summary form will be used to highlight selected topics.

6.4.1 Basic information

The Hydra Technology Company is a manufacturer of hydraulic components and systems. It is a major supplier to several producers of light- and medium-sized construction equipment as well as some farm equipment manufacturers. In a recent update of their product and manufacturing systems plan, a requirement was established to improve the production capability for cylinder type parts. These parts represent a middle-level volume in the company's product line, and involve moderate but increasing variation in customer specifications. Manufacturing lot sizes normally vary from 10 to 90 units.

In addition to reducing unit costs, the plan specifies improvements in quality, reduction in lead time, and more flexibility in meeting customer product specifications and quantities as essential to future competitiveness. A project team, which includes representatives from several departments and functions (marketing and sales, engineering design, manufacturing, plant maintenance, and so on), is being used to expand the requirement in more detail and to develop and analyse project alternatives. The project team is to develop a recommended course of action and to provide the analysis results on which the recommendation is based to company management. The summary discussion in this section is related to the economic analysis portion of the effort.

Assume in step 1 of the engineering economic analysis process (recognition and formulation of the problem) that the project team expanded in detail the requirement in the company plan, and developed additional

TABLE 6.3 *The basic information about the present operation and feasible alternatives*

Item	Present system	Changed system	FMS
Number of machine tools	23	17	6
Average order processing time	14 days	9 days	3 days
Operating employees	30	22	8
Floor space	9800 ft^2	7900 ft^2	2750 ft^2

information to support subsequent steps in the process. Also, assume in step 2 (development of the feasible alternatives) that the project team decided upon two alternatives for detailed analysis.

The first alternative involves changes to the current operation. The primary changes are in materials handling equipment, including one robot application and some other automation; in final assembly, including a new robot operation; and in milling operations, including two new computer numerical-controlled machines.

The second alternative is the development and implementation of a new flexible manufacturing system (FMS) for cylinder type parts. The design being considered is a linear flow type FMS with three machining centres which uses a cart design for transportation. Four pallets can be accommodated at the loading station, and eight pallets at the buffer station.

6.4.2 Economic analysis (steps 3, 4, and 5)

Some basic information about the present operation and the two feasible alternatives is shown in Table 6.3.

An actual dollar (A$) analysis was used by the project team to include inflation and price changes in the estimated cash flows. Annual differences with present operations in costs and revenues were first estimated in Year 0 (base year) real dollars for each alternative. Then, estimated effective annual escalation rates were used to develop the cash flows in actual dollars for each year of the analysis period.

The analysis period selected by the project team was seven years. This was based on the new production equipment (primarily automated) and related items. The requirement to produce cylinder type parts, based on the updated company plan, is considered to extend indefinitely. Thus, current capital investment to meet this need should be based on the expected useful life of the project assets in their primary use. This was estimated by the project team to be seven years for both project alternatives.

TABLE 6.4 *Basic, economic data by cost and revenue structure category*

Category	Annual escalation rate (e_j) (%)	Year(s)	Annual difference* (Year 0 $)	
			Changed system	FMS
Net sales revenue	2.0	1	350 000	350 000
		2	425 000	620 000
		3–7	500 000	910 000
Investment (first) cost*	0.0	0	−980 000	−2 480 000
Operating costs	4.6	1–7	233 800	659 000
Maintenance costs	5.8	1	−92 600	−125 000
		2	−71 500	−102 400
		3–7	−28 000	−74 200
Other costs	5.0	1	−31 100	−124 000
		2	−5 600	−82 500
		3–7	14 300	−49 000

* The investment (first) cost category entry is not an annual difference but a one-time expenditure in Year 0 (the base year). The annual differences in the other cost and revenue categories are relative to continued operation of the present system.

The basic data for each category in the cost and revenue structure are summarized in Table 6.4. To reduce the detail in this discussion, the individual cash flows (e.g. personnel, material and contract cost) were compiled for each category (e.g. maintenance costs) and used in the economic analysis at that level to develop the net cash flow for each of the two project alternatives. A positive entry in Table 6.4 for a cost category represents a savings compared to present operations, and a negative entry indicates an increased cost.

The economic analysis on an after-tax basis of the two project alternatives is shown in Table 6.5. The following comments apply to the analysis:

1. Actual dollar entries in columns 1, 3, 4 and 5 are based on the real dollar estimates and the escalation rates in Table 6.4. That is, the actual dollar estimate for alternative k in year n for cash flow j is:

   $$A\$_{k,n,j} = R\$_{k,n,j}(1 + e_j)^n$$

   For example, the A$ estimate for the FMS alternative ($k = 2$) in year 3 of the analysis period ($n = 3$) for net increased sales revenue ($j = R$) is:

   $$A\$_{2,3,R} = \$910\,000(1.02)^3 = \$965\,700$$

   The investment cost (column 2) occurs in the base year (year 0) and was not escalated. The purchasing power of actual dollars and real dollars are the same in the designated base year.

TABLE 6.5 *Economic analysis of project alternatives*[1]

	(1)	(2)	(3)	(4)	(5)	(6)	(7)	(8)	(9)	(10)	(11)	(12)
End of year	Sales revenue	Investment cost	Operating cost	Maintenance cost	Other costs	A$ BTCF	Depreciation[2]	Taxable income	Income taxes[3]	A$ ATCF	$1/(1.05)^n$	R$ ATCF
Alternative 1: Changes to present system												
0	0	−980.0	0	0	0	−980.0	0	0	0	−980.0	1.0	−980.0
1	357.0	0	244.6	−98.0	−32.7	470.9	−196.0	274.9	−107.2	363.7	0.9524	346.4
2	442.2	0	255.8	−80.0	− 6.2	611.8	−313.6	298.2	−116.3	495.5	0.9070	449.4
3	530.6	0	267.6	−33.2	16.6	781.6	−188.2	593.4	−231.4	550.2	0.8638	475.3
4	541.2	0	279.9	−35.1	17.4	803.4	−112.9	690.5	−269.3	534.1	0.8227	439.4
5	552.0	0	292.8	−37.1	18.3	826.0	−112.9	713.1	−278.1	547.9	0.7835	429.3
6	563.1	0	306.2	−39.3	19.2	849.2	− 56.4	792.8	−309.2	540.0	0.7462	402.9
7	574.3	0	320.3	−41.5	20.1	873.2	0	873.2	−340.5	532.7	0.7107	378.6
PW (MARR = 9%) = $1543.4												
Alternative 2: New FMS												
0	0	−2,480.0	0	0	0	−2,480.0	0	0	0	−2,480.0	1.0	−2,480.0
1	357.0	0	689.3	−132.3	−130.2	783.8	−496.0	287.8	−112.2	671.6	0.9524	639.6
2	645.0	0	721.0	−114.6	− 91.0	1,160.4	−793.6	366.8	−143.1	1,017.3	0.9070	922.7
3	965.7	0	754.2	− 87.9	− 56.7	1,575.3	−476.2	1,099.1	−428.6	1,146.7	0.8638	990.5
4	985.0	0	788.9	− 93.0	− 59.6	1,621.3	−285.7	1,335.6	−520.9	1,054.4	0.8227	867.5
5	1,004.7	0	825.2	− 98.4	− 62.5	1,669.0	−285.7	1,383.3	−539.5	1,129.5	0.7835	885.0
6	1,024.8	0	863.1	−104.1	− 65.7	1,718.1	−142.9	1,575.2	−614.3	1,103.8	0.7462	823.7
7	1,045.3	0	902.8	−110.1	− 68.9	1,769.1	0	1,769.1	−689.9	1,079.2	0.7107	767.0

PW (MARR = 9%) = $2607.4

1. Entries in the table are in thousands (000s) of dollars and based on economic data in Table 6.4. This is an actual dollar (A$) after-tax analysis with conversion of the after-tax cash flows (ATCF) to real dollars (R$) in columns 11 and 12 to assist interpretation of results.
2. Depreciation based on modified ACRS (MACRS), five-year class life; cost basis equals investment costs.
3. Effective income tax rate (t) for the company is 39%.

2. The increases in net sales revenue for the two alternatives relative to continuing present operations were estimated to be different, and to vary by year. The differences between alternatives were based on the better quality, lead time and flexibility of the FMS that are expected to eventually occur. The estimated degradation in sales if present operations were continued is included in the difference estimates. However, for both alternatives, the full effect of improved competitiveness was not estimated to occur until the third year even though more production capacity increased sales in the first year. This reflects the project team's judgment that time and operating experience are required before the company can realize the full potential of the advanced systems. The lower annual escalation rate (2%) indicates expected price competition and not being able to escalate unit prices as fast as general inflation. The estimated annual general inflation rate (f) for the seven-year analysis period is 5% per year.
3. Operating cost savings were a major factor for both alternatives. The primary component in these savings is operating personnel. Floor space and other items make up the remainder. The savings due to less floor space were based on reduced maintenance, janitorial services, and related material and supply costs. Other items included in the operating cost savings ranged from reduced toolcrib support to elimination of order expediting with the FMS alternative after reasonable operating experience is achieved.
4. Both alternatives were estimated by the project team to have increased maintenance costs. The FMS requirements are significantly greater than those for alternative 1 (changes to present system). These increased costs reflect added personnel with new skills. Also, they include training costs in the first year plus some contract maintenance assistance during years 1 and 2.
5. The other costs category also shows increases for the FMS alternative in each year, and during the first two years for alternative 1. The increased costs were projected primarily for engineering design and manufacturing engineering. Work-in-process (WIP) inventory reduction benefits have already been factored into this cost category to offset some of the increases.
6. The depreciation is based on five-year class life assets under the modified accelerated cost receovery system (MACRS) established by the Tax Reform Act, 1986. The cost basis used was the investment (first) cost for each alternative.
7. The effective marginal income tax rate used for the company was 39%. This is based on a federal rate of 34% and a state rate of 7.5%. That is:

$$t = 0.34 + 0.075 - (0.34)(0.075) = 0.3895, \quad \text{or} \quad 39\%$$

TABLE 6.6 *The results of the analysis and comparison for two feasible alternatives*

Analysis method	Changed system	FMS	Increment
Present worth (i = 9%)	$1 543 400	$2 607 400	$1 064 000
Internal rate of return (IRR)	44.5%	33.7%	26.2%
Simple payback	Year 3	Year 3	Year 4
Discounted payback	Year 3	Year 4	Year 4

8. The real dollar (R$) after-tax cash flow (ATCF) in column 12 was calculated from the actual dollar ATCF using the general inflation rate (f = 5%). That is:

$$R\$_n = A\$_n[1/(1.05)]^n$$

9. The MARR criterion used by the project team was 9% (nominal) on an after-tax basis. The rate was set by company management as part of economic policy. This reflects its awareness of potential problems if the hurdle rate applied to new-technology projects is too high, as well as expected continuing reductions in commercial interest rates during the period of the project and in the marginal cost of capital to the company.

The results of the analysis and comparison of the two feasible alternatives are shown in Table 6.6. The last column provides analysis results for the additional increment of investment ($1 500 000) required for the FMS over changes to present operations.

The project team found the economic results favourable for both alternatives, with the FMS being preferred. This was based on its greater PW value at i = MARR = 9%. This economic choice is confirmed by the PW on the increment being greater than 0, and its IRR exceeding the MARR. Both alternatives have relatively short payback periods (simple and discounted) for new-technology projects.

In column 12 of Table 6.5, the after-tax cash flow for each alternative in real dollars has a general downward trend after year 3 (except year 5 for the FMS). This result in dollars with constant purchasing power shows a different picture than the after-tax cash flows in actual dollars which basically are level starting with year 3.

One important topic that we have not discussed, due to the summary form of the example, is accomplishing sensitivity analyses to examine the impact of changes in critical factors and assumptions. This is a very important phase of an engineering economic analysis in practice. In this example, two important areas that should be in the sensitivity analyses are the net increase in sales revenue and the operating cost savings. Of particular interest, is the impact of more conservative estimates in both areas on the economic results.

6.5 Summary

The engineering economic analysis process discussed in this article, with life cycle considerations appropriately included, provides a sound approach to the analysis of advanced systems in production and manufacturing. However, there are a number of potential problem areas which require scrutiny to mitigate their impact on the economic results. One of the most important actions to lessen these impacts is a sound company planning process relative to its products and the manufacturing systems required to produce them competitively. This is essential for a continuous improvement process, and to provide a sound foundation for new-technology projects and their analysis.

6.6 References

Cooper, R. (1988) The rise of activity-based costing – Part one: what is an activity-based cost system? *Cost Management*, Summer, 45–54.

DeGarmo, E.P., Sullivan, W.G. and Bontadelli, J.A. (1989) *Engineering Economy*, 8th edn, Macmillan, New York.

Gold, B. (1983) Strengthening managerial approaches to improving technological capabilities. *Strategic Management Journal*, **4**.

Hodder, J.E. (1986) Evaluation of manufacturing investments: a comparison of US and Japanese practices. *Financial Management*, Spring, 17–24.

Hodder, J.E. and Riggs, H.E. (1985) Pitfalls in evaluating risky projects. *Harvard Business Review*, Jan–Feb, 128–35.

Hundy, B.B. (ed.) (1985) *Proceedings of 3rd European Conference on Automated Manufacturing*.

Huthwaite, B. and Spence, G. (1989) The power of cost measurement in new product development. *National Productivity Review*, **8** (3), 239–48.

Johnson, H.T. (1989) Managing costs: an outmoded philosophy. *Manufacturing Engineering*, May, 43–6.

Kaplan, R.S. (1986) Must CIM be justified by faith alone? *Harvard Business Review*, Mar–Apr, 239–47.

Kochan, D. (ed.) (1986) *CAM: Developments in Computer Integrated Manufacturing*, Springer-Verlag.

Mital, A. and George, L.J. (1989) Economic feasibility of a product assembly line: a case study. *The Engineering Economist*, **35** (1), 25–38.

Mize, J.H. and Seifert, D.J. (1985) CIM – A global view of the factory. *Proceedings: Fall Industrial Engineering Conference*, Institute for Industrial Engineers, 173–7.

Piekarski, J.A. (1989) Justification of high-tech manufacturing equipment. *Cost*

Engineering, **31** (5), 8–13.

Reeve, J.M. (1989) Variation and the cost of activities for investment justification. *Proceedings: IIE Integrated Systems Conference*, Institute of Industrial Engineers, 593–8.

Sullivan, W.G. and Bontadelli, J.A. (1980) The industrial engineer and inflation. *Industrial Engineering*, Mar, 24–33.

Tayyari, F. and Parsei, H.R. (1989) Cost analysis of economic justification for automated technologies. *Proceedings: IIE Integrated Systems Conference*, Institute of Industrial Engineers, 584–7.

Van Horne, J.C. (1980) *Financial Management and Policy*, Prentice-Hall, New Jersey.

White, B.E. (1989) Justification of CIM systems – is special treatment necessary? *Proceedings: IIE Integrated Systems Conference*, Institute of Industrial Engineers, 588–92.

Part Two

Models and Techniques

7 Analytical Techniques for Justification of Manufacturing Systems

M. Jeya Chandra and Catherine M. Harmonosky

7.1 Introduction

As companies evalute advanced manufacturing systems, such flexible manufacturing systems (FMS) and computer-integrated manufacturing systems (CIMS), for possible implementation in their plant, an economic analysis is always considered. Unfortunately, most of the advantages of such systems, e.g. increased manufacturing flexibility and improved quality, are not included in traditional economic analysis techniques. This often results in an inaccurate assessment of the advanced system's potential contribution to the companies competitiveness. Yet, decisions about implementing technologically advanced systems are still made within a capital budgeting framework in which the expected value of the investment is based on risk and expected return.

Quite often, strategic and financial considerations dominate technical concerns as investment decisions move to higher management levels (Dean, 1987). This becomes a major concern for engineers in industry who currently must justify an advanced system, which may be technologically superb and excellent for long-term goals, to financial superiors by the same accounting and economic analysis standards from 50 years ago. Although calculating a rate of return or developing a cost-benefit analysis for these systems is very difficult (Kaplan, 1983), it is still a necessary evil in current corporate structure. Some authors suggest approaching the problem by attaching a rationale statement to the cost numbers, trying to capture and translate enthusiasm of technical experts into a form understandable by upper management (Dean, 1987). Information that might be translated is numerous and may include lead time, delivery performance and improved quality (Randhawa and Bedworth, 1985). A two-phase approach assigning ordinal values to qualitative descriptions of how well a proposed system meets desired qualitative traits has been reported (Parsaei and Wilhelm, 1989). Meredith and Hill (1987) discuss a managerial approach linking the level of integration to different justification techniques; however, they do not discuss specific means of quantification. Keller and Noori (1988) suggest evaluation and justification concentrating on only one of the potential benefits, inventory setup costs, as an extension of work done by Porteus (1985). In all of this work, the methods of quantifying the typically intangible system benefits are not discussed.

The issue of system flexibility is often cited as a major benefit of advanced manufacturing systems. Therefore, trying to incorporate costs associated with flexibility into financial justification seems promising. Flexibility might be used as the primary driver for cost measurement. There has been work in the area of measuring flexibility in FMSs considering vectors and information-theoretics (Ito, 1987; Kumar, 1986; Ito *et al.*, 1985) although these measures do not include costs or economic analysis.

Park and Son (1988) have developed a more comprehensive justification approach by including cost of flexibility in the traditional net present value calculation. They concentrated on four types of flexibility: equipment, product, process and demand. To obtain these flexibility cost estimates, they suggest using computer simulation, however, they do not detail their techniques or exactly how the simulation information was issued as a quantification mechanism for intangibles in the cost equation.

This paper addresses quantification of intangibles by using analytical techniques to estimate the cost benefits of system flexibility based on some types of flexibility suggested by Park and Son (1988). In addition, material handling costs, an often overlooked element in economic analysis, will also be included.

7.2 Analysis

In this section we will identify some of the intangible factors which are not usually quantified, and estimate their corresponding values using analytical methods. These intangible factors are related to the flexibility measures explained by Park and Son (1988) as equipment flexibility, product flexibility, process flexibility and demand flexibility. Park and Son associate some measurable costs with these flexibility measures. They suggest measuring equipment flexibility in terms of idle cost, which is the cost in opportunity resulting from underuse of equipment. Product flexibility is related to set-up costs because, in advanced manufacturing systems, reduced set-up times will allow shorter production runs. This results in the ability to manufacture a large variety of products and meet changing market demands. Process flexibility is quantified in terms of the waiting cost. The capability to make rapid changes in part processing reduces the work-in-process inventory, which is directly proportional to the waiting time of parts processed. Finally, the demand flexibility is related to the inventory cost of finished goods and raw materials. Adaptability to changes in demand rate eliminates the necessity of maintaining large amounts of raw material and finished goods inventories.

To quantify these flexibility measures, analytical methods must be identified to compute the machine utilization which can be used to compute the idle cost, the average waiting time of jobs to be processed, and the savings in cost due to inventories of finished goods and raw materials. In this paper, several different analytical methods will be used to quantify different types of costs considered for justification. The analytical method of mean-value analysis, a technique developed by Reiser and Lavenberg (1980) to analyse queueing methods, will be used to compute the machine utilizations and the average waiting times of jobs. Another analytical method, a simple lot size inventory model, will be used in this paper to measure inventory levels and associated costs. In addition, material handling costs will be analysed by using a Leontief input-output model to determine number of handling moves in a system. First, the analytical techniques will each be discussed in general, and then they will be applied specifically to the justification quantification problem.

7.2.1 Mean-value analysis

Mean-value analysis allows the mean values of certain performance measures of closed queueing networks to be computed without using the actual probability of arrival times or processing time distributions. The advantage of using mean-value analysis for quantifying intangible costs for justification is simplicity of computation and the small amount of numerical data needed for input. Often when systems are being proposed, detailed data concerning distributions of processing times and arrival rates or details concerning system operation and control are simply not known. Making a detailed system model, such as a simulation model, may not be possible or worth the time and effort given the lack of data. Because mean-value analysis needs only basic mean information about system inputs, it seems to be an appropriate tool to apply to the system justification issue.

The primary performance measures that are computed are the mean number of customers of different classes waiting in queues for servers at various nodes in the network. The mean waiting times can be computed from the mean number of customers waiting at various nodes using Little's formula (Reiser and Lavenberg, 1980). The server utilizations can be computed using the mean service times and the rate at which the customers arrive at the nodes for service. The mean-value analysis is based on several assumptions, the most important assumptions being: 1. the service times are exponentially distributed and 2. the routing of the customers among the different nodes is determined probabilistically. The results obtained from the mean-value analysis when these assumptions are violated are only approximate (Hildebrant, 1980).

Any manufacturing system like a job shop or a flexible manufacturing system can be modelled as a queueing network with the jobs as customers and the machines as servers. The jobs visit a number of processors like machines and inspection stations before exiting the system. The jobs have to wait in queues in front of machines when the machines are busy processing other jobs. The main assumptions needed for mean-value analysis results to be accurate are not typically true in the case of manufacturing systems. The processing times are deterministic and the routing of jobs is predetermined. Nevertheless, reasonably accurate estimates of the mean performance measures of the manufacturing systems can be obtained using the mean-value analysis (Hildebrant, 1980). Also, the manufacturing systems are not really closed queueing networks in which the same customers move around the nodes. However, the number of different job types that can be present in the system is usually limited and the number of pallets circulating in the system is often held constraint. Therefore, it possible to model such systems using closed queueing networks (Suri and Hildebrant, 1984). Some conditions under which the mean-value analysis does not work very well are: 1. job selection rules other than first-come-first-served for processing, and 2. significant congestion in the material transport and/or material storage systems (Suri and Hildebrant, 1984).

One of the main principles used in this technique is that an arriving customer 'sees' the system, with himself removed, in equilibrium. To illustrate this, a system consisting of three classes of customers, with N_1, N_2 and N_3 customers, respectively, is considered. Let the mean waiting time of a class 2 customer or the mean number of class 2 customers waiting be of interest. Then, when a class 2 customer arrives at the facility, the system is seen with N_1, $(N_2 - 1)$ and N_3 customers in classes 1, 2 and 3, respectively. This customer's mean waiting time in the queue is made up of the following mean times related to the system, with N_1, $(N_2 - 1)$ and N_3 customers in the corresponding classes:

1. The sum of the mean service times of the mean number of customers waiting in the queue that the arriving customer finds; and
2. The mean residual service time of the customer in service upon the arrival of the identified customer.

Little's formula is invoked, which states that:

The mean number of customers of class i waiting in queue =
The mean waiting time in the queue of a class i customer (Total number of customers of class i − Mean number of customers of class i in queue and service)/Mean inter-arrival time of class i customers

Using this formula, a recursive relation is formulated in the mean-value analysis, which is used to calculate the mean values of the system performance measures.

In Section 7.3, mean-value analysis is used to compute the number of customer's waiting, customer waiting time, and machine utilization and to estimate the associated costs. Next, an inventory model is discussed which can be used to compute the inventory costs as a function of the setup costs.

7.2.2 Inventory model

In this section, a simple lot size inventory model is considered to illustrate the calculation of inventory costs of raw materials and finished products. The following notation is used in the analysis:

r is the constant (uniform) demand rate of the final product.
p is the rate of production of the product.
N_m is the number of units of raw material required per unit of final product.
C_{1f} is the carrying cost of final product per unit per unit time.
C_{1m} is the carrying cost of raw material per unit per unit time.
C_{3f} is the setup cost of the final product per batch.
C_{3m} is the ordering cost of raw material per batch.
q is the batch size of final product.
t_f is the time between successive starts of manufacture of final product.

It is assumed that the lead times are zero and that the replenishment rate for the raw material is zero. The inventory graphs of the final product and the raw material are given in Fig. 7.1.

The total cost per unit of the raw material and the final product is given by

$$TC = \frac{C_{1f}}{2}\frac{q(p-r)}{p} + \frac{C_{3f}r}{q} + \frac{C_{1m}qn_m}{2} + \frac{C_{3m}r}{q}$$

$$= \frac{q}{2}\left[\frac{C_{1f}(p-r)}{p} + C_{1m}n_m\right] + \frac{r}{q}\left[C_{3f} + C_{3m}\right] \quad (7.1)$$

Differentiating with respect to q and setting the derivative equal to 0, we get the optimum batch size of the final product as

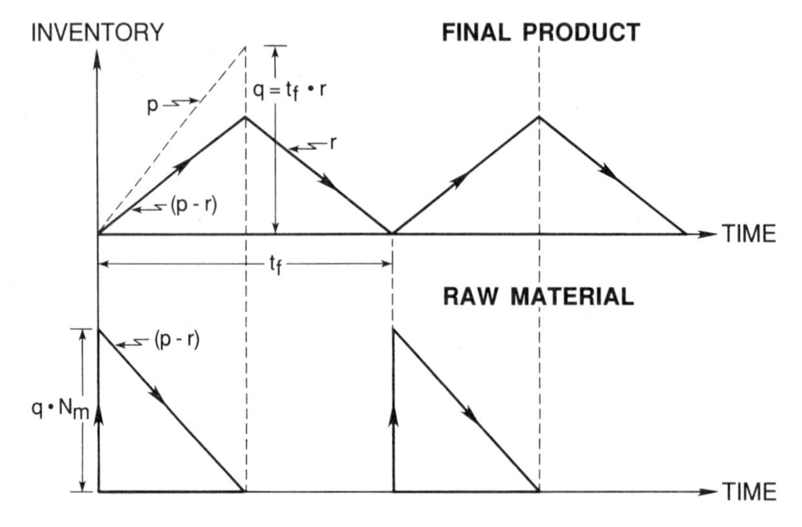

Fig. 7.1 *Inventory graphs.*

$$q^* = \sqrt{\left[\frac{2(C_{3f} + C_{3m})r}{C_{1f}(1 - r/p) + C_{1m}n_m}\right]} \tag{7.2}$$

Substituting this in the expression for the total cost, we get the optimum total cost per unit time as,

$$TC^* = \sqrt{\{2r(C_{3f} + C_{3m})[C_{1f}(1 - r/p) + C_{1m}n_m]\}} \tag{7.3}$$

It can be seen from the above expression that keeping the demand rate r, the inventory carrying costs C_{1f} and C_{1m} and the ordering cost of the raw material C_{3m} constant (typical constraints in reality), the total inventory cost can be minimized by reducing the set-up cost and the increasing production rate of final product. An advanced manufacturing system, such a flexible manufacturing system, will have a low set-up time, and hence, a low set-up cost and a high production rate. The optimum-cost expression quantifies the reduction in set-up time and the increase in the production rate.

7.2.3 Material handling cost

The reduction in the movement of materials in an advanced manufacturing system should be taken into consideration in the economic justification. We use the Leontief input-output model to estimate the average number of material transfers in any manufacturing system. This is based on the work of Chandra and Schall (1988).

The Leontief input-output model is a mathematical representation of a system in which a linear relation is assumed to describe the transfers among various parts of the system. In this model, the system is divided into sectors. The units which are transferred through the sectors of the system represent the activity of the system. There are internal and external flows, depending upon whether the flows occur from one sector to another sector or they occur from or into an external sector. The external portion of the model is designated as rest of the world (ROW). The general Leontief model can be represented as a set of linearly independent equations: one for each of the sectors. The machining centres along with inspection stations and the storage areas are nodes in the network representing the manufacturing system. Two nodes are connected by an arc if there is material flow between the two. Each node of the network corresponds to a sector in the Leontief model.

The structure of the equation is:

$$X_i = a_{i1}X_1 + a_{i2}X_2 + \ldots + a_{ij}X_j + \ldots + a_{im}X_m(-4) + y_i$$
$$\text{for} \quad i = 1, 2, \ldots, m,$$

in which $y_i \geqslant 0$, $X_i \geqslant 0$ for all i, $0 \leqslant a_{ij} \leqslant 1$ for all i and j, and

$$\sum_{i=1}^{m} a_{ij} = 1 \text{ for all } j$$

In equation (7.4): y_i is the amount of external input of activities into sector i; X_i is the total amount of activities through sector i; and a_{ij} is the proportion of the total amount of input of activities in sector j which flows out of j into sector i. This is also known as the technological coefficient.

The technological coefficients a_{ij} are computed from the actual amount of activities from each sector to all other sectors, known as transactions. If t_{ij} represent the total amount of activities that flow from sector j to sector i, then the technological coefficient is computed as:

$$a_{ij} = t_{ij}/X_j \tag{7.5}$$

In a manufacturing system, the transactions and the external input y_is can be estimated from the set of the components that will be processed by the system and their process planning.

Let $\boldsymbol{A}$ represent the $m \times m$ matrix consisting of the technological coefficients. Then equation (7.4) can be written as

$$\boldsymbol{X} = \boldsymbol{A}\boldsymbol{X} + \boldsymbol{Y}, \tag{7.6}$$

where $\boldsymbol{X}$ is an $m \times 1$ column vector containing the activities X_i, and $\boldsymbol{Y}$ is a $m \times 1$ column vector containing the external inputs y_is from (7.6). The vector $\boldsymbol{X}$ can be written as

$$\boldsymbol{X} = (\boldsymbol{I} - \boldsymbol{A})^{-1}\boldsymbol{Y} \tag{7.7}$$

where $\boldsymbol{I}$ is the identity matrix of size $m \times m$. The elements of $(\boldsymbol{I} - \boldsymbol{A})^{-1}$ are called interdependency coefficients and are denoted by β_{ij}. These coefficients represent the direct and indirect responses of sector i to one unit of input into sector j. These responses are:

1. Direct response of sector i to a total input of y_j units in sector $\beta_{ij} * y_j$.
2. Indirect response of sector i to a total input of y_j units in sector $j = (\beta_{ij} - a_{ij})\boldsymbol{Y}$. In addition, the sum $\Sigma_{i=1}^{m} \beta_{ij}$ for all j represent the average number of transfers a unit makes after it has entered through sector j and before exiting the system.

The estimates of performance measures concerning the manufacturing system that can be obtained from the Leontief input-output model are the following:

1. The total number of units processed at the machining centre i, as a result of accepting y_j units for processing at centre j, given by $\beta_{ij} * y_j$;
2. The total number of units processed at the machining centre i, as a result of all the units processed at all the centres within the system (this is equal to $\Sigma_j \beta_{ij} * y_j$); and
3. The average number of transfers (material handlings) within the system for one unit accepted for processing at centre j.

In this paper the information in 3. above is used to estimate the material handling costs. In the next section, a numerical example is presented to illustrate the use of all the analytical models discussed to this point.

7.3 Numerical example

Two systems are considered, an existing manufacturing system and a proposed flexible manufacturing system for replacement, schematically represented in Figs 7.2 and 7.4, respectively. The material transfers among the stores, machining centres and the inspection stations are represented by the directed areas connecting the sectors. The existing system has six receiving stores (Qs), eight machining centres (Ms, Ts, Ds), an inspection centre (I), and stores for receiving accepted finished units (F) and scrapped items (S). The corresponding technological coefficient matrix, containing the proportions of units transferred from any centre/stores to any other sector for the Leontief, is given in Fig. 7.3. The proposed flexible manufacturing system in Fig. 7.4 has two receiving stores (Q_1, Q_2), two intermediate stores (Q_3, Q_4), two flexible machining centres (FMS1, FMS2), an inspection station and stores for receiving accepted finished items and scrapped units. The corresponding technological coefficient matrix for the Leontief model is given in Fig. 7.5.

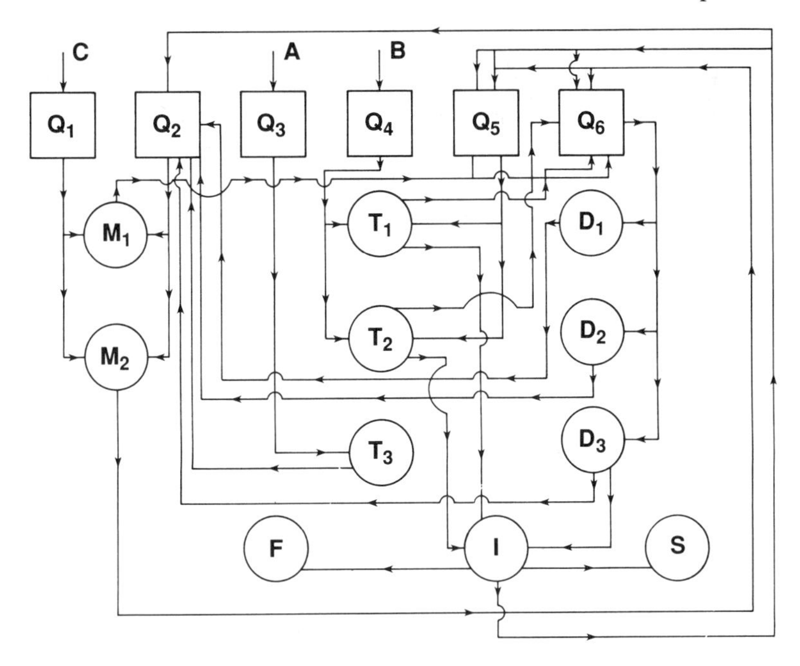

Fig. 7.2 *Flow diagram of existing facility.*

	Q_1	Q_2	Q_3	Q_4	Q_5	Q_6	T_1	T_2	T_3	M_1	M_2	D_1	D_2	D_3	I	S	F
Q_2 Q_1																	
Q_3										.67	.27				.03		
Q_4																	
Q_5									1.0			1.0	1.0	.09	.03		
Q_6							.38	.29		.25	.36				.03		
T_1		.55	.45														
T_2		.45	.55														
T_3	1.0																
M_1				.7	.3												
M_2				.3	.7												
D_1						.186											
D_2						.186											
D_3						.628											
I							.62	.71		.08	.37			.91			
S															.01		
F															.90		

Fig. 7.3 *Technological coefficient matrix.*

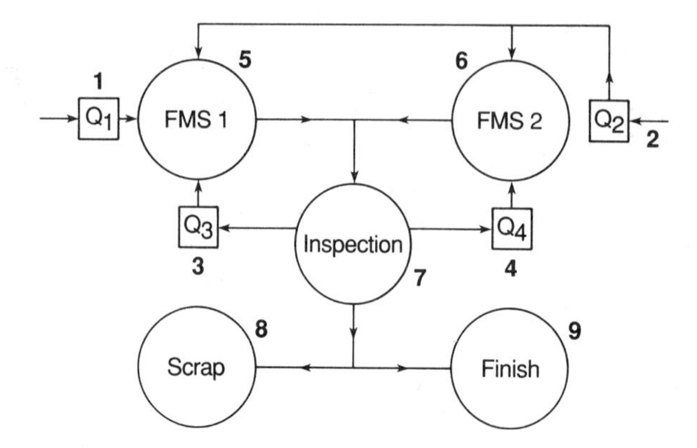

Fig. 7.4 *Flow diagram of FMS.*

	1 Q_1	2 Q_2	3 Q_3	4 Q_4	5 F_1	6 F_2	7 I	8 S	9 F
1 Q_1									
2 Q_2									
3 Q_3							.02		
4 Q_4							.02		
5 F_1	1.0	.28	1.0						
6 F_2		.72		1.0					
7 I					1.0	1.0			
8 S							.01		
9 F							.95		

Fig. 7.5 *Technological coefficient matrix of FMS.*

Three types of parts A, B and C are assumed to be processed in either system and the total amounts of these part types processed per day are 200, 150 and 300, respectively.

The information that follows is the output from the previously discussed analytical models applied to this numerical example. These computed measures relate to quantification of the performance improvements in an advanced manufacturing system due to increased system flexibility and decreased material handling needs, which leads to more accurate economic evaluation.

7.3.1 Mean-value analysis

The results from the mean-value analysis are given in Table 7.1. It can be seen from the results presented in Table 7.1 that the flexible machining

TABLE 7.1 *Results of mean-value analysis*

	Existing	FMS
I Average time in the system (minutes) (including processing)		
A	195	84
B	268	84
C	426	84
II Average number of parts in the system	27.5	13.6
III Average percentage idle time of the machining centres	41%	1%

TABLE 7.2

	Existing System	FMS
Production rate (p)	15 000 per year	25 000 per year
Set up time per batch	4 hours	20 minutes
Set up cost per hour	$60.00	$120.00

system results in lower idleness of machine centres, lower in-process inventories (average number of jobs in the system) and lower waiting times for the jobs. These idleness and inventory measures can be used to calculate cost values according to a company's information regarding operating costs and inventory holding costs.

7.3.2 Inventory of raw materials and finished goods

The following cost values are assumed for equations (7.2) and (7.3).

Demand rate (r) = 10 000 per year

Number of units of raw material required per unit of final product (N_m) = 5

Carrying cost of final product = $0.10/unit/year ($C_{1f}$)

Carrying cost of raw material = $0.05/unit/year ($C_{1m}$)

Ordering cost of raw material per batch (C_{3m}) = $100

From Table 7.2 the setup costs are calculated as:

Set-up cost of final product per batch in the existing system = $240 ($C_{3f}$)

Set-up cost of final product per batch in the FMS (C_{3f}) = $40.

TABLE 7.3

	Existing	FMS
Batch size (q^*)	4900	3000
Annual inventory cost per year (TC*)	$1390.00	$930.00

Using these values in equations (7.2) and (7.3) yields the results in Table 7.3. It can be seen that the proposed flexible manufacturing system results in lower inventory cost and smaller batch size.

7.3.3 Leontief input-output model

This model gave the following results concerning the average number of material transfers per part processed.

Existing system	6.45
FMS	3.13

The average number of material transfers multiplied by the total number of parts manufactured annually and the material handling cost per transfer yields the average material handling cost per year.

7.4 Conclusion

In this paper, analytical models were used to estimate costs related to some measures of flexibility, typically considered intangibles, which are essential for evaluating and justifying advanced manufacturing systems. The methods presented can provide quantitative information for use alone or for use in other models (Park and Son, 1988) for justification. Though simulation is a very valuable tool in estimating some of the performance measures of manufacturing systems, analytical techniques should be used whenever feasible. This is because simulation results are at best estimates and make performing sensitivity analysis difficult.

7.5 References

Carrie, A.S. and Banerjee, S.K. (1984) Approaches to implementing manufacturing systems. *Omega*, **12**, 251–9.

Chakravarty, A.K. and Shtub, A. (1985) New technology investments in multistage production systems. *Decision Sciences*, **16**, 248–64.

Chandra, M.J. and Schall, S.O. (1988) Economic justification of flexible manufacturing systems using the Leontief input-output model. *The Engineering Economist*, **34** (1), 27–50.

Dean, J.W., Jr (1987) *Deciding to Innovate: How Firms Justify Advanced Technology*, Ballenger Publishing, Cambridge.

Hildebrant, R.R. (1980) Scheduling flexible machining systems using mean value analysis. *Proceedings of the IEEE-Conference on Decision and Control*, 701–6.

Ito, Y. (1987) Evaluation of FMS: state of the art regarding how to evaluate system flexibility. *Robotics and Computer-Integrated Manufacturing*, **3**, 327–34.

Ito, Y., Ohmi, T. and Shima, Y. (1985) An evaluation method of flexible manufacturing system – a concept of flexibility evaluation vector and its application. *Proceedings of the 25th International Machine Tool Design and Research Conference*, 89–95.

Kaplan, R.S. (1983) Measuring manufacturing performance: a new challenge for managerial accounting research. *The Accounting Review*, **58** (4), 242–52.

Keller, G. and Noori, H. (1988) Justifying new technology acquisition through its impact on the cost of running an inventory policy. *IIE Transactions*, **20**, 284–91.

Kumar, V. (1986) On measurement of flexibility in flexible manufacturing systems: an information-theoretic approach. *Proceedings of the Second ORSA/TIMS Conference on Flexible Manufacturing Systems: Operations Research Models and Applications*, 131–43.

Meredith, J.R. and Hill, M.M. (1987) Justifying new manufacturing systems: a managerial approach. *Sloan Management Review*, **28** (4), 49–61.

Park, C.S. and Son, Y.K. (1988) An economic evaluation model for advanced manufacturing systems. *The Engineering Economist*, **34**, 1–26.

Parsaei, H.R. and Wilhelm, M.R. (1989) A justification methodology for automated manufacturing technologies. *Computers and Industrial Engineering*, **16**, 363–72.

Porteus, E.L. (1985) Investing in reduced setups in the EOQ model. *Management Science*, **31**, 998–1010.

Randhawa, S.U. and Bedworth, D. (1985) Factors identified for use in comprising conventional and flexible manufacturing systems. *Industrial Engineering*, **17** (6), 40–4.

Reiser, M. and Lavenberg, S.S. (1980) Mean-value analysis of closed multichain queueing networks. *Journal of the Association for Computing Machinery*, **27** (2), 313–23.

Suri, R. and Hildebrant, R.R. (1984) Modeling flexible manufacturing systems using mean-value analysis. *SME Journal of Manufacturing Systems*, **3**, 27–38.

8 Simulation *v.* Analytical Models for the Economic Evaluation of Flexible Manufacturing Systems: An Option-Based Approach

Giovanni Azzone and Umberto Bertelè

8.1 Introduction

One of the main obstacles to a wider diffusion of flexible manufacturing systems (FMSs) is their economic justification (Kalkunte *et al.*, 1986; Gold, 1988). Traditionally, investments in automation have been justified by reduction in direct labour costs. The same justification, however, applies only to some extent as far as FMS is concerned. Other advantages, such as better quality standards, improved timeliness in responding to market demands and, above all, flexibility, are the critical aspects involved in the decision to introduce an FMS. However, these aspects appear somehow intangible, and difficult to include in a financial analysis.

This paper deals with the problem of justifying capital investment in an FMS. Particularly, it outlines the impact of the flexibility of an FMS on cash flow in uncertain environments.

The paper is structured in three parts. The first part suggests a conceptual framework and some guidelines to be followed in order to include the impact of manufacturing flexibility in discounted cash flow (DCF) analysis. The second part compares briefly four different evaluation models based on simulation techniques and/or sensitivity analysis. Differences among these models and conditions under which each model can effectively be employed are discussed. The last part describes a new

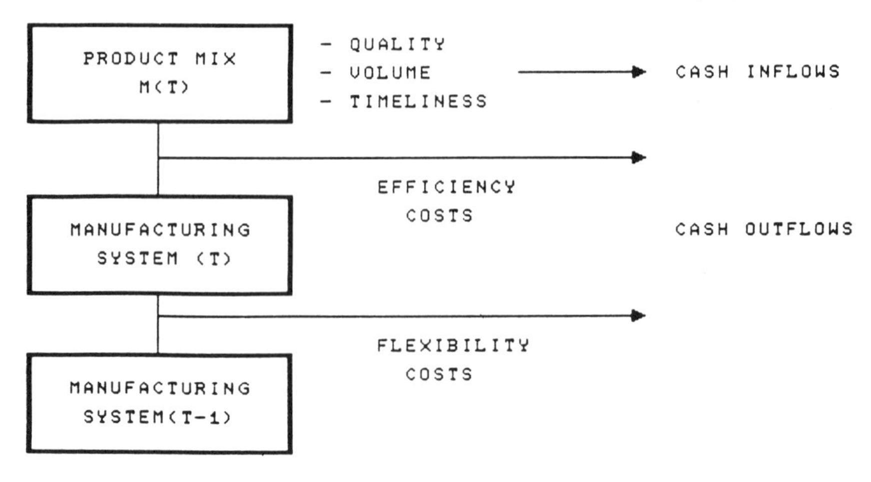

Fig. 8.1 *Value creation in an uncertain environment.*

evaluation model which is based on an extension of option theory applicable to real assets. Major advantages and disadvantages of this model, as compared to simulation-based models, are also discussed.

8.2 The economic effect of flexibility in manufacturing

It is quite logical that flexibility can improve the value of a company in today's competitive environment which is characterized by short product life cycles and frequent mix changes. Nevertheless, often, companies do not include even a rough evaluation of this impact in cash-flow analysis, when calculating the net present value of an investment in a manufacturing system. As a consequence, the resulting evaluation can be completely misleading for decision makers.

The following is given to illustrate the measurement of the value of flexibility. Fig. 8.1 describes the process in which a manufacturing system creates value in a turbulent environment.

The purpose of the manufacturing system is to produce a number of parts M and each part can be described at a given time t in terms of the following:

1. Demand distribution;
2. Quality standard; and
3. Target timeliness.

In a turbulent environment, the number of parts the manufacturing system must process will be a variable changing over time, and hence, it can be written as a function of time $[M(t)]$.

The product mix affects cash inflows in an economic justification analysis. If an FMS increases quality standards or timeliness, it increases cash inflows by way of a premium fetched or an increase in market share.

Production costs, on the contrary, define cash outflows directly. This aspect is considered in more detail below. Cash outflows at a given time t, in fact, can be seen as the sum of two different terms. The first term is the cost of producing the customer's requirements at that time. This depends on the efficiency of the resources used. The second term is the cost needed to adapt the manufacturing system to new requirements. This would arise when either the acquisition of new machines or the conversion of available ones occurs.

Hereafter, the former term of the cash outflows is referred to as **efficiency cost**, and the latter as **flexibility cost**. The importance of these two terms obviously depends on environmental turbulence. The more frequently a company changes its product mix, the more important is its need for flexibility. Today, the importance of flexibility is increasing in almost all industries.

While a traditional automation acts mainly on efficiency costs, an FMS acts mainly on flexibility costs. There is no logical reason as to why the flexibility costs should not be included in cash-flow analysis, when the less important efficiency costs are being included.

8.3 Simulation models

According to the conceptual scheme of Fig. 8.1, the measurement of the economic impact of the flexibility of an FMS requires three steps. First, the product mix dynamics over the project life must be forecasted by a deterministic or stochastic method. Second, the flexibility cost of the FMS (the cost to adapt it to possible mix changes) must be appraised. Finally, the mix dynamics must be simulated over the project life in order to evaluate the actual efficiency costs and flexibility costs.

Figure 8.2 illustrates the structure of a model for economic evaluation of FMS developed by Azzone and Bertele' (1989). Here, mix change is represented by the frequency of introduction of new products, their life cycle and the distribution of demands. The mix dynamics, linked to the efficiency costs and flexibility costs, determine the costs, revenues and investments which are to be used in the economic analysis.

As far as we know, there are four models that have been developed on the basis of these conceptual guidelines. These are models developed by Azzone and Bertele' (1989), Hundy and Hamblin (1988), Hutchinson and Holland (1982) and Miltenburg (1987). The main difference among these models is in the kind of mix changes that are considered. As a result, each model describes the flexibility of an FMS in a different way. Rather

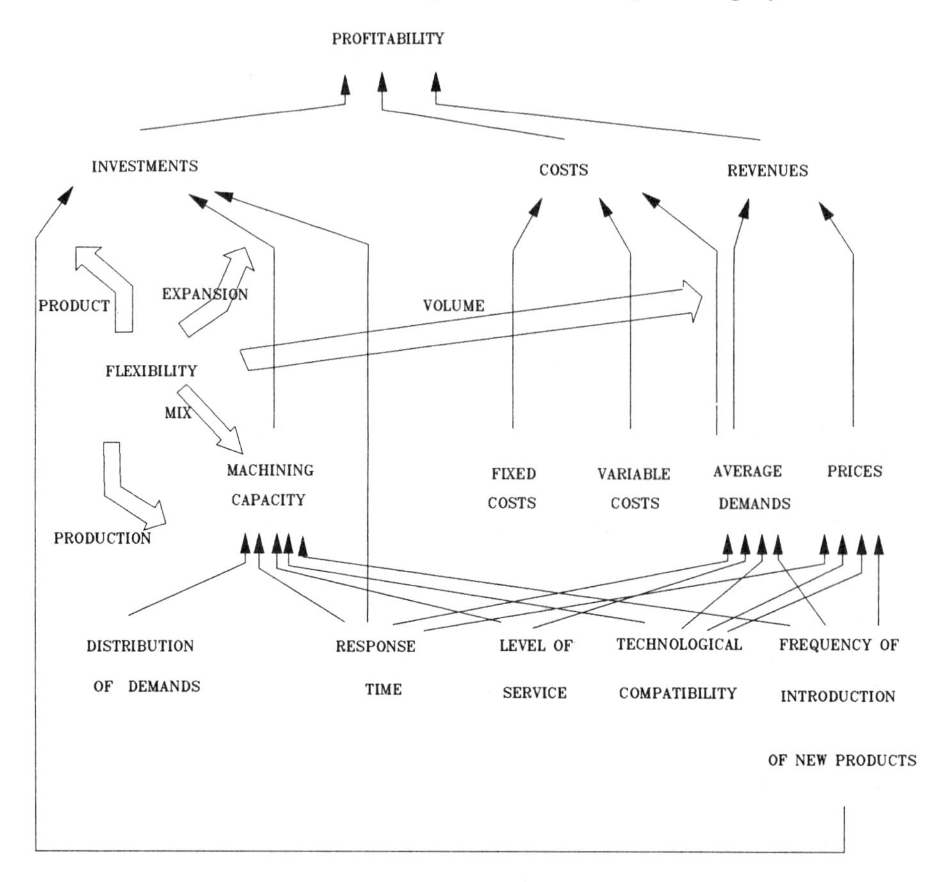

Fig. 8.2 *A model for the economic evaluation of FMS (adapted from Azzone and Bertele', 1989).*

than describing these models in detail, it would be better to suggest how a company can choose the model that best suits its requirements.

The flexibility of a manufacturing system can be defined generically as the ability to respond to perturbances at low cost and within a short time. This definition, however, is too wide to allow any measurement of the **adaptation cost**. This is due to the fact that a manufacturing system has to deal with many different kinds of perturbances, caused by situations varying from the introduction of a new product to the expansion of current product mix.

Hence, in order to determine the applicability of the four models, one needs to understand how flexibility can be divided into basic terms and, consequently, how they can be incorporated into each model.

8.3.1 A multidimensional definition of flexibility

Various classifications of flexibility have been suggested, among others by Browne *et al.* (1984), Barad and Sipper (1988), and Swamidass (1988). Here a taxonomy suggested by Azzone and Bertele' (1988) has been adopted. The criterion used to separate flexibility into basic terms is the size of the perturbance that must be dealt with, i.e. the impact the perturbance has on the manufacturing system.

8.3.2 'Small perturbance' *v.* 'large perturbance'

To respond to a perturbance, either a change in the operating conditions of the manufacturing system (hence a change in operating costs) or a set of structured modifications (such as introduction of new machines) may be needed. Accordingly, the type of perturbance may be referred to as 'small' or 'large', respectively. Besides, the effect of a perturbance on the manufacturing system depends not only on the size of the perturbance, but also on the capacity of the system to limit the impact of the perturbance.

Three types of flexibility can therefore be defined:

1. The cost of responding to a small perturbance, i.e. the cost of changing the operating conditions of the system while its structure stays unchanged;
2. The capacity to limit the impact of a perturbance, i.e. the range of mix changes that can be coped with without changing the structure of the system; and
3. The cost of responding to a large perturbance, i.e. the cost of modifying the manufacturing system.

8.3.3 Small perturbances

Small perturbances are caused by two factors:

1. Quantitative change in market demand (an increase or a decrease in demand); and
2. Qualitative change in demand (a new product is required or an old one is to go off production).

Correspondingly, two basic flexibilities may be defined:

1. Volume flexibility; and
2. Product flexibility.

Volume flexibility

This is measured by the system's capability to limit the effect of change in market demand on the operating margin.

Product flexibility

This is measured by the cost of introducing a new product into the market. Product flexibility is inversely proportional to the cost of introducing a new product.

8.3.4 Capacity to reduce the impact of perturbances

In this case also, a distinction between quantitative and qualitative changes needs to be made. Two types of flexibility can be defined for the purpose, they are:

1. Mix flexibility; and
2. Production flexibility

Mix flexibility

This is the ability to manage quantitative changes in market demand without adding any machines to the manufacturing system. It is measured by the set of product mixes which the system can produce in a time unit.

Mix flexibility depends on:

1. The available overcapacity;
2. The number of products sharing a resource; and
3. The setup times.

Production flexibility

This is the ability to accommodate qualitative changes in product mix without modifying the manufacturing system. It results from the availability of resources that are not allocated fully and is measured by the set of parts that the manufacturing system can process.

8.3.5 Large perturbations

These are the perturbations which require structured modifications such as acquisition of new machines, due to the introduction of new products or an increase in volume.

Flexibility	Hutchinson Holland	Miltenburg	Hundy Hamblin	Azzone Bertele'
Volume	*		*	*
Product	*	*	*	*
Mix				*
Production		*		*
Expansion	*	*	*	*

Fig. 8.3 *Elementary flexibilities considered on four evaluation models.*

In this regard, we can define **expansion flexibility** as the ability to reduce the cost of expanding the system. It is determined by the modularity of the resources employed by the system.

Figure 8.3 shows the classification of the four models according to the basic flexibilities which they consider. It must be noted that when the number of basic flexibilities considered in a model increases, the cost and time needed for the evaluation process will also increase. A company should, therefore, choose a model which considers only those types of flexibility that are relevant to its environment.

8.4 Analytical models: using option theory

Simulation-based models allow a good evaluation of the economic effect of flexibility in manufacturing. However, there are some limitations of these models that need to be pointed out. The first and obvious problem is the time and cost requirement of the evaluation process. The cost requirement of the simulation models is higher compared with that of traditional static DCF analysis. This higher cost is justified only when flexibility is of real importance to the company.

The second and less obvious problem is related to the reliability of results. In fact, the evaluation process described above assumes implicitly that all possible mix changes will actually be introduced. As a result, the **value of flexibility** is determined by multiplying the reduction in the adaptation cost by the number of planned mix changes. But this approach does not consider the role of the management (Kensinger, 1988).

In fact, management will not accept those product mix changes for which the expected cash flow is lower than the cost needed to adapt the manufacturing system to the change. Thus, product mix changes are introduced only when:

$$CF \geqslant AC$$

where: CF is the discounted cash flow due to product mix change; and AC is the adaptation cost.

8.4.1 An option-based model

A different approach which considers the impact of the management decision is introduced below.

The concept here is that a lower adaptation cost for a product mix change does not imply that the change must be implemented, but it rather gives the company an option to make the change.

A number of authors (such as Kester, 1984; Brealey and Myers, 1988) suggested the use of extended forms of the analytical expressions developed for stock options (Black and Scholes, 1973) to deal with the investments in real assets. However, the application of these analytical expressions is quite limited. The reason for this is that a number of basic assumptions, such as total divisibility, are unacceptable for investments in manufacturing systems (for a thorough review of these limitations, see Azzone *et al.*, 1989).

It is therefore better to develop a new model which is only aimed at evaluating changes allowed by a manufacturing system. Let us assume that the expected cash flow of a mix change is described by a random variable CF, as shown in Fig. 8.4.

If AC is the adaptation cost for the product mix change, the change will actually occur only if expected cash flow is higher than the adaptation cost.

The correct value of a product mix change, therefore, will be the maximum of 0 and $CF - AC$.

There are two impacts of an improved flexibility, i.e. of a reduction in the adaption cost. First, by reducing AC, the number of product mix changes that creates the value increases. So, there is a higher probability for the number of successful changes. Second, the net value of each acceptable product mix change ($CF - AC$) increases.

Hence, the correct value of flexibility is given by:

$$V = \int_{AC}^{+\infty} (CF - AC)p(CF)\mathrm{d}CF \tag{8.1}$$

where: V is the expected impact of flexibility on cash flow; and $p(CF)$ is the probability distribution of the discounted cash flow.

If it is assumed that CF is distributed according to a normal distribu-

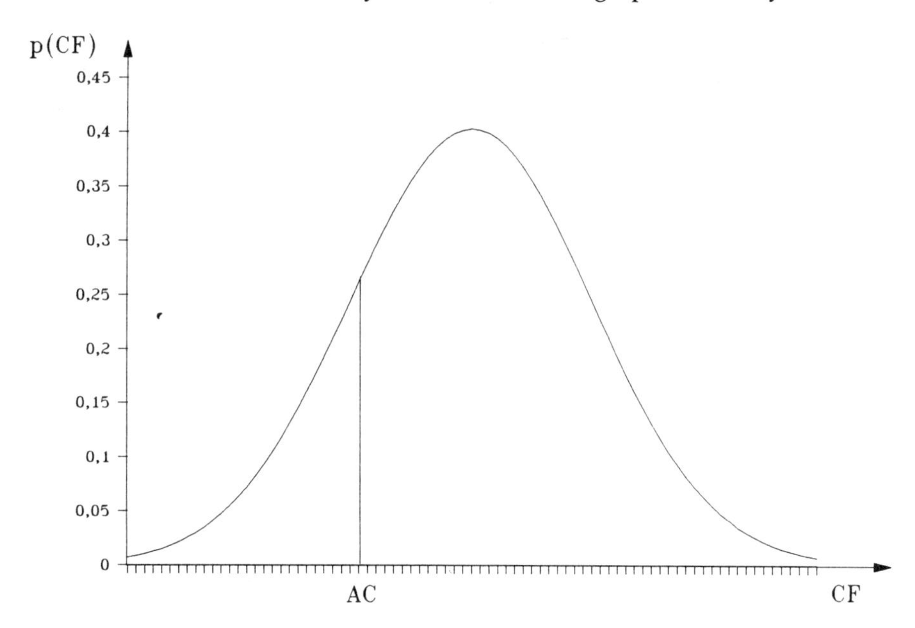

Fig. 8.4 *Economic valuation of an option.*

tions $N(a;s)$, then the following equation can be derived (Azzone *et al.*, 1989):

$$V = \frac{s}{\sqrt{(2\pi)}} \, e^{\frac{-(a - AC)^2}{2s^2}} + (a - AC)\left[1 - F\left(\frac{AC - a}{s}\right)\right] \qquad (8.2)$$

where: $F(x)$ is the standard normal cumulative distribution function.

This model is based on a closed analytical expression. Hence the computational effort required to solve the equation is minimal.

The following illustrates how the equation (8.2) can be used for determining the economic value of the basic flexibilities described above. These flexibilities, in fact, are linked to three of options, *viz.* product option, operating option and expansion option, as shown in Fig. 8.5.

Product option

The first type of option is the introduction of a new product. Here, *CF* represents the discounted cash flow that the new product will generate over its economic life. The adaptation cost *AC* is the investment needed for the introduction of the product. As mentioned earlier, this adaptation cost is reduced by an increase in product flexibility and production flexibility. The option to introduce the new product will actually be decided if the cash flow of the new product is higher than the adaptation cost.

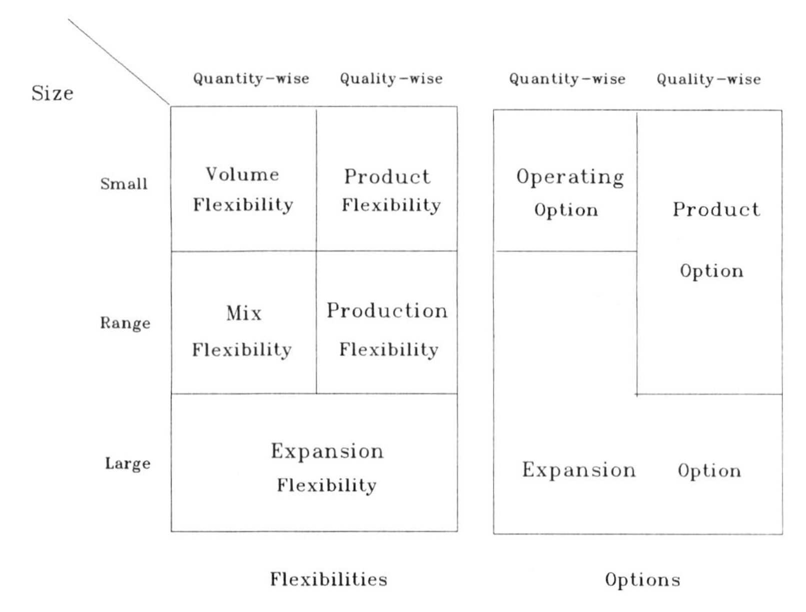

Fig. 8.5 *Elementary flexibilities and corresponding options.*

Operating option

At any time, a company can temporarily shut down or completely abandon its manufacturing system. The decision of shutting down the plant will be taken if the contribution margin is lower than the additional fixed costs that are needed to make production. This aspect can be seen (Kensinger, 1988) as an option to stop production. Here, *CF* is represented by the contribution margin of the current mix multiplied by market demand, while *AC* is the part of the fixed manufacturing costs that are not sunk, i.e. the cost to adapt the manufacturing system to production.

Expansion option

Finally, the third type of option is the expansion of the manufacturing system. The option to expand the system will be taken if demand exceeds available capacity (which is linked to mix flexibility), and if demand is surplus the additional capacity.

In order to apply equation (8.2) and find the economic value of this option, the following should be considered:

1. The total contribution margin of excess demand (unit contribution margin × demand, *CF*); and

2. The investment in additional capacity (that depends on expansion flexibility, AC).

8.4.2 A sample application

In order to understand how the option-based model can be applied and to compare its results with that of simulation models, a simple case study is considered. A company's evaluation of the replacement of a transfer line with an FMS is considered for this purpose.

The increase in product flexibility could lead to a reduction in the cost of introducing a new product from \$100 000 to \$50 000. The company assumes the probability of demand of the new product to be 0.5. Based on records and market analysis, the company also estimates that the cash flow from each new product opportunity can be described by a normal distribution N (120 000; 40 000).

Simulation models determine the yearly effect of FMS product flexibility by multiplying the reduction in the adaptation cost (\$100 000 – \$50 000) by the probability of demand for the new product (0.5). This leads to a reduction of \$25 000 in the cost of introducing the new product.

Actually, the new product will be introduced only if its cash flow is more than \$100 000 in the transfer line, and more than \$50 000 in the FMS. As a consequence, by applying equation (8.2) the value of product option would be \$66 240 for the FMS and \$26 400 for the transfer line. If, as mentioned above, the new product is demanded by the customer with a probability of 0.5, then the correct impact of the increase in product flexibility is \$19 840 a year. This amount should be included in cash flow analysis.

8.5 Option-based models v. *simulation models*

The option-based models overcome both limitations of the simulation-based models. They allow, in fact, a quick evaluation process based on closed analytical expressions and take into account also the role of management. However, the option-based models have some disadvantages when compared with the simulation models.

The main disadvantage is the joint effect between different options. For instance, the cost of introducing a new product depends not only on product flexibility but also on available capacity. In fact, if there is no idle capacity, then there will be a need to expand the system and then the expansion flexibility should be considered. In the three options described above it is considered that each kind of perturbation is independent. So far, no model which can take into account the interaction among perturbations has been introduced.

So it can be inferred that although there is no specific model available to evaluate the economic effects of flexibility in manufacturing, whatever is available can be used as a set of tools to tackle the present needs.

Simulation models, in particular, should be preferred in complex situations where different kinds of elementary flexibilities and their joint effects are important, and must be evaluated. Option theory, on the contrary, can give better results at a lower cost when only single dimensions of flexibilities are involved.

8.6 Conclusions

In this paper, a discussion of the economic evaluation of FMS and, in particular, the impact of flexibility on that evaluation was conducted. One of the main purposes of the paper is to show that there is no conceptual reason as to why flexibility should not be included in cash-flow analysis.

Two different approaches were presented and compared: simulation models and option-based models that can be used to evaluate the impact of flexibility.

More generally, the relevance as to when it is worthwhile to use one of these methods instead of traditional static DCF analysis, was discussed.

It was pointed out that there is no conceptual reason that prevents the inclusion of flexibility in cash flow analysis. However, this does not mean that companies must follow this approach under all circumstances. In fact, only when the company operates in turbulent environments, would the effort of the evaluation process be 'economically' justified. It is predicted that in the future the importance of flexibility in industries will increase and, as a consequence, more companies would need better techniques for measuring the value of flexibility.

8.7 References

Azzone, G. and Bertele', U. (1988) Synergetic effects and double counting risks in evaluating strategic investments in manufacturing flexibility. *8th Strategic Management Society Annual International Conference*, Amsterdam, October.

Azzone, G. and Bertele', U. (1989) Measuring the economic effectiveness of flexible automation: a new method. *International Journal of Production Research*, **17**, 5.

Azzone, G. Bertele', U. and Condini, M. (1989) L'utilizzo della teoria delle opzioni per la valutazione degli investimenti in tecnologie produttive: problemi e prospettive. WP 89-022. Dipartimento di Elettronica, Politecnico di Milano.

Barad, M. and Sipper, D. (1988) Flexibility in manufacturing systems: definitions and Petri net modelling. *International Journal of Production Research*, **26**, 2.

Black, F. and Scholes, M. (1973) The pricing of options and corporate liabilities. *Journal of Political Economy*, 81.

Brealey, R. and Myers, S. (1988) *Principles of Corporate Finance*, McGraw-Hill, **2**, (2), 114–17.

Browne, J., Dobois, D., Rathmill, K., Seithi, S. and Stecke, K.E. (1984) Classification of flexible manufacturing systems. *The FMS Magazine*, 2.2, Singapore.

Gold, B. (1988) Charting a course to superior technology evaluation. *Sloan Management Review*, pp. 19–39.

Hutchinson, G.K. and Holland J.R. (1982) The economic value of flexible automation. *Journal of Manufacturing Systems*, **6**, (2), 117–36.

Kalkunte, M.V., Sarin, S.C. and Whilelm, W.E. (1986) Flexible manufacturing systems: a review of modeling approaches for design, justification and operation. In *Flexible Manufacturing Systems: Methods and Studies*. Elsevier Amsterdam.

Kensinger, J.W. (1988) The capital investment project as a set of exchange options. *Managerial Finance*, **14**, 2/3.

Kester, C.W. (1984) Today's options for tomorrow growth. *Harvard Business Review*, **62**, (4), 153–60.

Swamidass, P.M. (1988) Manufacturing flexibility. *Operations Management Association Monograph*, No. 2, Waco, Texas.

9 Multivariate Learning Curve Model for Manufacturing Economic Analysis

Adedeji B. Badiru

9.1 Introduction

Learning, in the context of operations management, refers to the improved efficiency obtained from repetition of a production operation. Workers learn and improve by repeating operations. Learning is time-dependent and externally controllable. Several research studies have confirmed that human performance improves with reinforcement or frequent repetitions (Yelle, 1979). Several publications have documented the historical aspects of the evolution of the learning curve (Hirchman, 1964; Conley, 1970; Yelle, 1979; Belkaoui, 1986). Learning curves appear to be the most discussed subject in cost analysis in industry.

Reductions in operation processing times achieved through learning curve effects can translate directly to cost savings for manufacturers and improved morale for employees (Badiru, 1988). Learning curves are essential for setting production goals and monitoring progress (Richardson, 1978). The effect of learning should be of importance to all manufacturers. If the effect of learning is not considered, costs and operation times may be overestimated and result in waste and inefficiency (Knecht, 1974; Abernathy and Wayne 1974; Yelle, 1980).

Learning curves present the relationship between production cost (or time) and production volume based on the effect of learning. For example, an early study by Wright (1936) disclosed an '80% learning' effect which indicates that a given operation is subject to a 20% productivity improvement each time the production quantity doubles. With this type of information, a learning curve can serve as a predictive tool for obtaining time estimates for tasks that are repeated within a production cycle (Chase and Aquilano, 1981). Learning curves are applicable to all aspects of manufacturing planning and control. Several terms have been used to describe the learning phenomenon over the years. Some of the terms synonymous with 'learning curve' include:

1. Manufacturing progress function;
2. Cost-quantity relationship;
3. Cost curve;
4. Product acceleration curve;
5. Improvement curve;
6. Performance curve;
7. Experience curve; and
8. Efficiency curve.

The topic of interest in this paper is the manufacturing progress function. However, the term 'learning curve' will be used throughout for the sake of consistency of terminology. Since production time and cost are related inherently, both terms are often used interchangeably in learning curve analysis. For consistency, cost is used as the basis for discussion in this paper.

The paper presents a survey of the various learning-curve models available for performing manufacturing economic analysis. The paper also extends the previous developments in learning-curve analysis to a framework for developing a multivariate learning-curve model. Such a model is needed to account for the multivariate influence on learning and, thus, facilitate a wider scope of learning-curve analysis for manufacturing operations.

9.2 Learning-curve models

The conventional learning-curve model presents several limitations in practice. For example, the model is not completely defined because it does not include an error term to account for lack of perfect fit for historical data. Since the first formal publication of learning-curve theory in 1936 (Wright, 1936), there have been numerous alternative propositions concerning the geometry and functional form of the learning curve (Baloff, 1971; Jewell, 1984; Kopcso, 1983; Smunt, 1986; Towill and Kaloo, 1978; Yelle, 1983). Some of the most notable versions of the learning curve include:

1. The log-linear model (Wright, 1936);
2. The S-curve (Carr, 1946);
3. The Stanford-B model (Asher, 1956);
4. DeJong's learning formula (DeJong, 1957);

5. Levy's adaptation function (Levy, 1965);
6. Glover's learning formula (Glover, 1966);
7. Pegels' exponential function (Pegels, 1969);
8. Knecht's upturn model (Knecht, 1974);
9. Yelle's product model (Yelle, 1976); and
10. Multiplicative Power Model (Waller and Dwyer, 1981)

9.2.1 The log-linear model

The log-linear model (Wright, 1936) is often referred to as the conventional learning-curve model. This model states that the improvement in productivity is constant (constant slope) as output increases. There are two basic forms of the log-linear model: the average-cost function and the unit-cost function.

Average-cost model

The average-cost model is more popular than the unit-cost model. It specifies the relationship between the cumulative average cost per unit and cumulative production. The relationship indicates that cumulative cost per unit will decrease by a constant percentage as the cumulative production volume doubles. The model is expressed as:

$$C_x = C_1 x^b$$

$$= C_1 x^{\left(\frac{\log p}{\log 2}\right)}$$

where: C_x is the cumulative average cost of producing x units;
C_1 is the cost of the first unit;
b is the learning-curve exponent (i.e. constant slope of the learning curve on log-log paper); p is the percentage learning rate; x is the cumulative production count; $b = (\log p)/(\log 2)$; and $p = 2^b$.

When linear graph paper is used, the log-linear learning curve is a hyperbola of the form shown in Fig. 9.1. On log-log paper, the model is represented by the following straight-line equation:

$$\log C_x = \log C_1 + b \log x$$

where b is the constant slope of the line. The expression for p is derived by considering two production levels where one level is double the other.

Level I: x_1

Level II: $x_2 = 2x_1$

Then:

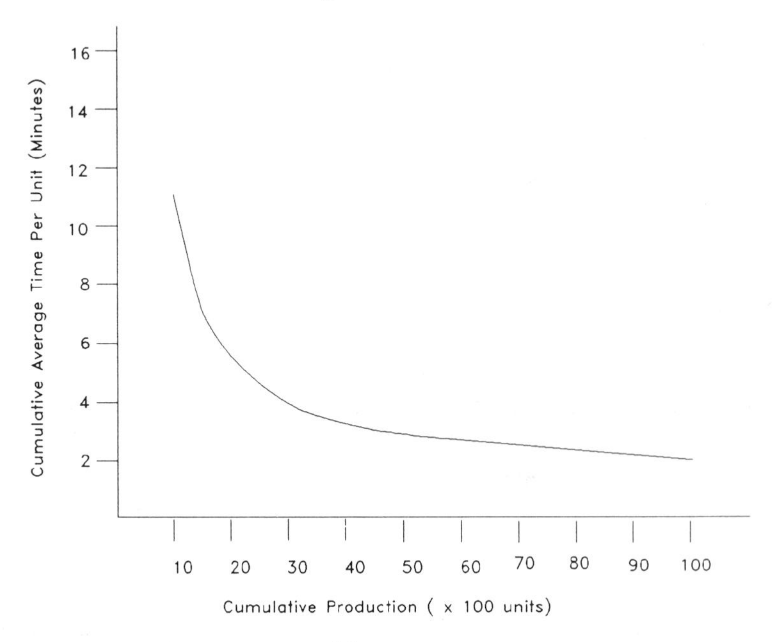

Fig. 9.1 *Log-linear learning-curve model.*

$$C_{x1} = C_1(x_1)^b$$
$$C_{x2} = C_1(2x_1)^b$$

The percentage productivity gain is then computed as:

$$p = \frac{C_1(2x_1)^b}{C_1(x_1)^b}$$
$$= 2^b$$

For example, assume that 50 units are produced at a cumulative average cost of \$20 per unit. We want to compute the learning percentage when 100 units are produced at a cumulative average cost of \$15 per unit. That is,

At first production level: units = 50; cost = \$20

At second production level: units = 100; cost = \$15.

Using the log relationship, we obtain the following simultaneous equations:

$$\log 20 = \log C_1 + b \log 50$$
$$\log 15 = \log C_1 + b \log 100$$

Subtracting the second equation from the first yields

$$\log 20 - \log 15 = b \log 50 - b \log 100$$

That is:

$$b = \frac{\log 20 - \log 15}{\log 50 - \log 100}$$
$$= \frac{\log(20/15)}{\log(50/100)}$$
$$= -0.415$$

Therefore:

$$p = (2)^{-0.415}$$
$$= 0.75$$
$$= 75\% \text{ learning rate}$$

In general:

$$b = \frac{\log C_{x1} - \log C_{x2}}{\log x_1 - \log x_2}$$

where: x_1 is the first production level; x_2 is the second production level; C_{x1} is the cumulative average cost per unit at the first production level; and C_{x2} is the cumulative average cost per unit at the second production level.

Expression for total cost

Using the basic cumulative average cost function, the total cost of producing x units is computed as:

$$TC_x = (x)C_x$$
$$= (x)C_1x^b$$
$$= C_1x^{(b+1)}$$

Expression for unit cost

The unit cost of producing the xth unit is given by:

$$UC_x = C_1x^{(b+1)} - C_1(x - 1)^{(b+1)}$$
$$= C_1[x^{(b+1)} - (x - 1)^{(b+1)}]$$

Expression for marginal cost

The marginal cost of producing the xth unit is given by:

$$MC_x = \frac{d[TC_x]}{dx}$$
$$= (b + 1)C_1x^b$$

Let us consider another scenario. In a production run of a certain high-precision component of a jet engine, it was observed that the cumulative hours required to produce 100 units are 100 000 hours with a learning curve effect of 85%. For project-planning purposes, an industrial engineer needs to calculate the number of hours spent in producing the 50th unit.

Solution

Following the notation we used before, we have the following information:

$$P = 0.85$$
$$X = 100 \text{ units}$$
$$C_X = 100\,000 \text{ hours/100 units} = 1000 \text{ hours/unit}$$

Now:

$$0.85 = 2^m$$

Therefore:

$$m = -0.2345$$

Also:

$$1000 = C_1(100)^m$$

Therefore:

$$C_1 = 2944.42 \text{ hours}$$

Thus:

$$C_{50} = C_1(50)^m$$
$$= 1176.50 \text{ hours}$$

That is, the cumulative average hours for 50 units is 1176.50 hours. Therefore, the cumulative total hours for 50 units = 58 824.91 hours.

Also:

$$C_{49} = C_1(49)^m$$
$$= 1182.09 \text{ hours}$$

That is, the cumulative average hours for 49 units is 1182.09 hours. Therefore, the cumulative total hours for 49 units = 57 922.17 hours.

Consequently, the number of hours for the 50th unit are given by:

$$C_{50} - C_{49} = 58\,824.91 \text{ hours} - 57\,922.17 \text{ hours}$$
$$= 902.74 \text{ hours}$$

The unit-cost model

The unit-cost model is expressed in terms of the specific cost of producing the xth unit. The unit-cost formula specifies that the individual cost per unit will decrease by a constant percentage as cumulative production doubles. The functional form of the unit-cost model is the same as for the average-cost model except that the interpretations of the terms are different. It is expressed as:

$$UC_x = C_1 x^b$$

where: UC_x is the cost of producing the xth unit; C_1 is the cost of the first unit; x is the cumulative production count; b is the learning curve exponent = (log p)/(log 2); and p is the percentage learning rate = 2^b.

From the unit-cost formula we can derive expressions for the other cost basis. For the discrete case, the total cost of producing x units is given by:

$$TC_x = \sum_{i=1}^{x} UC_i$$
$$= C_1 \sum_{i=1}^{x} (i)^b$$

The cumulative average cost per unit is given by:

$$Y_x = \frac{TC_x}{x}$$
$$= \frac{1}{x} \sum_{i=1}^{x} UC_i$$
$$= \frac{C_1}{x} \sum_{i=1}^{x} (i)^b$$

The marginal cost is found as follows:

$$MC_x = \frac{\mathrm{d}[TC_x]}{\mathrm{d}x}$$
$$= \frac{\mathrm{d}\left[C_1 \sum_{i=1}^{x} (i)^b\right]}{\mathrm{d}x}$$

For the continuous case, we have the following corresponding expressions:

$$
\begin{aligned}
TC_x &= \int_0^x UC(z)\,\mathrm{d}z \\
&= C_1 \int_0^x z^b \mathrm{d}z \\
&= \frac{C_1 x^{(b+1)}}{b+1} \\
Y_x &= \left(\frac{1}{x}\right)\frac{C_1 x^{(b+1)}}{b+1} \\
MC_x &= \frac{\mathrm{d}[TC_x]}{\mathrm{d}x} \\
&= \frac{\mathrm{d}\left[\dfrac{C_1 x^{(b+1)}}{b+1}\right]}{\mathrm{d}x} \\
&= C_1 x^b.
\end{aligned}
$$

9.2.2 The S-curve

Carr (1946) proposed an S-shaped learning-curve function based on an assumption of a gradual start up. The function has the shape of the cumulative normal distribution function for the start-up curve and the shape of an operating characteristics function for the learning curve. The gradual start up is based on the fact that the early stages of production are typically in a transient state with changes in tooling, methods, materials, design, and even the workers. The basic form of the S-curve formula is:

$$MC_x = C_1[M + (1 - M)(x + B)^b]$$

where: M is the incompressibility factor (a constant); and B is the equivalent experience units (a constant).

Assumptions about at least three out of the four parameters (M, B, C_1 and b) are needed to solve for the fourth one. Alternatively, the coefficients of the S-curve can be determined by fitting a cubic curve on a log-log plot. An example of such a cubic function is:

$$\log MC_x = A + B(\log x) + C(\log x^2) + D(\log x^3)$$

9.2.3 The Standard-B model

An early study commissioned by the US Defense Department at the Stanford Research Institute (Asher, 1956) led to the development of the

Stanford-B model, which was found to be more representative of World War II data. The model is represented as:

$$Y_x = C_1(x + B)^b$$

where: Y_x is the direct cost of producing the xth unit; C_1 is the cost of the first unit when $B = 0$; b is the slope of the asymptote; and B is a constant, which is the equivalent units of previous experience at the start of the process, that is the number of units produced prior to first unit acceptance ($1 < B < 10$).

It is noted that when $B = 0$, the Stanford-B model reduces to $Y_x = C_1x^b$, which is the conventional log-linear model. The Boeing Company found that the Stanford-B model was the best model for the manufacturing of the Boeing 707. Hoffman (1950) found that the inclusion of the B parameter resulted in smaller sums of squared deviations.

9.2.4 DeJong's learning formula

DeJong (1957) presented a power function which incorporates parameters for the proportion of manual activity in a task. When operations are controlled by manual tasks, the time will be compressible as successive units are completed. If, by contrast, machine cycle times control operations, then the time will be less compressible as the number of units increases. DeJong's formula introduces an incompressible factor, M, into the log-linear model to account for the man: machine ratio. The model is expressed as:

$$MC_x = C_1[M + (1 - M)x^{-b}]$$

where M is the incompressibility factor (constant).

When $M = 0$, the model reduces to the log-linear model, which implies a completely manual operation. In completely machine-dominated operations, $M = 1$. In that case, the unit cost reduces to a constant equal to C_1, which suggests that no cost improvement is possible in machine-controlled operations. This represents a condition of high incompressibility. Regrettably, no significant published data are available on whether or not DeJong's model has been used successfully to account for the degree of automation in any given operation. With the increasing move towards automation in industry, this certainly is a topic for urgent research.

9.2.5 Levy's adaptation function

Recognizing that the log-linear model does not account for leveling off of production rate and the factors that may influence learning, Levy (1965) proposed the following learning cost function:

$$MC_x = \left[\frac{1}{\beta} - \left(\frac{1}{\beta} - \frac{x^b}{C_1}\right)k^{-kx}\right]^{-1}$$

where: β is the production index for the first unit; and k is the constant used to flatten the learning curve for large values of x.

The flattening constant k forces the curve to reach a plateau instead of continuing to decrease or turning in the upward direction.

9.2.6 Glover's learning formula

Glover (1966) presented a model which incorporates a work-commencement factor. The model is based on a bottom-up approach which uses individual worker learning results as the basis for plant-wide learning-curve standards. The functional form of the model is expressed as:

$$\sum_{i=1}^{n} y_i + a = C_1\left(\sum_{i=1}^{n} x_i\right)^m$$

where: y_i is the elapsed time or cumulative quantity; x_i is the cumulative quantity or elapsed time (y and x are used interchangeably); a is the commencement factor; n is the index of the curve ($= 1 + b$); and m is the model parameter.

If Σy_i is used to represent the total elapsed time, then x_i will be 1. That is:

$$\sum_{i=1}^{n} x_i = n$$

and

$$\sum_{i=1}^{n} y_i - a = C_1 n^m$$

9.2.7 Pegels' exponential function

Pegels (1969) presented an alternative algebraic function for the learning curve. His model, a form of an exponential function, is represented as:

$$MC_x = \alpha a^{x-1} + \beta$$

where α, β and a are parameters based on empirical data analysis.

The total cost of producing x units is derived from the marginal cost as follows:

$$TC_x = \int (\alpha a^{x-1} + \beta)\,\mathrm{d}x$$

$$= \frac{\alpha a^{x-1}}{\ln(a)} + \beta x + c$$

where c is a constant to be derived after the other parameters are found. The constant can be found by letting the marginal cost, total cost and average cost of the first unit all be equal. That is, $MC_1 = TC_1 = AC_1$. This yieds:

$$C = \alpha - \frac{\alpha}{\ln(a)}$$

Pegels' exponential model assumes that the marginal cost of the first unit is known. Thus:

$$\begin{aligned} MC_1 &= \alpha + \beta \\ &= y_0. \end{aligned}$$

Pegels also presented another mathematical expression for the total labour cost in start-up curves (Pegels, 1976). He expressed the total cost as:

$$TC_x = \frac{a}{1 - b} x^{1-b}$$

where: x is the cumulative number of units produced; and a, b are empirically determined parameters.

The expressions for marginal cost, average cost and unit cost can be derived as shown earlier for other models.

9.2.8 Knecht's upturn model

Knecht (1974) presents a modification to the functional form of the learning curve to express analytically the observed divergence of actual costs from those predicted by learning curve theory when units produced exceed 200. This permits the consideration of non-constant slopes for the learning curve model. If UC_x is defined as the unit cost of the xth unit, then it approaches 0 asymptotically as x increases. To avoid a zero limit unit cost, the basic functional form is modified. In the continuous case, the formula for cumulative average costs is derived as:

$$\begin{aligned} C_x &= \int_0^x C_1 z^b \mathrm{d}z \\ &= \frac{C_1 x^{b+1}}{(1 + b)} \end{aligned}$$

This cumulative cost also approaches zero as x goes to infinity. Knecht alters the expression for the cumulative curve to allow for an upturn in the learning curve at large cumulative production levels. He suggested the functional form:

$$C_x = C_1 x^b e^{cx}$$

where c is a second constant. Differentiating the modified cumulative average cost expression gives the unit cost of the xth unit as shown below:

$$\begin{aligned} UC_x &= \frac{d}{dx}[Y_1 x^b e^{cx}] \\ &= Y_1(bx^{b-1}e^{cx} + cx^b e^{cx}) \\ &= Y_1 x^b e^{cx}\left(c + \frac{b}{x}\right) \end{aligned}$$

A model similar to Knecht's model is the plateau model described by Baloff (1971). The plateau model assumes that production cost reaches a steady state at which point cost levels off.

9.2.9 Multiplicative power model

Waller and Dwyer (1981) discuss alternative models for parametric cost analysis. One of the cost estimating relationships they proposed is of the form:

$$\text{Cost} = b_0 x_1^{b_1} x_2^{b_2} \ldots x_n^{b_n} \varepsilon$$

where: b_0 is the model coefficient; x_i is the ith independent variable ($i = 1, 2, \ldots, n$); b_i is the exponent of the ith variable; and ε is an error term.

The model above can be fitted by using logarithmic transformation and standard multiple linear regression technique. The model was said to have been fitted successfully for missile-tooling and test-equipment cost. Another cost-estimating model presented by Waller and Dwyer is of the form:

$$\text{Cost} = c_1 x_1^{b_1} + c_2 x_2^{b_2} + \ldots + c_n x_n^{b_n} + \varepsilon$$

where c_i ($i = 1, 2, \ldots, n$) is the coefficient of the ith independent variable.

9.2.10 Yelle's combined-product learning curve

Yelle (1976) proposed a learning-curve model for products by aggregating and extrapolating the individual learning curves of the operations making up a product. The model, which is similar to one of the cost-estimating relationships presented by Waller and Dwyer (1981), is expressed as:

$$C_x = k_1 x_1^{b_1} + k_2 x_2^{b_2} + \ldots + k_n x_n^{b_n},$$

where: C_x is the cost of producing the xth unit of the product; n is the number of operations making up the product; and $k_i x_i^{b_i}$ is the learning curve for the ith operation.

Observations drawn from the mathematical expression above were then used to fit a least-squares function of the log-linear form for the product linear curve. As pointed out by Howell (1980), Yelle's model contains several deficiencies. Some of the noted shortcomings are:

1. A learning curve formulated by aggregating several different learning curves (with different slopes) will not necessarily be a straight line on a linear plot.
2. A learning curve extrapolated from different learning curves will not necessarily be a straight line.
3. A product-specific learning curve seems to be a more reasonable model than an integrated product curve.

For example, an aggregated learning curve with 96.6% learning rate obtained from individual learning curves with the respective learning rates of 80%, 70%, 85%, 80% and 85% does not appear to represent reality. If this type of synergism is always possible, then one can always improve the learning rate for any operation by decomposing it into smaller integrated operations. Figure 9.2 shows comparative plots of some of the models discussed above on a log-log scale.

9.3 Multivariate learning curve model

The learning rate of employees is often influenced by other factors that are within the control of the organization. The conventional learning-curve model is developed as a function of production level only. But, there are other factors apart from time that can influence how fast, how far and how well a production operator learns within a given time horizon. The overall effect of learning is influenced by several factors including such things as:

1. Skill level;
2. Level of experience;
3. Level of prior training;
4. Amount of concurrent training;
5. Design changes;
6. Methods improvement;
7. Material substitutions;
8. Changes in tolerance levels;
9. Complexity of the task;
10. Degree of external interference affecting the task and/or operator;

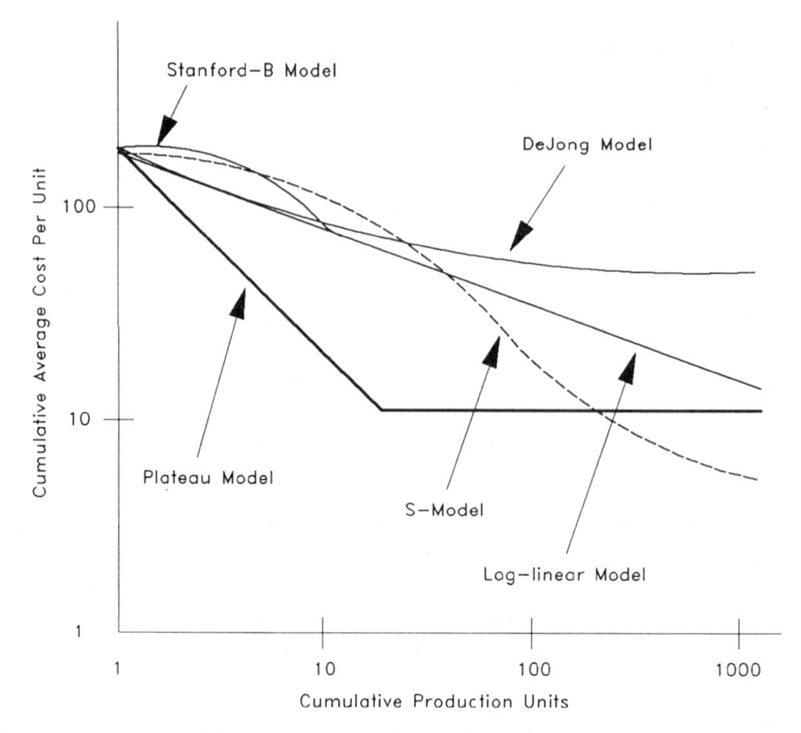

Fig. 9.2 *Comparison of learning-curve models on a log scale.*

11. Level of competence with available tools; and
12. Prior experience with related job functions (transfer of skill).

9.3.1 Model formulation

Conway and Schultz (1959) first suggested the need for multivariate generalized progress function. They point out that there are other factors that influence cost in learning-curve analysis. For example, they present a hypothetical response surface relating cost to production rate and cumulative production volume. This paper extends that idea to an analytical model development. To account for the multivariate influence on learning, a model is proposed here to facilitate a wider scope of learning-curve analysis for manufacturing operations. The general form of the multivariate model is given as:

$$C_x = K \prod_{i=1}^{n} c_i x_i^{b_i},$$

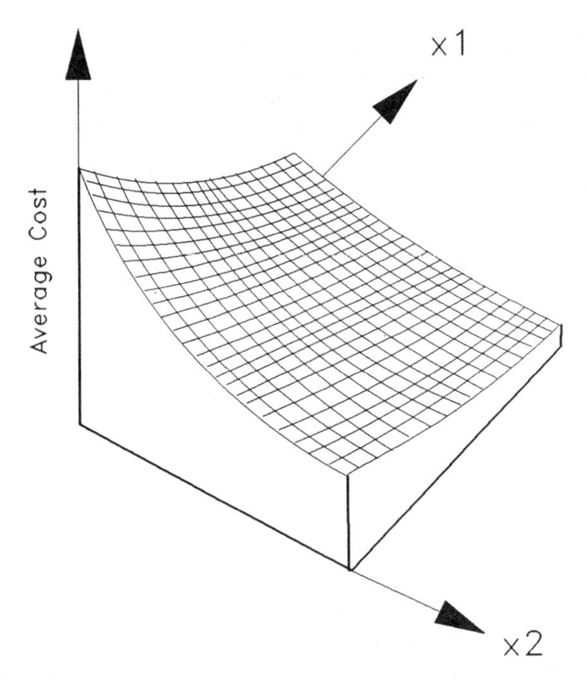

Fig. 9.3 *Generic response surface for a two-factor learning-curve model.*

where: C_x is the cumulative average cost per unit for a given set of factor values; K is a model parameter, which equals the cost of the first unit of the product; x is the vector of specific values of independent variables (factors); x_i is the specific value of the ith factor; n is the number of factors in the model; c_i is the coefficient for the ith factor; and b_i is the learning exponent for the ith factor.

Figure 9.3 shows a generic example of the response surface for a two-factor ($n = 2$) learning-curve model involving production volume (x_1) and time (x_2). For ease of analysis, production volume is assumed to be a continuous variable. Including time as a second variable in the learning-curve model is reasonable because most of the other factors that can influence learning can be expressed as time-dependent variables.

9.3.2 A two-factor model

With a two-factor model the expected cost of production can be estimated not only on the basis of volume-dependent learning, but also on the basis of time. Previous learning-curve models have considered the increase in productivity due to increases in production levels. But not much con-

TABLE 9.1 *Multivariate learning curve data*

Cumulative average cost	Cumulative production units	Elapsed time (h)
120	10	11
95	20	54
80	40	100
65	80	220
55	160	410
40	320	660
32	640	810
25	1280	890
20	2560	990
19	5120	1155

sideration has been given to the fact that production level itself is time-dependent and influenced by other factors. Carlson (1973) pointed out that multislope and curvilinear learning-curve models have not received the recognition they deserve, probably due to the 'straight-line' syndrome of the log-linear model. This paper attempts to rectify that long-standing neglect. For a two-factor model, we present a specific mathematical expression for the learning curve:

$$C_{x_1x_2} = Kc_1x_1^{b_1}c_2x_2^{b_2}$$

where: C_x is the cumulative average cost per unit for a given set of factor values; K is the model parameter; x_1 is the specific value of the first factor; x_2 is the specific value of the second factor; c_i is the coefficient for the ith factor; and b_i is the learning exponent for the ith factor.

The set of data in Table 9.1 is used to illustrate how to fit a multivariate learning-curve function. It should be noted that more data cycles should be used in fitting any model that will be used for operational analysis. The more data sets used, the more accurately will the fitted surface represent the relationship among the factors. The data used here are for illustrative purposes only. The two-factor model is represented in logarithmic scale to facilitate the curve-fitting procedure as shown below:

$$\begin{aligned}\log C_x &= [\log K + \log(c_1c_2)] + b_1 \log x_1 + b_2 \log x_2 \\ &= \log a + b_1 \log x_1 + b_2 \log x_2\end{aligned}$$

where a represents the combined constant in the model.

Figure 9.4 shows a 3-D plot of the raw data in Table 9.1, while Fig. 9.5 shows a 3-D plot of the log-transformed data. Using the multiple regression function in STATGRAPHICS software, the following model was fitted:

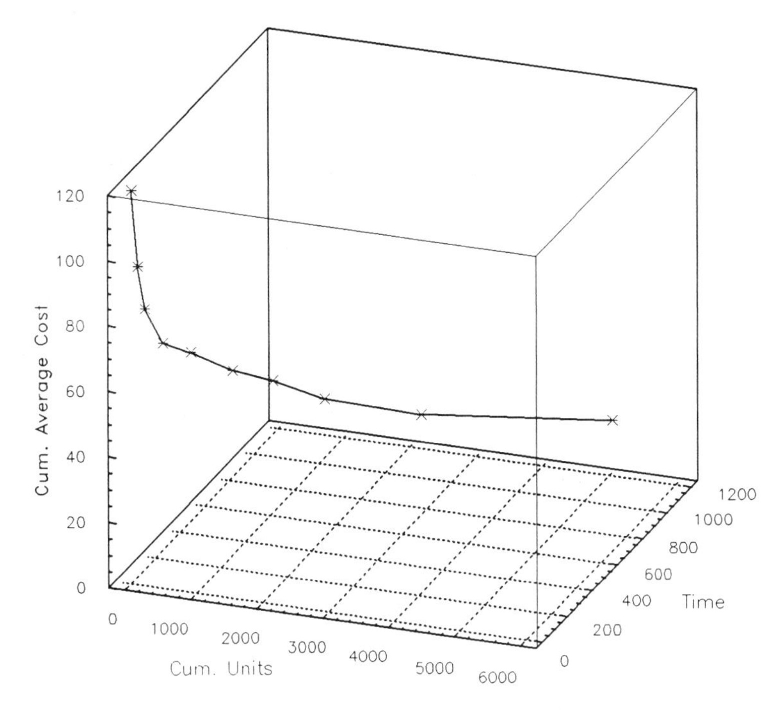

Fig. 9.4 *3-D plot of multivariate learning-curve data.*

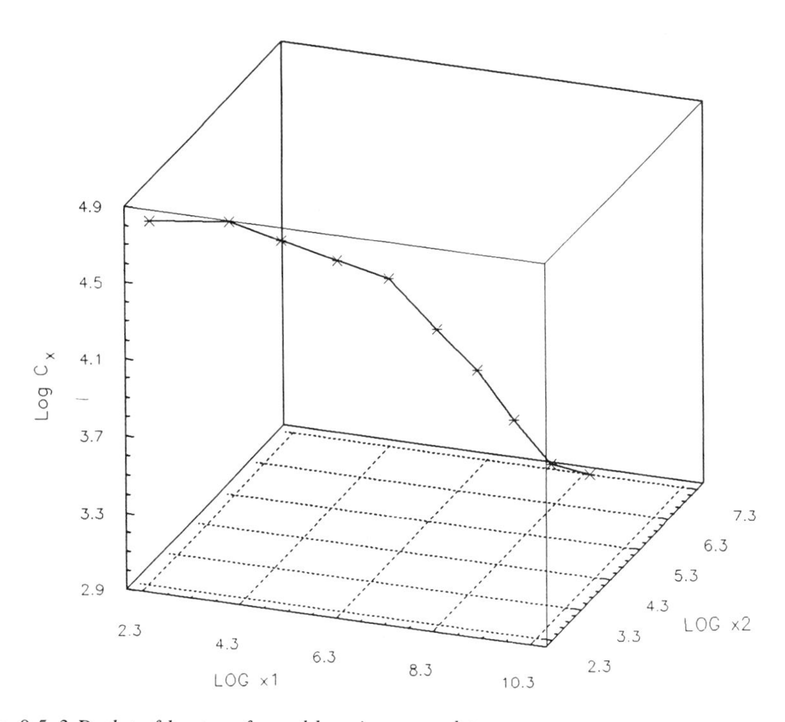

Fig. 9.5 *3-D plot of log-transformed learning-curve data.*

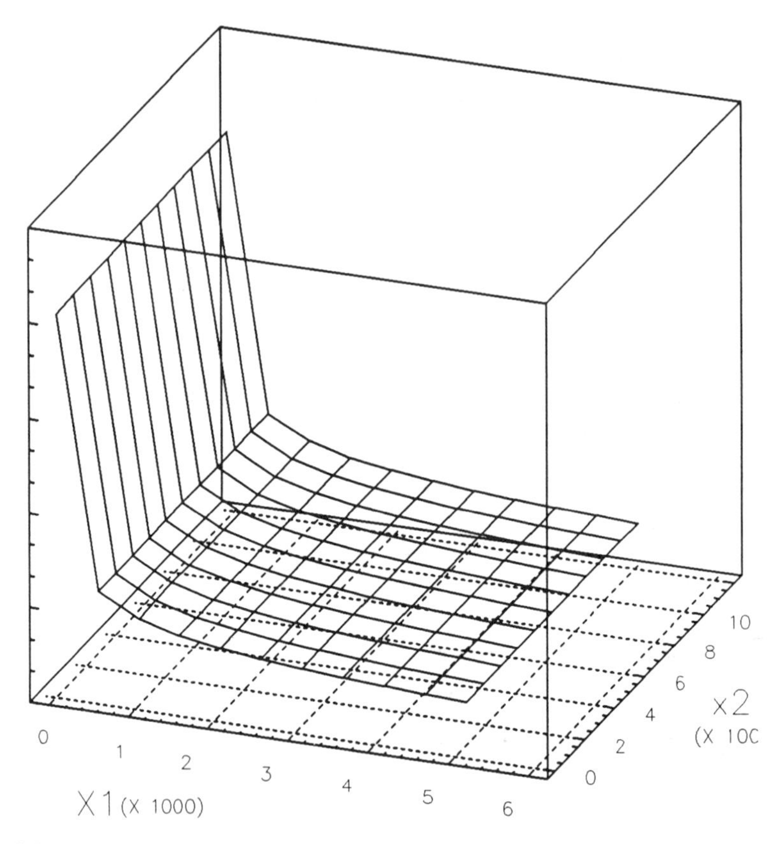

Fig. 9.6 *Response surface of the fitted multivariate learning-curve model.*

$$\log C_x = 5.5130 - 0.3140 \log x_1 + 0.0020 \log x_2$$

which transforms to the multiplicative model given by:

$$C_x = 247.8937 x_1^{-0.314} x_2^{0.002}$$

where $a = (Kc_1c_2) = 247.8937$ and $\log(a) = 5.5130$. If two of the constants K, c_1 or c_2 are known, then the third can be computed. The constants may be determined empirically from the analysis of historical data. Figure 9.6 shows the response surface for the fitted multiplicative multivariate learning-curve model. Diagnostic statistical analyses indicate that the model is a good fit for the data. The 95% confidence intervals for the parameters in the model are shown in Table 9.2. Given a production level of 6000 units and an elapsed production time of 1300 hours, the multivariate model indicates an estimated cumulative average cost per unit shown below:

TABLE 9.2 *95% Confidence Interval for Model Parameters*

Parameter	Estimate	Lower limit	Upper limit
log a	5.5130	5.2986	5.7273
b_1	−0.3140	−0.3762	−0.2518
b_2	0.0020	−0.0829	+0.0869

$$C_{(6000,\ 1300)} = (247.8937)(6000^{-0.314})(1300^{0.002})$$
$$= 16.37$$

9.3.3 Manufacturing economic analysis

Results of multivariate learning-curve analysis are important for various types of economic analysis in the manufacturing environment. The declining state of manufacturing in the United States has been a subject of much discussion in recent years. A reliable methodology for cost analysis of manufacturing technology for specific operations is essential to the full exploitation of the recent advances in the available technology. Manufacturing economic analysis is the process of evaluating manufacturing operations on a cost basis. In manufacturing systems, many tangible and intangible, quantitative and qualitative factors intermingle to compound the cost-analysis problem. Consequently, a more comprehensive evaluation methodology such as a multivariate learning curve can be very useful.

Break-even analysis

The conventional break-even analysis assumes that variable cost per unit is constant. On the contrary, learning-curve analysis recognizes the potential reduction in variable cost per unit due to the effect of learning. Due to the multiple factors involved in manufacturing, multivariate learning-curve models should be investigated and adopted for break-even cost analysis.

Make-or-buy decision

Make-or-buy decisions can be enhanced by considering the effect of learning on items that are manufactured in-house. Make-or-buy analysis involves a choice between the cost of producing an item and the cost of purchasing it. Multivariate learning curves can provide the data for determining accurately the cost of producing an item. A make-or-buy analysis can be coupled with break-even analysis to determine at which production level it will become cost effective to make a product versus buying it.

Manpower scheduling

A consideration of the effect of learning in the manufacturing environment can lead to a more accurate analysis of manpower requirements and the accompanying schedules. In integrated production, where parts move sequentially from one production station to another, the effect of multivariate learning curves can become even more applicable. The allocation of resources (Liao, 1979) during production scheduling should not be made without consideration for the effect of learning.

Production planning

The overall production-planning process can benefit from multivariate learning-curve models. Pre-production planning analysis of the effect of multivariate learning curves can identify areas where better and more detailed planning may be needed. The more pre-production planning that is done, the higher the potential for lowering the production cost of the first unit, which has a significant effect on learning-curve analysis.

Labour estimating

Carlson (1973) showed that the validity of log-linear learning curves may be suspect in many labour-analysis problems. For manufacturing activities involving operations in different stations, several factors interact to determine the learning rate of workers. Multivariate curves can be of use in developing accurate labour standards in such cases. Multivariate learning-curve analysis can complement conventional work-measurement studies.

Budgeting and resource allocation

Budgeting or capital rationing is a significant effort in any manufacturing operation. Multivariate learning-curve analysis can provide a management guide for allocating resources to production operations on a more equitable basis. The effects of learning can be particularly useful in zero-base budgeting policies (Badiru, 1988). Other manufacturing cost analysis where a multivariate learning-curve analysis could be of use include bidding (Yelle, 1979), inventory analysis, productivity improvement programmes (Towill and Kaloo, 1978), setting goals (Richardson, 1978) and lot sizing (Kopcso, 1983).

The learning-curve phenomenon has been of interest to researchers and practitioners for many years. The variety of situations to which learning curves are applicable has necessitated the development of various functional forms for the curves. The conventional view of learning curves considers only one factor at a time as the major influence on productivity improvement. But in today's integrated manufacturing environment, it is obvious that several factors interact to activate and perpetrate productivity improvement. This paper presents a general multivariate learning-curve model. The model is useful for various aspects of manufacturing economic analysis such as break-even analysis, make-or-buy decisions, manpower scheduling, production planning, budgeting, resource allocation, labour estimating and cost optimization.

9.4 References

Abernathy, W.J. and K. Wayne (1974) Limits of the learning curve. *Harvard Business Review*, **52**, 109–19.

Asher, H. (1956) Cost-quantity relationships in the airframe industry. *Report No. R-291*, July, The Rand Corporation, Santa Monica, California.

Badiru, A.B. (1988) *Project Management in Manufacturing and High Technology Operations*, Wiley, New York.

Baloff, N. (1971) Extension of the learning curve:- some empirical results. *Operations Research Quarterly*, **22** (4), 329–40.

Belkaoui, A. (1986) *The Learning Curve*, Quorum Books, Westport, Connecticut.

Carlson, J.G.H. (1973) Cubic learning curves: precision tool for labor estimating. *Manufacturing Engineering and Management*, **71** (5), 22–5.

Carr, G.W. (1946) Peacetime cost estimating requires new learning curves. *Aviation*, **45**, 76–7.

Chase, R.B. and Aquilano, N.J. (1981) *Production and Operations Management*, Irwin, Homewood, Illinois.

Conley, P. (1970) Experience curves as a planning tool. *IEEE Spectrum*, **7** (6), 63–8.

Conway, R.W. and Schultz, A., Jr (1959) The manufacturing progress function. *Journal of Industrial Engineering*, **1**, 39–53.

DeJong, J.R. (1957) The effects of increasing skill on cycle time and its consequences for time standards. *Ergonomics*, November, 51–60.

Glover, J.H. (1966) Manufacturing progress functions: an alternative model and its comparison with existing functions. *International Journal of Production Research*, **4** (4), 279–300.

Hirchmann, W.B. (1964) Learning curve. *Chemical Engineering*, **71** (7), 95–100.

Hoffman, F.S. (1950) Comments on the modified form of the aircraft progress functions. *Report No. RN-464*, The Rand Corporation, Santa Monica, California.

Howell, S.D. (1980) Learning curves for new products. *Industrial Marketing Management*, **9** (2), 97–9.

Jewell, W.S. (1984) A generalized framework for learning curve reliability growth models, *Operations Research*, **32** (3), May–June, 547–58.

Karger, D.W. and Bayha, F.H. (1977) *Engineered Work Measurement* (3rd edn), Industrial Press, New York.

Knecht, G.R. (1974) Costing, technological growth and generalized learning curves. *Operations Research Quarterly*, **25** (3), 487–91.

Kopcso, D.P. (1983) Learning curves and lot sizing for independent and dependent demand. *Journal of Operations Management*, **4** (1), 73–83.

Levy, F.K. (1965) Adaptation in the production process. *Management Science*, **11** (6), B136–54.

Liao, W.M. (1979) Effects of learning on resource allocation decisions. *Decision Sciences*, **10**, 116–25.

Pegels, C.C. (1969) On startup or learning curves: an expanded view. *AIIE Transactions*, **1** (3), 216–22.

Pegels, C.C. (1976) Start up or learning curves – some new approaches. *Decision Sciences*, **7** (4), 705–13.

Richardson, W.J. (1978) Use of learning curves to set goals and monitor progress in cost reduction programs. *Proceedings of 1978 IIE Spring Conference*, 235–9.

Smunt, T.L. (1986) A comparison of learning curve analysis and moving average ratio analysis for detailed operational planning. *Decision Sciences*, **17** (4), 475–95.

Towill, D.R. and Kaloo, U. (1978) Productivity drift in extended learning curves. *Omega*, **6** (4), 295–304.

Waller, E.W. and Dwyer, T.J. (1981) Alternative techniques for use in parametric cost analysis. *Concepts – Journal of Defense Systems Acquisition Management*, **4** (2), 48–59.

Wright, T.P. (1936) Factors affecting the cost of airplanes. *Journal of Aeronautical Science*, **3** (2), 122–8.

Yelle, L.E. (1976) Estimating learning curves for potential products. *Industrial Marketing Management*, **5** (2/3), 147–54.

Yelle, L.E. (1979) The learning curve: historical review and comprehensive survey. *Decision Sciences*, **10** (2), 302–28.

Yelle, L.E. (1980) Industrial life cycles and learning curves: interaction of marketing and production. *Industrial Marketing Management*, **9** (4), 311–18.

Yelle, L.E. (1983) Adding life cycles to learning curves. *Long Range Planning*, **16** (6), 82–7.

10 The Economics of Variance Reduction in Sequential Manufacturing

Robert G. Batson and Jessica O. Matson

10.1 Introduction

Manufactured goods are the result of numerous fabrication and assembly steps that occur in a sequence, as determined by process planners. The sequential nature of assembly of separately fabricated parts, and the associated problems of tolerance stack-up and tolerance allocation justifiably have received great attention in the manufacturing and quality-control literature (Evans, 1974, 1975; Spotts, 1978; Greenwood and Chase, 1987; Chase and Greenwood, 1988). Tolerance analysis of an existing assembly, or tolerance allocation for a proposed assembly often points toward one part – specifically one feature of that part – whose variation due to the current (proposed) manufacturing process is too large. If this variance can be reduced, immediate improvement in quality and cost of assembly would be realized. Numerous approaches have been proposed over the past 20 years to determine the minimum-cost allocation of the dimensional tolerance of an assembly to the tolerances of its components (Peters, 1970; Speckhart, 1972; Spotts, 1973; Wilde and Prentice, 1975; Oswald and Huang, 1977; Trucks, 1987). However, none address the more fundamental manufacturing question of how economically to achieve the tolerance that has been allocated to a part.

In today's manufacturing environment, especially in high-precision manufacturing, it is important to understand how best to control and reduce the variance introduced by sequential fabrication steps on an individual part. In this article, we explore the economics of variance reduction in a sequential manufacturing process. Examples of such processes abound, and include many processes composed of machining, chemical or electro-chemical subprocesses. The focus of this paper is therefore on achieving process capability via economically justified actions to improve the capability of the subprocesses that affect a given quality characteristic, such as the diameter of a hole or the thickness of a plated part. We do not address the scheduling, sequencing or queueing time of parts; however, mathematical models do play an important role in our economic analysis.

A basic assumption in our models is that subprocesses affect some key quality characteristic, e.g. a dimension or a hardness, according to a simple (processing rate) × (exposure time) model. Certainly, many well-known chemical subprocesses (e.g. etching, anodizing and plating) follow such a model, as do paint and metal spraying. The high-precision machining techniques utilizing laser or electron beams to cut holes or slots are described by the volume of material vaporized per unit time. Perhaps not so obviously, many machining subprocesses also fit this simple model. For example, the quality characteristic 'depth of cut,' executed for a given time T, is dependent on the feed rate of the drill or lathe. Machining process capability is a 'much neglected and poorly understood problem that deserves much greater attention' (Black, 1987).

Emphasis on manufacturing quality requires us to examine the elements of variance introduced by sequential subprocesses and to find economical fabrication approaches to variance reduction. This article provides both a theoretical basis and an algorithm for selecting variance-reducing actions in a manufacturing process that is in statistical control, but not capable relative to some key quality characteristic. An experimental method for determining the necessary data on means and variances of the subprocess rates and times is suggested. The relationship of our work to Taguchi's loss function (Taguchi, 1986; Ross, 1988) is also explained.

10.2 A process model

In sequential manufacturing of discrete parts, an individual part generally requires a variety of machining and/or chemical subprocessing steps before reaching its finished state. For the purpose of exposition, we focus on a dimensional quality characteristic. A finished part dimension may not be affected by each subprocess; knowing which subprocesses do actually change the dimension is an important part of overall process specification (Juran, 1988). Each finished dimension is the result of several subprocesses which remove or add material, including an initial subprocess (e.g. casting or drilling) that creates the part feature whose dimension is subsequently modified by the several effective subprocesses. Three key assumptions about these subprocesses used in this paper are:

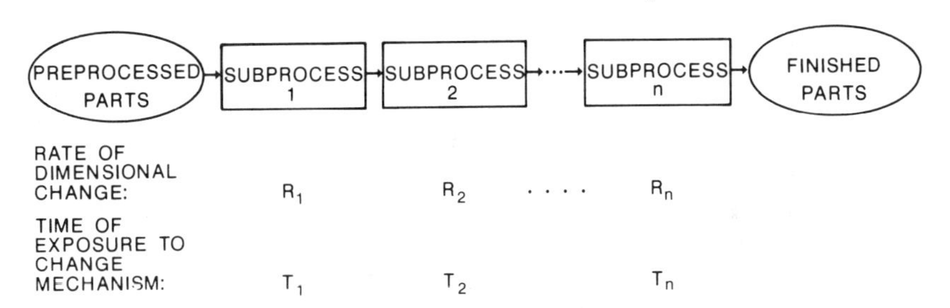

Fig. 10.1 *A sequential manufacturing process.*

1. Subprocess effects are independent.
2. Subprocess i will be in effect for time T_i and has no effect on the dimension for times outside that interval.
3. The rate of change R_i in dimension per unit processing time is constant – no acceleration or deceleration occurs during the processing time T_i and therefore the overall effect of subprocess i is a linear function of time for the time intervals T_i we might consider.

The simple process model used in this paper is illustrated in Fig. 10.1. Some examples that conform to this model include plating lines, anodizing and etching processes for aluminum parts, and certain machining processes. Faulty thinking leads one to treat the rates R_i and times T_i as deterministic quantities, whereas in reality these process variables are seldom subject to perfect control and exhibit random behaviour. This is especially true for human controllers; however, even automatic controllers for process variables are known to maintain the rate (or time) only within the capability of their design. Hence, a key conceptual step in our process model is to treat the subprocess rates and times as random variables that are in statistical control.

We now proceed from the schematic model (Fig. 10.1) to several mathematical models which describe the individual and combined effects of the n subprocesses upon the random variable F of final dimension. Let X_0 be the random variable of starting dimension, X_i be the change in dimension during subprocess i, and F be the random variable of final dimension. Then, using E for expectation and V for variance, we have the following identities:

$$F = X_0 + X_1 + X_2 + \ldots + X_n,$$

and so, by independence of the X_i:

$$E(F) = E(X_0) + \sum_{i=1}^{n} E(X_i)$$

and

$$V(F) = V(X_0) + \sum_{i=1}^{n} V(X_i)$$

We will use T to represent the **target mean** (specification centreline) for the random variable F and, when necessary, will use V to represent a **target variance** for F in order to produce a high percentage of parts in specification. We repeat that we are assuming each subprocess is in statistical control, and therefore so is F.

If T and $E(F)$ do not agree, but $V(F)$ is acceptable, then the usual approach to centre the process would be to adjust $E(X_0)$, the mean of the starting dimension, and to leave the subprocesses alone. Using the prime notation to indicate the resetting of a mean or variance, setting

$$E(X_0)' = E(X_0) + [T - E(F)]$$

accomplishes the centreing. Even if certain $E(X_i)$ were to shift for purposes of reducing $V(F)$, to be discussed later, it is still possible to centre the process on T by resetting the centreline on X_0. Letting

$$E(X_0)' = E(X_0) + [T - E(F)] + \sum_{i=1}^{n} [E(X_i) - E(X_i)']$$

implies that

$$\begin{aligned} E(F)' &= E(X_0)' + \sum_{i=1}^{n} E(X_i)' \\ &= E(X_0) + [T - E(F)] + \sum_{i=1}^{n} E(X_i)' + \sum_{i=1}^{n} [E(X_i) - E(X_i)'] \\ &= E(X_0) + \sum_{i=1}^{n} E(X_i) + [T - E(F)] \\ &= E(F) + [T - E(F)] = T \end{aligned}$$

We therefore focus our attention in the remainder of this paper upon how to select, based on subprocess capability and economic considerations, the best action(s) to reduce $V(F)$ – assuming it is too wide. From the identity $V(F) = \sum_{i=0}^{n} V(X_i)$, it is clear that to reduce $V(F)$ we must reduce one or more of the $V(X_i)$. Two key questions arise:

Q1. *Process Level.* Which $V(X_i)$ should be reduced first and if more than one $V(X_i)$ must be reduced, what order should be followed?

Q2. *Subprocess Level.* Given that $V(X_i)$ is to be reduced, should one attempt to better control the rate of processing, the time of processing, and are there any cross-over points where the most economic variance reducing action switches from control of rate (time) to control of time (rate)?

Fig. 10.2 *A process requiring reduction in variance of final dimension.*

In the next section, we show that these two questions are not separable, and that simple economic models do provide answers to both questions in a manner that can be implemented by manufacturing or quality engineers.

In examining a subprocess with rate of processing R_i and time of processing T_i, it is well known that $E(X_i) = E(R_i)\,E(T_i)$; the following formula (a proof appears in Batson, 1989) for the variance of the product of independent random variables is extremely useful:

$$\begin{aligned} V(X_i) &= V(R_i T_i) \\ &= V(R_i)\,E^2(T_i) + V(T_i)\,E^2(R_i) + V(R_i)\,V(T_i) \end{aligned}$$

This formula leads to several interesting observations regarding control of the variance of subprocess i, $V(X_i)$. For example, the rate of change of $V(X_i)$ with respect to $V(R_i)$ is $E^2(T_i) + V(T_i)$. The rate of change of $V(X_i)$ with respect to $V(T_i)$ is $E^2(R_i) + V(R_i)$. $V(X_i)$ is indeed sensitive not only to $V(R_i)$ and $V(T_i)$, but also $E(R_i)$ and $E(T_i)$. These relationships are exploited in the next section.

10.3 Economic analysis

Suppose we want to find the most economic sequence of actions on the subprocesses in order to reduce $V(F)$ from its current (or predicted) level to some target level V as shown in Fig. 10.2. By 'actions' we mean, of course, subprocess variance-reduction actions that might include better operator training, timing or measurement devices, or perhaps an increase in automation. We do not consider elimination of a subprocess as a viable variance-reduction action in this analysis, although this alternative should always be examined in variance-reduction studies because each step in sequential manufacturing can do nothing but add to variance of the final product. We do permit the possibility of reduction of $V(X_i)$ all the way to zero in our baseline analysis. A later variation will treat the case where there is a lower (control technology) limit on reduction in $V(R_i)$ or $V(T_i)$, and hence on reduction in $V(X_i)$.

In order to answer questions 1 and 2 posed in the previous section, we let: CR_i be the cost of reducing $V(R_i)$ by one unit; and CT_i be the cost of reducing $V(T_i)$ by one unit; and we assume that $E(T_i)$ and $E(R_i)$ are fixed

$(i = 1, \ldots, n)$ at levels that yield $E(X_i) = E(R_i)E(T_i)$ at levels assuring that $E(F)$ is on target. This assumption is justified by the Taguchi loss function, explained at the end of this section. Thus, we assume adjustment of means has been completed and although the process is centred, its capability is unacceptable ($V(F)$ is too large).

Two simple rate of change formulae are:

$$\Delta V(X_i) = \Delta V(R_i)[E^2(T_i) + V(T_i)] \tag{10.1}$$

and

$$\Delta V(X_i) = \Delta V(T_i)[E^2(R_i) + V(R_i)] \tag{10.2}$$

where, in (10.1), we are considering reducing $V(X_i)$ through reduction in $V(R_i)$, and, in (10.2), through reduction in $V(T_i)$. Using alternative (10.1), we would achieve a reduction in $V(X_i)$ of $E^2(T_i) + V(T_i)$ for each CR_i cost expended; similarly for alternative (10.2), we would achieve a reduction in $V(X_i)$ of $E^2(R_i) + V(R_i)$ for each CT_i cost expended. Based on this benefit-cost argument, we choose alternative (10.1) – reduction in $V(R_i)$ – if:

$$\frac{E^2(T_i) + V(T_i)}{CR_i} > \frac{E^2(R_i) + V(R_i)}{CT_i}$$

and alternative (10.2) if the inequality is reversed, and remain indifferent in the unlikely event of equality. Let us settle the cross-over issue raised in Q2 of Section 10.2. Suppose $V(R_i)$ is chosen for reduction based on the benefit-cost comparison above, and has been reduced to $V(R_i)'$, but not all the way to zero. One would never switch off to reduction in $V(T_i)$ at this point because:

$$\frac{E^2(T_i) + V(T_i)}{CR_i} > \frac{E^2(R_i) + V(R_i)}{CT_i} > \frac{E^2(R_i) + V(R_i)'}{CT_i}$$

Note also that if we choose alternative (10.1) and reduce $V(R_i)$ all the way to zero, $V(X_i)$ is not yet zero because in that case:

$$V(X_i) = \mathrm{Var}(T_i)E^2(R_i)$$

Similarly, if alternative (10.2) is chosen and $V(T_i)$ is reduced to zero, we still have: $V(X_i) = \mathrm{Var}(R_i)E^2(T_i)$.

Only by driving $V(R_i)$ and $V(T_i)$ both to zero can $V(X_i)$ be forced to zero. A bit of thought leads to the recognition that taking such drastic action on subprocess i might not be the most economic alternative when the process is viewed as a whole. We accommodate this very real possibility in the **economic decision algorithm** which is described next, which selects the optimal sequence of variance-reduction actions from among the $2n$ control possibilities, since there are n subprocesses in our process model. To select the first variance-reduction action, define:

$$m_1 = \max_i \left\{ \max \left[\frac{E^2(T_i) + V(T_i)}{CR_i}, \frac{E^2(R_i) + V(R_i)}{CT_i} \right] \right\}$$

We will use the notation $V(R_{i1})$, or $V(T_{i1})$, for the respective rate or time that was selected as the first (most economic) to be reduced. Suppose that $V(R_{i1})$ is selected by the maximization of benefit-to-cost ratios above. Reducing $V(R_{i1})$ to zero reduces $V(X_{i1})$ to $V(T_{i1})E^2(R_{i1})$.

Now if the reduction in $V(X_{i1})$ succeeded in reducing $V(F)$ below V, that is if:

$$V(R_{i1})[E^2(T_{i1}) + V(T_{i1})] > V(F) - V$$

then the variance-reduction procedure stops. Otherwise, we must return to the benefit-cost ratios and compute

$$m_2 = \max_{i \neq i1} \left\{ \max \left[\frac{E^2(T_i) + V(T_i)}{CR_i}, \frac{E^2(R_i) + V(R_i)}{CT_i} \right] \right\}$$

and find the second-best variance-reducing alternative. From among the $2(n - 1)$ ratios above, one is selected as the maximum which yields $m_2 \leqslant m_1$. Before declaring this rate R_{i2} (or time T_{i2}) as the next to be driven to zero, one must compare m_2 with the effect of completing the reduction of $V(X_{i1})$ all the way to zero. That is, continuing the above supposition that $V(R_{i1})$ was selected in the m_1 step, then choose to reduce $V(T_{i1})$ to zero only if:

$$\frac{E^2(R_{i1})}{CT_{i1}} > m_2$$

in which case redefine m_2 as $E^2(R_{i1})/CT_{i1}$.

Otherwise, implement the variance reduction action indicated by m_2 – reduction in R_{i2} (or T_{i2}). It should be obvious that if $V(T_{i1})$ is selected in the m_1 step, then next choose to reduce $V(R_{i1})$ to zero only if:

$$\frac{E^2(T_{i1})}{CR_{i1}} > m_2$$

in which case redefine m_2 as $E^2(T_{i1})/CR_{i1}$.

Should this comparison with m_2 lead to the conclusion that it is most economic to reduce $V(X_{i1})$ to zero, then obviously any subsequent decisions about reduction in $V(F)$ would not involve subprocess X_{i1}. In fact, this would be true even if $V(R_{i1})$ or $V(T_{i1})$, or both, were bounded below and could not be driven to zero. The case where reduction in $V(R_{i1})$ or $V(T_{i1})$ could continue beyond some bound, but at a higher cost than the originally stated rates of $CR_{i1}(CT_{i1})$, is addressed as a variation below.

The algorithm just described behaves as follows:

1. Each iteration j requires the comparison of a collection of benefit-cost ratios to select the maximum m_j.
2. At step j, there are $2n - j$ benefit-cost ratios to be compared.
3. The benefit-cost ratios for subprocess i start out with the form

$$\frac{E^2(T_i) + V(T_i)}{CR_i} \quad \text{and} \quad \frac{E^2(R_i) + V(R_i)}{CT_i}$$

but revert (respectively) to

$$\frac{E^2(T_i)}{CR_i} \quad \text{if } V(T_i) \to 0$$

and

$$\frac{E^2(R_i)}{CT_i} \quad \text{if } V(R_i) \to 0$$

4. Once a $V(R_i)$ or $V(T_i)$ is driven to zero (or its lower technological limit), its benefit-cost ratio is set to zero and it is never considered in any subsequent maximization. This is why there is one less benefit-cost ratio to be compared at each iteration.
5. The algorithm continues until $V(F)$ has been reduced below V.

We first illustrate the baseline algorithm.

10.3.1 An example application

Suppose the variance in a final dimension F is 600 mm^2, where it has been determined that the variance of parts entering the process is $V(X_0) =$ 122 mm^2. The process consists of three subprocesses and the desired reduction in variance of F due to the process is from 478 mm^2 to 250 mm^2. The data for the process are:

$$\begin{array}{lll}
E(R_1) = 2 & E(R_2) = 5 & E(R_3) = 6 \\
V(R_1) = 2 & V(R_2) = 1 & V(R_3) = 4 \\
CR_1 = \$5 & CR_2 = \$5 & CR_3 = \$5 \\
E(T_1) = 10 & E(T_2) = 2 & E(T_3) = 3 \\
V(T_1) = 2 & V(T_2) = 1 & V(T_3) = 5 \\
CT_1 = \$10 & CT_2 = \$5 & CT_3 = \$10
\end{array}$$

One can easily verify that:

$$V(X_1) + V(X_2) + V(X_3) = 212 + 30 + 236 = 478$$

In order to determine the optimal sequence of variance-reducing actions, we begin by computing m_1:

$$m_1 = \max\left\{\max\left[\frac{102}{5}, \frac{6}{10}\right], \max\left[\frac{3}{5}, \frac{26}{5}\right], \max\left[\frac{14}{5}, \frac{40}{10}\right]\right\}$$

$$= \frac{102}{5}$$

which was

$$\frac{E^2(T_1) + V(T_1)}{CR_1}$$

Hence, first reduce $V(R_1)$ to zero. Next, we compute m_2:

$$m_2 = \max\left\{\max\left[\frac{3}{5}, \frac{26}{5}\right], \max\left[\frac{14}{5}, \frac{40}{10}\right]\right\}$$

$$= \frac{26}{5}$$

which was

$$\frac{E^2(R_2) + V(R_2)}{CT_2}$$

We compare m_2 with the benefit-cost ratio

$$\frac{E^2(R_1)}{CT_1} = \frac{4}{10}$$

to determine if driving $V(T_1)$ to zero would be preferred to reduction of $V(T_2)$. Because

$$m_2 = \frac{26}{5} > \frac{4}{10}$$

the algorithm chooses to reduce $V(T_2)$ to zero. This happens to be the stopping point for this example because the two actions determined by m_1 and m_2 have resulted in a reduction in variance of 204 + 26 = 230, so that $V(F)$ now stands at 478 − 230 = 248. The total expense of this reduction was 2($5) + 1($5) = $15.

Note that even though reducing $V(T_3)$ to zero at step two would have reduced $V(F)$ by 200 (reduced $V(X_3)$ from 236 to 36), this action was not chosen due to economics – reduction in (VT_3) reduces $V(X_3)$ at a rate of 4 mm^2 per dollar whereas reducing $V(T_2)$ results in a more favourable 5.2 mm^2 per dollar. Reduction in $V(T_3)$ would have been chosen as the next variance reducing action if a reduction below 248 mm^2 for the process was desired. Since driving $V(T_3)$ all the way to zero would have left only 48 mm^2 variance due to the process (an order of magnitude reduction from the original 478 mm^2) it is likely that the action to reduce $V(T_3)$ would not have required the \$50 expense to drive it completely to zero, but rather a partial reduction.

We now examine several variations on the baseline economic decision algorithm.

10.3.2 Variation 1 (cost breaks)

Suppose that the cost of reducing $V(X_i)$ is quantified by the following piece-wise linear functions.

$$\begin{aligned}
&\text{If } 0 \leqslant \Delta V(R_i) \leqslant A_1, && \text{then cost} = CR_i^{(1)} \text{ per unit reduction in } V(R_i) \text{ in this range;}\\
&\text{If } A_1 \leqslant \Delta V(R_i) \leqslant A_2, && \text{then cost} = CR_i^{(2)} \text{ per unit reduction in } V(R_i) \text{ in this range;}\\
&\quad\vdots && \quad\vdots\\
&\text{If } A_{k-1} \leqslant \Delta V(R_i) \leqslant V(R_i), && \text{then cost} = CR_i^{(k)} \text{ per unit reduction in } V(R_i) \text{ in this range;}
\end{aligned}$$

where $CR_i^{(1)} \leqslant CR_i^{(2)} \leqslant \ldots \leqslant CR_i^{(k)}$. Also,

$$\begin{aligned}
&\text{If } 0 \leqslant \Delta V(T_i) \leqslant B_1, && \text{then cost} = CT_i^{(1)} \text{ per unit reduction in } V(T_i) \text{ in this range;}\\
&\text{If } B_1 \leqslant \Delta V(T_i) \leqslant B_2, && \text{then cost} = CT_i^{(2)} \text{ per unit reduction in } V(T_i) \text{ in this range;}\\
&\quad\vdots && \quad\vdots\\
&\text{If } B_{s-1} \leqslant \Delta V(T_i) \leqslant V(T_i), && \text{then cost} = CT_i^{(s)} \text{ per unit reduction in } V(T_i) \text{ in this range.}
\end{aligned}$$

Suppose in the baseline algorithm that $V(R_i)$ was selected for reduction based on the computation on m_1 where each benefit-cost ratio uses the appropriate $CR_i^{(1)}$ or $CT_i^{(1)}$ in its denominator. Then, rather than reducing $V(R_i)$ all the way to zero, it is only reduced by amount A_1 to the level $V(R_i) - A_1$. The denominator of the benefit-cost ratio associated with reduction in $V(R_i)$ would then be increased from $CR_i^{(1)}$ to $CR_i^{(2)}$ because further decrease in $V(R_i)$ would occur in the range $[A_1, A_2]$ at a cost of $CR_i^{(2)}$ per unit reduction.

At the end of each benefit-cost maximization iteration, the cost in the denominator of the subprocess rate (time) selected for reduction is increased to its next level, unless the $V(R_i)(V(T_i))$ has reached zero (or some technological limit), in which case that particular variance is no longer a candidate for reduction and drops out of the maximization comparison. Note also, that in the process of reducing $V(R_i)$ from its original value to levels of:

$$
\begin{aligned}
&V(R_i) - A_1, \\
&V(R_i) - A_2, \\
&\quad\vdots \qquad \vdots \\
&V(R_i) - A_{k-1}
\end{aligned}
$$

the numerator of the benefit-cost ratio associated with the reductions in $V(T_i)$ would have to be adjusted prior to the next maximization step. For example, if $V(R_i)$ is reduced to $V(R_i) - A_y$, the benefit-cost ratio for reduction in $V(T_i)$ becomes:

$$\frac{E^2(R_i) + V(R_i) - A_y}{CT_i^{(x)}}$$

where x is the current cost slope associated with further reduction in $V(T_i)$.

10.3.3 Variation 2 (only $V(T_i)$ adjustable)

In many real situations one would encounter, the processing rates would have variances that were essentially functions of the equipment or materials used and could not be adjusted. The challenge for the manufacturing engineer would then be to reduce $V(F)$ strictly through better control of subprocess times T_i. The resultant (simplified) decision process would be to select $V(T_i)$ for reduction based on:

$$m_1 = \max_i \left[\frac{E^2(R_i) + V(R_i)}{CT_i} \right]$$

Assuming $V(T_{i1})$ can be driven to zero, then select the second best $V(T_i)$ for reduction by:

$$m_2 = \max_{i \neq i1} \left[\frac{E^2(R_i) + V(R_i)}{CT_i} \right]$$

Obviously, cost breaks in reduction of some or all of $V(T_i)$ could be applied in this case. Also note that because we assume $V(R_i)$ cannot be adjusted, the smallest that $V(F)$ can be made is:

$$V(F) = V(X_0) + \sum_{i=1}^{n} V(R_i)\, E^2(T_i)$$

and this assumes each $V(T_i)$ has been reduced to zero, i.e. deterministic control through automation. Finally, note that it might be possible to continue the reduction in $V(F)$ by reduction in one or more $E(T_i)$. Of course, such a shift in $E(T_i)$ creates a shift in $E(X_i) = E(R_i)\, E(T_i)$, and in order to keep $E(F)$ on target T, some sort of compensatory shift in other

$E(T_i)$ or $E(X_0)$ would be required. In most manufacturing situations, one is not free arbitrarily to reallocate dimensional means because of equipment limitations, limitations of gauges, effects on throughput rates, and so on.

10.3.4 Variation 3 (Taguchi loss function)

Consider the dimensional quality characteristic F in the context of Taguchi's quadratic loss function (Taguchi, 1986; Ross, 1988) for deviations from the target value T:

$$\text{Loss} = \$k\,(f - T)^2 \qquad \text{for a given part with } F = f$$
$$\text{Loss} = \$k\{V(F) + [E(F) - T]^2\} \qquad \text{for the population of parts with distribution } F$$

where k is a constant determined by economic analysis of the costs of imperfect ($f \neq T$) parts to the society which uses the product containing the part. From the loss function, the benefit of reducing $V(F)$ by one unit is \$$k$, regardless of which subprocess is improved, and regardless of whether it is $V(T_i)$ or $V(R_i)$ that is reduced. In essence, there would have been a \$$k$ in each benefit-cost numerator above, so the decisions recommended by our baseline algorithm and Variations 1 and 2 are consistent with Taguchi philosophy, so long as $E(F)$ is considered to be fixed on T.

Now if $E(F) \neq T$, it can be shown that a reduction of $|E(F) - T|$ by x units results in a loss reduction of $\$kx[2|E(F) - T| - x]$, which is quadratic in x. Such reductions would, of course, be achieved by shifts in $E(T_i)$ or $E(R_i)$, which were held constant during the variance-reduction algorithm. Our economic-decision algorithm can be used in conjunction with an algorithm that computes the most economic shift in subprocess means in order to determine a sequence of actions that reduce the loss due to $V(F) > 0$ and $E(F) \neq T$, toward zero, in an optimal trajectory. At any point in the loss reduction sequence, we would compare the cost per unit loss reduction for the best variance-reducing action (from our algorithm) with the best mean-shift action (using the above quadratic formula) and select the most effective action.

We leave the details of such an enhancement for a later paper. Manufacturing experience points toward a predominance of instances where a shift of $E(F)$ all the way to T is feasible, and is quite inexpensive relative to any variance reduction activity. Thus, even in instances where there is some loss due to $E(F)$ not being on target, the economic-decision algorithm of this paper is a vital tool for planning the sequence of actions on the process to reduce loss in an economically optimal way.

10.4 Conclusions

The determination of the mean and variance of the respective processing rates and times for each subprocess appears at first to be an onerous task. However, in the course of process-capability analysis it is natural to collect data on such rates and times in order to develop a better understanding of the process. Often the rates (at least the range of rates) is set by the equipment type or an internal process specification. More than likely, a data-collection experiment with suitable randomization is needed to establish the mean and variance of a processing rate, be it a chemical process or a machining process. Certainly, the best way to learn about variation in processing times is to perform either a time study on each part processed or to collect random samples on time (work-sampling study). By setting up a comprehensive data-collection procedure to measure the 'before' dimension, 'after' dimension, and processing time for each subprocess a part encounters, one can very rapidly build up a database on processing rates (material addition or removal), processing times, and individual and overall processing effects. Experimental design techniques, either classical (Montgomery, 1984; Box, *et al.*, 1978) or Taguchi orthogonal arrays (Taguchi, 1986; Ross, 1988) may be used along with randomization either to block or to average out the effects of variables such as production mix, day of week, machine or processing tank utilized.

The methods described in this paper can be applied to various fabrication processes and are especially appropriate for decisions concerning control of processes that today are manual or only semi-automatic. The tool permits the systematic selection of subprocesses to be improved by better equipping or training of the operator, or by purchase of automatic control equipment. Expenditures are rank ordered by the extent of capability improvement they offer, and quality-improvement teams have a rational basis to decide on whether first to work to reduce the variance in processing time or rate. Finally, the method presented can be used to determine the specification on the starting dimension X_0 of a critical quality characteristic. Often, manufacturing management mistakenly insists that X_0 must have target T, or that the specification range on X_0 should be the same width as that on the finished dimension F.

Such misperceptions must be corrected with data on the current subprocess capabilities, which reveal each subprocess's contribution to growth in variance as the parts are processed. Economic justification for quality improvement through subprocess control is provided by our methodology, as well as the proper sequence of actions to take based on benefit-cost. Our methodology is compatible with Taguchi's quadratic loss function, in fact it provides a cost function for optimal variance reduction which may be compared with the cost of adjusting the mean.

10.5 References

Batson, R.G. (1989) Cost risk analysis: a state-of-the-art review. *Estimator*, Spring, 21–49.

Black, J.T. (1987) Machining processes. In *Production Handbook*, 4th edn (ed. J.A. White) Wiley, New York, 418–37.

Box, G.E.P., Hunter, W.G. and Hunter, J.S. (1978) *Statistics for Experimenters*, Wiley, New York.

Chase, K.W. and Greenwood, W.H. (1988) Design issues in mechanical tolerance analysis. *Manufacturing Review*, **1** (1), 50–9.

Evans, D.H. (1974) Statistical tolerancing: the state of the art, part 1. *Journal of Quality Technology*, **6**, 188–95.

Evans, D.H. (1975) Statistical tolerancing: the state of the art, part 2. *Journal of Quality Technology*, **7**, 1–12.

Greenwood, W.H. and Chase, K.W. (1987) A new tolerance analysis method for designers and manufacturers. *Transactions of the ASME, Journal of Engineering for Industry*, **109**, 112–16.

Juran, J.M. (1988) *Quality Control Handbook*, 4th edn, McGraw-Hill, New York.

Montgomery, D.C. (1984) *Design and Analysis of Experiments*, Wiley, New York.

Oswald, P.F. and Huang, J. (1977) A method for optimal tolerance selection. *Transactions of the ASME, Journal of Engineering for Industry*, **99**, 558–65.

Peters, J. (1970) Tolerancing the components of an assembly for minimum cost. *Transactions of the ASME, Journal of Engineering for Industry*, **92**, 677–82.

Ross, P.J. (1988) *Taguchi Techniques for Quality Engineering*, McGraw-Hill, New York.

Speckhart, F.H. (1972) Calculation of tolerance based on a minimum cost approach. *Transactions of the ASME, Journal of Engineering for Industry*, **94**, 447–53.

Spotts, M.F. (1973) Allocation of tolerances to minimize the cost of assembly. *Transactions of the ASME, Journal of Engineering for Industry*, **95**, 762–4.

Spotts, M.F. (1978) Dimensioning stacked assemblies. *Machine design*, **50**, 60–3.

Taguchi, G. (1986) *Introduction to Quality Engineering* Asian Productivity Organization/UNIPUB, White Plains, New York.

Trucks, H.E. (1987) *Design for Economical Production*, 2nd edn, Society of Manufacturing Engineers, Dearborn, Michigan.

Wilde, D. and Prentice, E. (1975) Minimum exponential cost allocation of sure-fit tolerances. *Transactions of the ASME, Journal of Engineering for Industry*, **97**, 1395–8.

Part Three

Costing and Investment Methods

11 Memo to global competitors: it is time to replace cost accounting with process based information

H. Thomas Johnson and Richard W. Sapp

11.1 Introduction

Today's global competition demands that companies identify their sources of competitive advantage and monitor and control the customer-focused processes within their organizations. In general this means that to be successful over the long term, a firm must sell products that satisfy customers by meeting global standards for quality, flexibility, service, dependability and price; and those products must cost less than customers are willing to pay for the value received. Why have most companies not been able to keep up with world-class standards of competitiveness and profitability in recent years?

One reason is the deeply entrenched practice of using cost-accounting information to manage business performance. For more than 60 years, managers have used cost information derived from transaction-based financial-accounting systems to judge the impact of their decisions on company profits. Driven by the procedures and cycles of the organization's financial reporting system, the cost information is too late, too aggregated and too distorted to be relevant for world-class competitiveness. Such cost-information systems do not give an accurate picture of what companies must do to achieve and sustain competitive advantage in today's global economy. With increased emphasis on meeting quarterly or annual earnings targets, managerial accounting information systems focus too narrowly on producing a monthly earnings report. And despite the considerable resources devoted to computing a monthly or quarterly income figure, this figure fails to measure the actual increase or decrease in economic value that has occurred during this period. Consequently:

1. Management accounting reports are of little help to operating managers attempting to reduce variation delay and excess in processes.
2. Management accounting provides no information on customer satisfaction.
3. Managers' horizons contract to the short-term cycle of their monthly profit-and-loss statement.
4. Managers think and act locally rather than globally.

If you don't know what it takes to satisfy customers and you don't know if your processes are capable and in control, you are not likely to be competitive in the long run.

US manufacturers need new management information giving signposts that point to sources of competitiveness and profitability in the global economy today and tomorrow. This new information will not come from patching up or adding to traditional management accounting systems. Vigorous global competition, rapid progress in product and process technology, wide fluctuations in currency exchange rates and raw material prices demand excellence from management-accounting systems. In contrast, traditional management-accounting information reinforces behaviour that impairs business performance in the long run. It must be replaced, entirely, by information that points managers towards the right things to do to satisfy customers.

11.2 Why traditional management accounting impedes competitiveness and profitability in today's global economy

Considering that virtually all management-accounting practices in use today had been developed by 1925, it is not surprising that traditional management-accounting reinforces the belief that a business achieves higher profits by managing costs: holding administrative overhead costs in line, holding direct and indirect factory costs to competitors' levels, and abandoning high-cost products that earn inadequate margins. This was the formula for success in the non-global, locally focused firm which approached its business with a mindset remarkably similar to its competitors. However, for the emerging global competitor facing non-traditional production technologies, unintended consequences usually result from its efforts to increase profit by managing its businesses using accounting costs of operations and products.

11.2.1 Managing costs of operations

In running factory operations, managing cost usually means being 'efficient' – that is, minimizing unit direct costs and using systems of incentives or controls to hold indirect costs in check. Most US manufacturers since World War II have believed the best way to be efficient is to achieve economies of scale, by producing output in large lots.

Large lots reduce unit costs of output, but they also reduce flexibility and may diminish quality. However, companies cushion against the uncertainties of quality and inflexibility, presumably at costs that do not exceed the benefits of large lots, by holding economic amounts of inventory.

This scale-economy strategy for optimizing costs impedes competitiveness by influencing how a company organizes work in its factories. This influence can be seen by studying the customary way manufacturers manage purchasing and set-up costs.

Purchasing

Most manufacturers periodically receive large shipments of components which they store and release to assembly lines as needed. Until very recently, US companies have tended to reject proposals to receive daily component shipments directly to the lines. The reason for rejection is that it is too costly. Daily shipments to assembly lines might jeopardize discounts suppliers normally give for large-volume shipments, and interruption of supplies, even for a day, might cause costly shutdowns. Lost discounts and possible costs of interruption are usually deemed to outweigh the cost of storing components shipped in large lots.

To manage the cost of large-scale shipments, companies purchase components in economic lot sizes that minimize the total cost of purchase and storage. Their goal is to reduce unit component costs by achieving the largest possible economies from large-scale orders. To increase the economies of large orders even more in the future, many companies are currently investigating the use of automated storage and retrieval systems to reduce storage costs.

Set-up costs

Manufacturers usually fabricate machined parts for storage, and release those parts from inventory according to schedule. They manage the cost of setting up machines in much the same way as they manage the cost of component shipments. Parts are fabricated in economic lot sizes to minimize the total cost of set up and storage. The governing strategy, as with purchasing, is to minimize unit costs of fabricated parts by increasing the size of economic lots. Plans to automate storerooms will increase the economic size of production batches just as it will increase the economic size of purchased component shipments. In fact, the anticipated reduction in unit costs of larger batches of purchased and fabricated parts is often a major part of the savings that justify proposals to invest in automated inventory and retrieval systems.

United States plant managers have followed the above strategies to control unit costs of purchased and fabricated parts for many years. However, today's global standards of competitiveness compel plants to do more than just hold unit costs in line. To meet foreign competition, for example, they must respond faster to changes in the market and give more attention to customer quality than ever before. They also face increased pressure to slash total costs, to meet competitors' sharply lower prices.

Plant managers frequently respond to this recent pressure for improved flexibility, quality and cost by arguing that limits exist to what they can achieve. More flexibility, for instance, could be achieved with more frequent set-ups or more frequent component deliveries. And higher quality might be had by paying vendors higher prices for components, by fabricating parts in smaller batches (which would mean more set-ups), or by employing more inspectors. However, these steps to improve flexibility and quality always entail higher cost. But higher cost is precluded by competitive price pressures.

It is evident why plant managers usually claim it is impossible to improve flexibility or quality without incurring higher costs. Over time the strategy of cutting unit costs by increasing scale and throughput has increased the size of purchased and fabricated batches. Increased batch sizes, in turn, increase the size of inventories, the size of machines, the space utilized, the distances parts move, lead times, and the number of undetected defects. All those increases imply more costs – added costs that will be made up, presumably, by the lower unit costs of larger batches. Cutting batch sizes to improve flexibility and quality won't cause those added costs to go away. Indeed, cutting batch sizes will only raise the unit cost of items in a batch. Therefore, managing costs by pursuing economies of scale creates conditions that cause plant managers to believe improved flexibility or quality come only at higher cost.

Reinforcing the trade-offs caused by pursuing large-lot scale economies is the practice US manufacturers follow of measuring plant performance with standard cost variances. In general, variances calculate the difference between actual costs incurred for a period and standard costs **earned** (also termed **absorbed** or **applied**) by the direct labour hours charged to units of output produced during the period. The ratio of earned to actual costs is often referred to as an **efficiency ratio**, with a goal of 1:1.

Designed originally to value inventories for financial reporting purposes, standard cost rates for attaching costs to products have become one of the primary management-accounting tools US manufacturing companies use to evaluate performance and control the behaviour of plant managers. United States plant managers live and die by their skill at managing variances.

In practice, pressure to minimize standard-cost variances causes plant managers to produce output both in larger lots and in larger amounts, regardless if what they produce is saleable. Pressure to 'earn' direct labour hours encourages long runs of large batches. Moreover, pressure to 'earn' overhead encourages overproduction – even at the cost of overtime at the end of an accounting period.

11.2.2 Managing product costs

Companies use financial product cost information to monitor the profit impact of changes in product mix and input prices. Product costs presumably help people make decisions or take actions that will enhance profitability. Unfortunately, financial product-cost information tends to have the opposite effect.

Financial product costs, compiled by accountants to value inventories and to report the cost of goods sold, misdirect managers' efforts to make profitable marketing decisions. Manufacturers usually try to enhance profits by expanding product lines with high accounting gross margins. Ironically, this strategy tends to erode a company's profitability over time. This happens in part because financial product costs omit those costs incurred outside the factory itself, such as costs for distribution or general administration. But even greater distortion occurs because financial-accounting procedures distribute overhead costs in proportion to the output volumes of products.

The bases accountants customarily use to distribute overhead cost to products – direct labour or direct materials – vary more or less with output volume. Consequently, low-volume, complex products such as sophisticated new lines or custom work are charged with far less than their share of indirect costs (for setup, engineering change orders, special parts orders, and so forth), while high-volume, standard products bear far more than their share of indirect costs. Therefore, complex, sophisticated, low-volume products show higher gross margins than they actually earn in the long run. A company that seeks higher profits by proliferating its lines of such products will find its real indirect costs increase far more than the financial product costs indicate. The result is higher than expected growth of overhead and an unexpected slump in profitability.

An example of the unfortunate effects of using financial product cost information to judge products' costs comes from the semiconductor industry. In the early days of the industry plants made low-density chips – say, 16K capacity. They did not incur nearly the overhead they would later incur when making high-density chips in the 256K range. Processes to make low-density chips are more labour-intensive and require less sophisticated clean rooms or assembly machines than processes to make high-density chips.

Those overhead costs came to haunt many plants that eventually built high-density chip lines alongside existing low-density lines. The usual cost-accounting system, by distributing overhead over direct labour hours with a single plantwide rate, charged proportionately more of those overhead costs to the low-density, direct labour intensive chips. The high-density chips that caused the overhead didn't bear their full share of the cost.

Competitive market prices, however, tend to follow actual cost. Hence, prices of low-density 'commodity' chips eventually fell relative to prices of the newer, more sophisticated high-density chips. Companies who relied on their cost systems for product-cost information perceived declining margins for the commodity chips and rising margins for the new high-density items. In the late 1970s and early 1980s, many large semiconductor manufacturers with diverse product lines (hence more distorted product costs) surrendered what they thought were unprofitable lines to focused (often foreign) competitors who were not misled by distorted cost signals.

11.3 Managing processes: a management information framework for the future

Managing with information from financial-accounting systems impedes business performance today because traditional cost-accounting data do not track sources of competitiveness and profitability in the global economy. Cost information, *per se*, does not track sources of competitive advantage such as quality, flexibility and dependability; furthermore, traditional product cost information, based on plantwide indirect cost allocations, is increasingly inaccurate and unreliable in today's technological environment.

A promising tool for tracking competitiveness and profitability is process-based information (PBI). PBI focuses on the resource-consuming activities a business performs such as order processing, raw material and WIP handling, incoming parts inspection, production scheduling and set-ups. These kinds of processes are the basis for establishing competitiveness and flexibility. Businesses, therefore, need information about how each process consumes the firm's resources and whether these activities deliver value to the customer or not. PBI is both a mindset and a tool to create this critical linkage between processes, resource consumption and customer value creation.

As a mindset, a PBI approach to management information attempts to focus managers' attention on the underlying causes of cost and profits. It comprises any relevant information about the processes across the entire chain of value – design, engineering, sourcing, production, distribution, marketing and after-sale service.

PBI, as a tool, can be used to identify and manage waste. Some activities consume resources but do not add value to the customer. Managing waste within PBI can be approached in five steps:

1. Chart the flow of processes throughout the organization.
2. Identify sources of customer satisfaction in every process.
3. Eliminate any work that contributes no value to the customer.
4. Identify and remove all causes of delay, excess and variation in all processes.
5. Track specific indicators of waste.

Identifying generators of delay, excess and variation calls for the cooperation of everyone in the organization. No process should escape attention. Once you identify causes of waste, the entire arsenal of new management methods associated with just-in-time, total quality control and employee involvement should be used to remove them. The presence of any non-value work limits a company's ability to be as competitive as possible.

Specific indicators of waste include the lapsed time to do something, the distance parts are moved, space occupied by production, number of part numbers and set-up time. Tracking such indicators of waste helps companies achieve a goal of continuous improvement in identifying and removing waste which significantly enhances a company's competitive position. One of the most insidious causes of waste is the scale economy (e.g. producing output in large lots) mindset that most US firms exhibit.

11.3.1 Managing competitive operations

Driven to be cost-competitive by achieving scale economies, US manufacturers tend to overlook the impact small-sized lots can have on dimensions of competitiveness such as flexibility and quality, not to mention cost. Smaller batches cut lead times, which improves flexibility, and they enable workers to inspect their own work, which improves quality. Financial cost information does not convey that impact. Indeed, cost records in traditional US manufacturing firms will associate smaller batches with higher unit costs – one basis for the idea that flexibility and quality trade off for higher cost.

However, the impact of small batches is reflected in the improved market position and profitability of companies that have adopted just-in-time (JIT) purchasing and production programmes. What enables a JIT operation's costs to fall, while a traditional manufacturing operation's costs rise, as batch sizes fall?

One reason is the different way they manage set-up activities. The traditional operation takes set-up time for granted and manages set-up cost by setting up large lots as few times a day as possible. The JIT operation manages set-ups by cutting set-up time so that as many set-ups as possible, on different lots, can be run during a day. Besides making smaller batches feasible, shorter set-up times also reduce total costs by forcing production processes to come closer together, by reducing in-process inventory buffers, and by shrinking lead times. Short set-up times, by making small lots and fast changeover feasible (not large lots to spread set-up costs over many units) are the key to global competitiveness.

Businesses need quite different information to manage set-up time than they use to manage set-up costs. Managing time entails eliminating non-value work – that is, waste. Essentially, this means finding and eliminating any cause of delay, excess or variation. In the case of set-ups activity, causes of waste might include inadequate worker training, poor tool conditions, and awkward design of machines. Lists of such causes have no end, and are often uncovered by repetitive problem-solving. In addition, businesses need information confirming progress at eliminating waste from work. With set-ups this would be information on set-up times and set-up frequency. Information confirming progress at removing waste motivates people to improve on past accomplishments continuously.

Managing costs with large lots and inventory cushions is not the way to be competitive in the global economy. From now on, most world-class manufacturers will produce small lots in great variety and they will maintain minimal inventories. Their performance will exceed what most US manufacturers until now have achieved in terms of quality, flexibility, dependability, service and even cost. United States manufacturers who persist in managing costs the old way will not survive against this new breed of competitor.

11.3.2 Identifying profitable products

Financial product costs, compiled to value inventory for external reporting purposes, tend to give a distorted view of what products cost. This paradox, as we noted above, arises from the procedure accountants follow to distribute manufacturing overhead costs to products. This paradox can be resolved by compiling product costs from information about a company's activities. Activity-based product-cost information, being developed in several large US companies, tracks total cost and long-term profitability over a product's entire life cycle.

Unlike financial product-cost information, activity-based product costs do not start with the assumption that the direct labour or direct material in a product cause overhead costs to be incurred. Instead, they presume that products incur overhead costs by requiring resource-consuming activities such as design, engineering, production set-ups, distribution, marketing and service. Activities themselves cause overhead costs by triggering the consumption of resources. Costs of products differ, then, according to their actual requirements for activities, not according to volume-based allocations of over-aggregated indirect costs.

In the microchip example cited earlier, activity-based product costs would trace the costs of clean rooms and computer-controlled assembly machines to those particular chips that required their use. Costs of such overhead items would no longer be spread indiscriminately over all products according to their direct labour content.

Activity-based product costs inform managers more reliably than financial costs about differences in products' long-term costs. With activity-based product costs, companies that attempt to improve profitability by selling more of what they think are high-margin products will not suffer the surprises that await companies who pursue such a strategy using financial product-cost information.

Activity-based product costs are linked by some people to the process-based operating information discussed earlier. However, strategic decisions made with activity-based product costs will not automatically lead to competitiveness-enhancing modifications of operating processes.

To see why, consider once more the case of set-ups. By tracing costs of set-ups to products that cause set-ups, activity-based product costs show higher costs of low-volume products and lower costs of high-volume products than traditional financial product costs show. By considering only activity-based product-cost information, profit-motivated managers might be encouraged to reduce sales of low-volume products and emphasize markets for high-volume products. However, long runs and large-sized batches cannot raise profitability in the long run. Long-run competitiveness and profitability in the global economy will require fast changeover and small lots. Activity-based product costs, *per se*, don't tell managers how to achieve faster set-ups and smaller batches. They tell managers what different products cost.

11.4 Putting process-based information to work

To be profitable in the long run, manufacturers must stop doing two things traditional management accounting encourages them to do. They must stop managing the costs of work that will disappear if they do what it takes to be competitive in the global economy, and they must stop basing business decisions on financial product-cost information.

Taking the following steps will put process-based information to work:

1. For every product sold, identify the processes that must be performed to satisfy the customer. Be sure to perform those processes. This is the key idea behind the concept of quality function deployment.
2. In all processes, identify sources of delay, excess, and variation that cause waste and impede continuous flow. This step opens the door to continuous improvement. There is no end to finding waste and interruptions in human work.
3. Embark on programmes to cut lead time and to improve flow. Here is where all the ideas popularly associated with Japanese manufacturing techniques are applied.
4. Track indicators of time (and waste) to confirm the success of these programmes and to motivate people to do more. This step yields the 'charts on the wall' that are so familiar to workers (and involved managers) in Japanese factories.
5. Calculate product costs by adding up the costs of activities it takes to design, engineer, make, distribute, sell and service each product. To do this, use the full arsenal of techniques associated with the concept of activity-based product costing.

Following these steps will not be easy without rejecting the mindset that says 'Optimize costs by economizing with large scale.' Instead, understand that competitiveness and profitability come to those who improve continuously at cutting lot sizes and changeover time. Remember: becoming competitive doesn't follow from doing what was once thought to be necessary to be profitable. Rather, being profitable comes from doing what world-class manufacturers know it takes to be competitive.

How rapidly the US regains lost ground in world markets depends to a large degree on how soon US manufacturers take this message to heart and act. Obviously corporate managers is not solely responsible for the declining competitiveness of the US in world markets. National economic policy must also share the blame. But company managers themselves can do much to reverse the trend. And unlike government policy-makers, company managers can act quickly and decisively. The time to begin is now.

12 Joint Cost Allocation to Multiple Products: Cost Accounting *v.* Engineering Techniques

Fariborz Tayyari and Hamid R. Parsaei

12.1 Introduction

In many manufacturing situations, several different types of product are produced from processing the same input materials and/or the same production process or department. Some common examples of joint products are:

1. The various products produced through processing a cow in a meat-packing plant, such as steaks, ground beef, beef stew, cow-hide and bones;
2. The various products produced by cracking crude oil, such as gasoline, heating oil and kerosene;
3. The variety of metals extracted and produced simultaneously in a mining operation, such as gold, copper and nickel.

12.1.1 Terminologies

To better understand the concepts and problems of joint cost allocations, the reader should be familiar with some special terminologies related to this subject. In a production process of this nature, one input (i.e. raw material) is used to produce two or more different products either in fixed proportion or in variable proportion. Since the content of this section deals with joint cost allocations, the following two definitions offered by Williams and Kennedy (1983) with some modifications are used to clarify the meaning of joint costs:

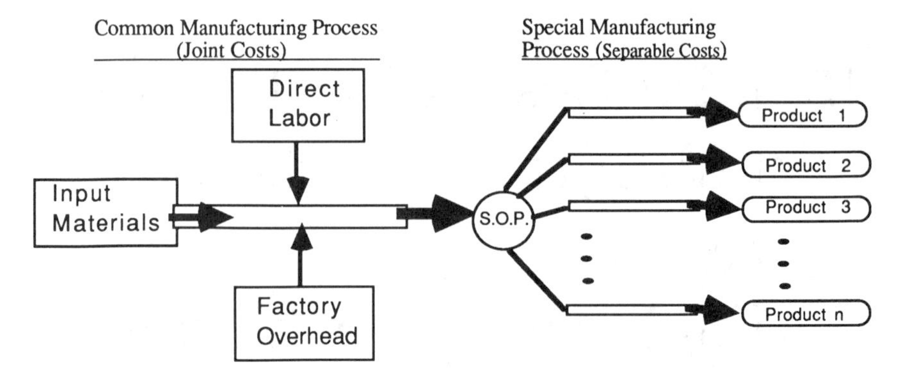

Fig. 12.1 *A typical multiple-product flow diagram showing the production of n different products being produced from processing a common input material which can be identified at the split-off point (SOP).*

1. Joint products are those products necessarily produced in fixed proportions through a single production process.
2. Common products are those products that are produced in common through processing the same input materials and process facility in variable proportions.

Figure 12.1 provides some useful information. As shown, some input (raw) materials are initially put into the production process. The processing of the materials requires the incurrence of some direct labour cost and factory overhead. Two or more identifiable products (i.e. products 1, 2, . . . , *n*) emerge from this production process. The emerging point, where the products are identifiable and separated from each other, is called the **split-off point** (SOP). The total costs (i.e. material, direct and overhead costs) incurred during this initial production process up to the split-off point are referred to as **joint costs**. Any costs incurred beyond the split-off point are called **separable costs**, **additional processing costs**, or **special manufacturing costs**.

The manufactured multiple products are usually referred to as **joint products** and **by-products**, depending on the market value of the products as compared to each other. A product with a significant contribution to the market value of all products is regarded as a joint product. On the other hand, a by-product is an output of a manufacturing process which adds a relatively small amount to the total market value of all products. Accountants, however, have struggled to reach agreement on using standard terminologies for these two types of products (Manes and Smith, 1965). The terms joint products, major products, co-products and main products may be used interchangeably; the terms by-products and minor products may be used as synonyms; and the terms scrap, waste, spoiled and defective products are used for other purposes.

12.2 Methods of joint cost allocation

12.2.1 Cost-accounting techniques

For years, cost accountants have traditionally used two basic methods for allocating joint costs to major (joint) products, namely **physical-measure** and **relative-market-value** methods (e.g. Myer, 1972; Davidson *et al.*, 1981; Needles *et al.*, 1981; Gray and Ricketts, 1982; Kaplan, 1982). They also use two methods, within these methods, to account for by-products, namely **zero-cost** and **net-realizable-value** methods.

Hardy *et al.* (1981) have criticized these models for their shortcoming in a fair allocation when demand for products differs from year to year, which results in changes in inventory size of each product category. They have demonstrated that the **sales-to-production ratio** is a suitable alternative for joint cost allocation when economic fluctuations affect the demand for the joint products of a company.

Manes and Smith (1965), based on arguments made by Stigler (1952), believed that maximizing profit where fixed proportions of the output products are involved, it is not necessary to allocate joint costs to the joint products. This statement, however, can be rejected by the requirements of cost-based pricing, inventory costing and planning decisions. The demand is usually a price-driven function and, hence, underpricing or overpricing will not be of any advantage in profit maximization. That is, maximum profit can be obtained if price is right; and to set a right price, the manufacturing unit cost must be right.

The disagreement among accountants in joint cost allocation has made their techniques somewhat arbitrary and thus unjustified, as far as economists are concerned (Manes and Smith, 1965).

Physical-measure method

Under the physical-measure method, the joint costs are allocated to each product in proportion to the number of units of the products produced. This is the simplest method of joint cost allocation and reasonable in situations where the market (sales) values of the products are about the same.

Market-value method

Under the market-value method, the joint costs are allocated to the products in proportion to their market (sales) values at the split-off point. The joint products may not all be salable at the split-off point and may require further special processing before they are completed and can be sold. The special processing costs are often called separable costs because they are indeed separable and can easily be assigned to each of the joint products. In such a case, there is usually no information on the market values of the products at the split-off point, which are approximated by deducting the separable costs of the joint products from their corresponding final sales values. The results are called **relative market values** and used as a reasonable basis for the joint costs allocation. The latter technique is referred to as the **relative-market-value method**.

Accounting for by-products

As an alternative, the income from the sales of by-products is regarded as other income, and all the joint costs are assigned to the main (joint) products. This technique is called **zero-cost method** of accounting for by-products. Under this method the only costs assigned to the by-products are their separable costs.

A second alternative of accounting for by-products is to assign to each by-product an amount of the joint costs equivalent to its net realizable (or sales) value at the split-off point. As for relative market value of a joint product, the net-realizable value of a by-product at the split-off point is estimated by deducting its separable costs (including further processing, handling and/or disposal costs) from its sales value. This technique is called **net-realizable-value methods**, under which the costs assigned to by-products are deducted from the total joint costs and, then, the remaining joint costs are assigned to the main (joint) products according to whatever method is used for the joint cost allocation described previously. It should be noted that under the net-realizable-value method the expected income from the sales of by-products is zero.

TABLE 12.1

Product	Sales Price per unit	Production output (units)	Separable cost/unit
M1	$100.00	8 000	$17.50
M2	50.00	10 000	6.00
BP	5.00	1 000	1.00

Example

Suppose that a company produced two joint (main) products, M1 and M2, and a by-product, BP, through a single production process. The total joint costs are assumed to be $500 000. The information given in Table 12.1 is also available.

Under the net-realizable value method of accounting for the by-product,

TABLE 12.2 *Physical-measure method of joint-cost allocation*

Product	Product (units)	Proportion	Share of joint cost	Average joint cost/unit
M1	8 000	4/9	$220 444	$27.56
M2	10 000	5/9	275 556	27.56
TOTAL	18 000	1.00	$496 000	

TABLE 12.3 *Relative-market-value method of joint-cost allocation*

Product	Relative market value at SOP	Proportion	Share of joint cost	Average joint cost/unit
M1	$660 000	0.60	$297 600	$37.20
M2	440 000	0.40	198 400	19.84
TOTAL	$1 100 000	1.00	$496 000	

a total of (5.00 − 1.00) × 1000 = $4000 is deducted from the $500 000 joint costs, and then the remaining (i.e. $496 000) will be allocated to the joint products M1 and M2. But under the zero-cost method of accounting for the by-product the total $500 000 joint costs will be allocated to the joint products M1 and M2.

Assuming that the net-realizable-value method for the by-product is to be followed, the results of joint costs allocation to the main products using the physical-measure method are shown in Table 12.2, and the relative-market-value method in Table 12.3.

12.2.2 Engineering techniques

To overcome the difficulty in selecting the choice of technique, engineers and quantitative analysts have applied the non-linear programming to joint costs allocation as well as by-product identification (e.g. Manes and Smith, 1965; Weil, 1968; Hartley, 1971; Jensen, 1974; Hamlen *et al.*, 1980; Gangolly, 1981; Balachandran and Ramakrishnan, 1981; Kaplan, 1982; Schneider, 1986).

Formulating joint cost allocation

Let N be the set of n sequenced numbers representing the product indexes, and $X_i (i = 1, \ldots, n)$ be the sales quantity of the product i. The following assumptions are made for the purposes of formulating the joint cost allocation:

1. n products are simultaneously produced in fixed proportions. That is, processing each unit of the input material yields a_i units of the ith product.
2. The production process is subject to a limited amount of K units of the input material.
3. There exists enough information about the demand functions for the joint products.

Hence, there will be n constraints representing the relations of sales quantities (i.e. $X_i \leqslant a_i Y$, when Y units of input material are processed and a_i units of product i will be produced, where $i = 1, 2, \ldots, n$).

Letting P_i indicate the price of the ith product ($i = 1, \ldots, n$) and X_i the number of units of this product produced and sold, then P_i can be expressed as an inverse function of the demand quantity. This function can be denoted as:

$$P_i = g_i(X_i), \qquad i \in N \tag{12.1}$$

Let us assume that Y units of the input (raw) material are processed at a cost of C dollars per unit (material and other processing variable costs). If X_i units of product $i (i = 1, \ldots, n)$ are sold, the total revenue, $TR(\boldsymbol{X})$, and total cost, $TC(\boldsymbol{X})$, are written as:

$$\begin{aligned} TR(\boldsymbol{X}) &= \sum_{i=1}^{n} X_i P_i = \sum_{i=1}^{n} X_i g_i(X_i) \\ TC(\boldsymbol{X}) &= TR(Y) = CY \end{aligned} \tag{12.2}$$

Note that $\boldsymbol{X} = (X_i, \ldots, X_n)$ is the sales vector. The contribution margin is calculated as: $Z(\boldsymbol{X}, Y) = TR(\boldsymbol{X}) - TC(Y)$ (12.3)

Therefore, the decision problem is to maximize the contribution margin (which guarantees the maximum profits) subject to the availability and non-negativity constraints. That is:

$$\text{Maximize: } Z(\boldsymbol{X}, Y) = \sum_{i=1}^{n} X_i g_i(X_i) - CY \tag{12.4}$$

$$\begin{aligned} \text{Subject to: } & X_i \leqslant a_i Y, \quad i = 1, \ldots, n && \text{(sales constraints)} \\ & Y \leqslant K, && \text{(material constraint)} \\ & X_i, Y \geqslant 0, \quad i = 1, \ldots, n && \text{(non-negativity constraints)} \end{aligned}$$

This is a nonlinear programming problem that may be solved by the classical quantitative techniques in operations research. For example, the Lagrangean methods can be used to obtain an optimal solution. This method will be particularly simple when the demand-price functions are assumed to be linear. In such a case, the objective function would only involve linear and squared variables, and the problem, then, would be called a quadratic programming problem.

Kaplan (1982) has applied the Lagrangean method in some numerical examples with linear demand functions and relaxed the material constraint. By doing so, which seems to be practical, and assuming $P_i = b_i - m_iX_i$ (both b_i and m_i are non-negative constant coefficients of the price-demand function for product i, where $i = 1, 2, \ldots, n$), the above nonlinear programming problem is reduced to the following form:

$$\text{Maximize: } Z(\boldsymbol{X}, Y) = \sum_{i=1}^{n} X_i(b_i - m_iX_i) - CY, \quad (i = 1, \ldots, n) \tag{12.5}$$

Subject to: $a_iY - X_i \geqslant 0,$ $\quad i = 1, \ldots, n$ (sales constraints)

$X_i, Y \geqslant 0,$ $\quad i = 1, \ldots, n$ (non-negativity constraints)

Under the Lagrangean method, each constraint is assigned a non-negative Lagrangean multiplier λ_i, such that $\lambda_i(a_iY - X_i) = 0$, and add all $\lambda_i(a_iY - X_i)$ to the objective function to obtain the following form of nonlinear programming with non-negativity constraints only:

$$\text{Maximize: } L(\boldsymbol{X}, Y, \lambda) = \sum_{i=1}^{n}(b_iX_i - m_iX_i^2) - CY + \sum_{i=1}^{n}(a_iY - X_i) \tag{12.6}$$

Subject to: $X_i, Y, \lambda_i \geqslant 0, i = 1, \ldots, n$ (non-negativity constraints)

Now, the optimal solution can be obtained by using Kuhn–Tucker conditions or simply setting the first partial derivative of $L(\boldsymbol{X}, Y, \lambda)$ with respect to each variable (i.e. X_i, Y, λ_i) equal to zero and, then solving the following resulting equations simultaneously:

$$\frac{\partial L}{\partial X_i} = b_i - 2m_iX_i - \lambda_i = 0, \quad i \in N \tag{12.7a}$$

$$\frac{\partial L}{\partial Y} = -C + \sum_{i=1}^{n} a_i\lambda_i = 0, \quad i \in N \tag{12.7b}$$

$$\frac{\partial L}{\partial \lambda_i} = a_iY - X_i = 0, \quad i \in N \tag{12.7c}$$

$$X_i, Y, \lambda_i \geqslant 0 \tag{12.7d}$$

Given that all conditions (12.7a) through (12.7d) are satisfied, the results will be the optimal solution to the problem and appear in the following forms:

$$Y = \frac{-C + \sum_{i=1}^{n} a_i b_i}{2 \sum_{i=1}^{n} m_i a_i^2} \quad \text{(Optimal number of units of input material to be processed)} \tag{12.8a}$$

$$X_j = a_j \left(\frac{-C + \sum_{i=1}^{n} a_i b_i}{2 \sum_{i=1}^{n} m_i a_i^2} \right) \quad \text{(Production and sales units of product } j\text{)} \tag{12.8b}$$

$$\lambda_j = b_j + m_j a_j \left(\frac{C - \sum_{i=1}^{n} a_i b_i}{\sum_{i=1}^{n} m_i a_i^2} \right) \quad \text{(Joint cost assigned to each unit of product } j\text{)} \tag{12.8c}$$

In equation (12.7b) ($\Sigma a_i \lambda_i = C$) C is the joint cost allocated to the n products as λ_i dollars to each unit of product i.

However, if any of the λ_i is found to be negative, which violates the non-negativity conditions, it must be set equal to zero and the corresponding product be treated as a by-product, on a zero-cost basis. Then, the value $\lambda_i = 0$ is substituted in the equations in the equation sets (12.7a) and (12.7b), the corresponding equation in the equation set (12.7c) is eliminated, and the new equations are solved again. In such circumstances, some units of product i will remain unsold that should be discarded. Otherwise, selling all units of that by-product will reduce the price as well as the overall profits, unless the residuals are sold in a different market.

Example

Suppose that processing each unit of an input material costs \$80 and yields one unit of product 1 and four units of product 2. Assume also that the demand function for product 1 is $X_1 = 16 - P_1$ and for product 2 is $X_2 = 100 - P_2$. Then, the price functions can be written as:

$$P_1 = 16 - X_1$$
$$P_2 = 100 - X_2$$

and the problem is to maximize the contribution margin; which is presented as follows:

$$\text{Maximize: } Z(X_1, X_2, Y) = (16 - X_1)X_1 + (100 - X_2)X_2 - 80Y$$
$$\text{Subject to: } \quad X_1 \leqslant Y$$
$$X_2 \leqslant 4Y$$
$$X_1, X_2, Y \geqslant 0$$

The Lagrangean form of this problem is:

$$\text{Maximize: } L(X_1, X_2, Y, \lambda_1, \lambda_2) = (16 - X_1)X_1 + (100 - X_2)X_2 - 80Y + \lambda_1(Y - X_1) + \lambda_2(4Y - X_2)$$

$$\text{Subject to: } X_1, X_2, Y, \lambda_1, \lambda_2 \geqslant 0$$

The Lagrangean first-order conditions are:

$$\begin{aligned} 16 - 2X_1 - \lambda_1 &= 0 \\ 100 - 2X_2 - \lambda_2 &= 0 \\ -80 + \lambda_1 + 4\lambda_2 &= 0 \\ Y - X_1 &= 0 \\ 4Y - X_2 &= 0 \end{aligned}$$

Solving these equations simultaneously or using equations (12.8c) yields $\lambda_1 = -3.76$ and $\lambda_2 = 20.96$. Since $\lambda_1 = -3.76$ is negative and not acceptable, the value $\lambda_1 = 0$ will be assumed and the Lagrangean first-order conditions will be modified as follows:

$$\begin{aligned} 16 - 2X_1 &= 0 \\ 100 - 2X_2 - \lambda_2 &= 0 \\ -80 + 4\lambda_2 &= 0 \\ 4Y - X_2 &= 0 \end{aligned}$$

Now, solving these equations simultaneously results in the following solution:

Y = 10 units
X_1 = 8 units
X_2 = 40 units
λ_1 = \$0 (joint cost assigned to each unit of product 1; note that product 1 is a by-product)
λ_2 = \$20 (joint cost assigned to each unit of product 2)
P_1 = \$8 (sales price per unit of product 1)
P_2 = \$60 (sales price per unit of product 2)

12.3 Discussion

Cost accountants traditionally have been using rather arbitrary methods in joint cost allocation. The most common methods used by these professionals are physical measure and relative market value at the split-off point. As Kaplan (1982) has pointed out, using the results of these methods of joint cost allocation in decision analysis or evaluation of the profitability of the products may end up with misleading conclusions. The economic impact on the demands for the products is a complicated phenomenon and cannot be incorporated in simple procedures such as these methods, and hence is left out.

Many quantitative analysts have tried to overcome this deficiency and applied mathematical models as a remedy to the problem. For example, Kaplan (1982) has used the nonlinear programming approach to this type of cost allocation. He has applied an excellent methodology, but the literature search did not show any general model for the situation. The authors of this paper have attempted to take a step in generalizing this methodology.

The general model developed in this paper for joint cost allocation can easily be programmed for computer applications. It is of the same nature as that introduced by Kaplan (1982) through examples which reasonably incorporate the economic impacts on the demand quantities, and hence it can be used in price decisions and profitability analysis of jointly produced products.

12.4 References

Balachandran, B.V. and Ramakrishnan, R.T.S. (1981) Joint cost allocation: a unified approach. *The Accounting Review*, **LVI** (1), 85–96.

Davidson, S., Maher, M.W., Stickney, C.P. and Weil, R.L. (1981) *Managerial Accounting: An Introduction to Concepts, Methods, and Uses*, 2nd edn, The Dryden Press, Chicago, Illinois.

Gangolly, J.S. (1981) On joint cost allocation: independent cost proportional scheme (ICPS) and its properties. *Journal of Accounting Research*, **19** (2), 299–312.

Gray, J. and Ricketts, D. (1982) *Cost and Managerial Accounting*, McGraw-Hill, New York.

Hamlen, S.S., Hamlen, W.A. and Tschirhart, J. (1980) The use of the generalized Shapley allocation in joint cost allocation. *The Accounting Review*, **LV** (2), 269–87.

Hardy, J.W., Orton, B.B. and Pope, L.M. (1981) The sales to production ratio: a new approach to joint cost allocation, (professional notes and letters). *Journal of Accountancy*, **152** (4), 105–6, 108 and 110.

Hartley, R.V. (1971) Decision making when joint products are involved. *The Accounting Review*, **XLVI** (4), 746–55.

Jensen, D.L. (1974) The role of cost in pricing joint products: a case of production in fixed proportions. *The Accounting Review*, **XLIX** (3), 465–76.

Kaplan, R.S. (1982) *Advanced Management Accounting*, Prentice-Hall, Englewood Cliffs, New Jersey.

Manes, R.P. and Smith, V.L. (1965) Economic joint cost theory and accounting practice. *The Accounting Review*, **XL** (1), 31–5.

Myer, J.N. (1972) *Cost Accounting for Non-Accountants*, Hawthorn, New York.

Needles, B.E., Jr., Anderson, H.R. and Caldwell, J.C. (1981) *Principles of Accounting*, Houghton Mifflin, Boston.

Schneider, A. (1986) Simultaneous determination of cost allocations and cost-plus prices for joint products. *Journal of Business Finance & Accounting*, **13** (2), 187–95.

Stigler, G. (1952) *The Theory of Price*, rev. edn, p. 129. (Cited in Manes and Smith, 1965).

Williams, D.J. and Kennedy, J.S. (1983) A unique procedure for allocating joint costs from a production process? *Journal of Accounting Research*, **21** (2), 644–5.

Weil, R.L., Jr. (1968) Allocating joint costs, (Communications). *The American Economics Review*, **LVIII** (5, Part 1), 1342–5.

13 A Totally Integrated Manufacturing Cost Estimating System (TIMCES)

Julius P. Wong, Ibrahim N. Imam, Ali Khosravi-Kamrani, Hamid R. Parsaei, and Fariborz Tayyari

13.1 Introduction

The use of computers in manufacturing has proven to be an effective approach to increase the productivity and to control the manufacturing related costs. Since 1960, computer-aided design and computer-aided manufacturing have been the subject of discussion in many engineering and trade publications. Due to the initiation of links between computer-aided design and manufacturing, the manufacturing firms have been able to develop a new and sophisticated manufacturing philosophy which is called flexible manufacturing systems (FMSs).

Computers are the foundations of CAD/CAM. Computers first appeared in manufacturing in the late 1960s to monitor machine functions, to perform data acquisition, and to assist in quality control (Rembold *et al.*, 1985). The power of the computer was utilized further by incorporating the storage capabilities and transferring data through communication lines. This led to the management information system and the introduction of distributed computing systems. Controllers were added to the machine tools and numerically control machines found their way into the manufacturing environment. Other manufacturing functions such as scheduling and planning began to utilize the power of the computer.

Computers are increasingly being used to control sophisticated machine tools, such as numerically controlled machines, industrial robots, automated storage and retrieval systems, and automated guided vehicles. The communication ports are being updated by engineers in order to achieve the overall goal, **computer-integrated manufacturing** (CIM). Proper communication, integration of manufacturing activities, and information management systems are the objectives behind the CIM philosophy which is the integration of manufacturing and management support functions. It is an integration of a business system with a computer-aided design system and a computer-aided manufacturing system. The efficient flow of data, database management and communications are vital for the existence of the system. An efficient and well-designed CIM system will combine manufacturing flexibility with efficient production management assistance to enhance the overall productivity of a manufacturing organization. Figure 13.1 illustrates an overall view of various components of a CIM system.

13.2 Background

Cost estimating is the task of determining and evaluating the costs involved in an engineering product, a project, or a system using scientific and engineering laws and methods (Ostwald 1984). It is a part of cost engineering which the American Association of Cost Engineers (AACE) has defined as: 'the area of engineering practice where engineering judgement and experience are utilized in the application of scientific principles and techniques to the problems of cost estimating, cost control and profitability' (Clark and Lorenzoni, 1987). Since the term indicates an estimate of the cost, the cost estimator should seek all possible ways to estimate a cost which falls within an acceptable range.

Clark and Lorenzoni (1987) have proposed several classifications for cost estimating. These include:

1. *Screening estimate*. The screening estimate allows the decision maker to decide on which way to go, and whether to accept such a project. The decision period in this phase is short due to the low level of complexity.
2. *Budget estimate*. Since the screening estimate does not give the detail required for budgetary decisions, the budget estimate is required to provide more detail, if desired.
3. *Definitive estimate*. The estimate from this phase is the most accurate and detailed. The decision period is much longer and more effort is required in order to reach the proper estimate. This decision period can be months or years depending on the project complexity and the degrees of accuracy required.

The subject of cost estimating usually deals with time, cost and performance. A list of potential objectives for cost estimating may include (Ostwald, 1984; Malstrom, 1984; Jelen, 1970):

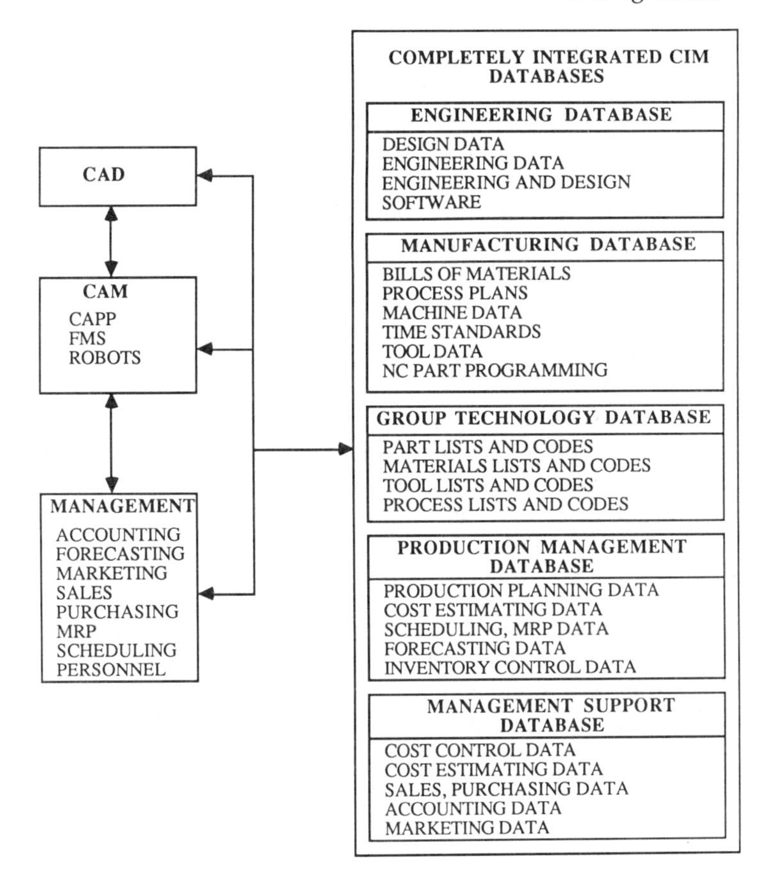

Fig. 13.1 *An overview of the required databases for a CIM system.*

1. Assist in submitting bids;
2. Revise quotations;
3. Assist in evaluating alternatives;
4. Control of manufacturing expenses;
5. Assist in make-or-buy decision;
6. Establish ground for selling price;
7. Provide and evaluate standards; and
8. Long-term planning.

Methods which are often used to estimate costs in the manufacturing environments are (Ostwald, 1984; Whealon, 1985; Torres, 1980):

1. The opinion method;
2. The conference estimating method;
3. The comparison method;

4. The unit estimate method;
5. The cost and time relationship;
6. The power law and sizing model method;
7. The probabilistic method;
8. The statistical models:
   (a) Estimating by confidence interval,
   (b) Estimating by tolerance interval, and
   (c) Estimating by prediction interval;
9. The simulation method;
10. The factor method; and
11. The detailed computerized method.

All the above methods are used only to estimate costs. It is desired to have these costs fall within an acceptable range. Therefore, the need for accuracy is evident. The errors of the estimation task must be reduced in order to achieve an optimal estimate. Estimating errors are often categorized as controllable and uncontrollable errors (Clark and Lorenzoni, 1987; Tyran, 1982). Controllable errors are often due to:

1. Failure to developed detailed data necessary for the cost estimate;
2. Errors in interpreting information;
3. Making wrong assumptions;
4. Use of poor documented data;
5. Failure to spend the time necessary for accurate estimations;
6. Poor analysis of the problem in hand; and
7. Lack of experience.

Uncontrollable errors are usually due to:

1. Unpredictable change in equipment;
2. Unexpected conditions such as fires, storms and industrial accidents;
3. Labour strike; and
4. Decline in productivity levels due to employee attitudes and morale.

If the management is not confident about the level of accuracy of the estimates, the cost estimate will not only be irrelevant, but also a burden on the company that cause unnecessary overhead expenses.

The cost estimate is the summation of various costs involved in the estimation of cost for a product project, or a system. These costs are classified into two groups;

1. Direct costs:
   (a) Direct material: materials which are an integral part of the finished product, and
   (b) Direct labour: costs that can be traceable directly to the making of the product;
2. Indirect costs:

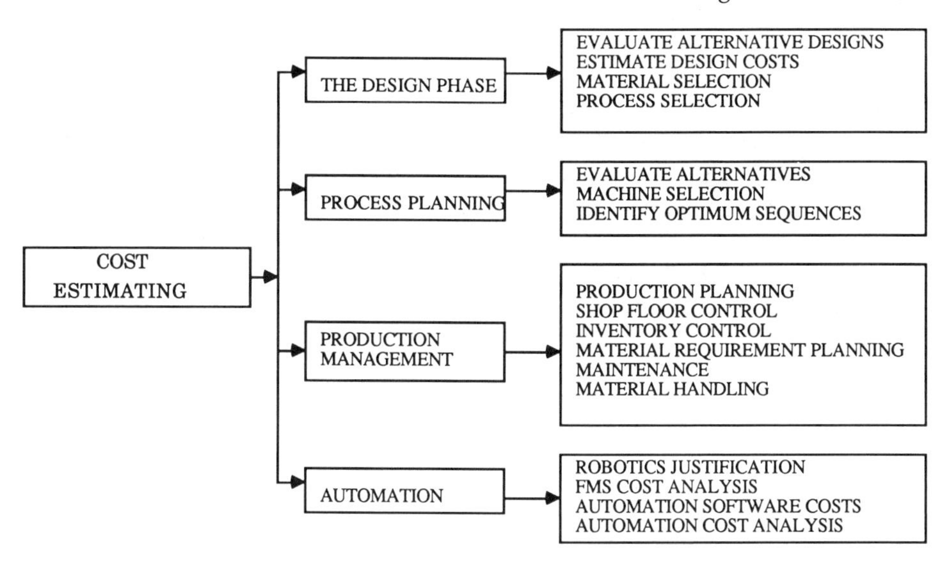

Fig. 13.2 *An illustration of the various roles of cost estimating in a computer-integrated manufacturing environment.*

(a) Manufacturing overhead: all manufacturing costs except direct material and labour costs,
(b) Indirect materials, and
(c) Administrative overhead.

The cost-estimating process plays an important role in the manufacturing, planning and control. It assists manufacturing firms in decision-making at various stages including: design and production planning, materials handling, facility layout, inventory control and shop-floor control. The role played by cost estimating in manufacturing is illustrated in Fig. 13.2.

13.3 Computer-aided cost estimating

Computer-aided cost Estimating (CACE) is the use of computers to estimate costs of products, projects or systems. The use of CACE increases the productivity of the estimator. The estimates will be easily revised and verified and data will be stored economically and readily accessible (Bacher, 1983). CACE is an important tool to:

1. Develop cost estimates in shorter time;
2. Improve estimate accuracy by minimizing the human interface;
3. Improve in cost data availability and security;

4. Evaluate more alternatives, and
5. Improve management moral.

Automated cost estimating was originally a translation from the manual cost-estimating methods to a computerized method that employs the computer for calculations and storage purposes. With the recent advances in database methodology, data structures and algorithm development, a new generation of cost-estimating programs is evolving. Since cost-estimating software requires a large amount of information from a variety of databases, it is essential that data structures, data retrieval and storage methods be carefully examined and to select the ones that ensure to efficiency of the cost-estimating software. Figure 13.3 is the flow diagram representing the steps required to estimate the cost of a product.

13.4 A review of the existing cost-estimating systems

A number of CACE systems are already available in industry. Some of these systems are presented in this section (Whealon, 1985; Goldberg, 1987; Casey, 1987).

13.4.1 CACE in manufacturing

The COSTIMATOR

This system was developed by Manufacturing Technologies, Inc. It can provide operation cost estimating on a variety of machine tools. The COSTIMATOR software has built-in information for approximately 100 machines, and can be expanded to a capacity of 10 000 machines. The COSTIMATOR also has information for approximately 200 materials and is expandable to 500. Machining parameters such as feeds and speeds are included in machinability data.

The COSTIMATOR uses a sonic digitizer to digitize information from drawings or actual parts. The digitizer consists of a sonic pinpoint that is depressed on the drawing part. A unit with precise microphones picks up the sound from the pen for recording the elapsed time. The time elapsed is then converted to actual measurements. This technique is used to emulate machine tool paths and automatically obtain machine data.

The system has extended capabilities to allow the cost estimator to ask 'what if' questions in order to determine potential savings. Changes in method, materials, and labour are incorporated automatically into the system and adjustments are made to all the parameters and subestimates affected by the change.

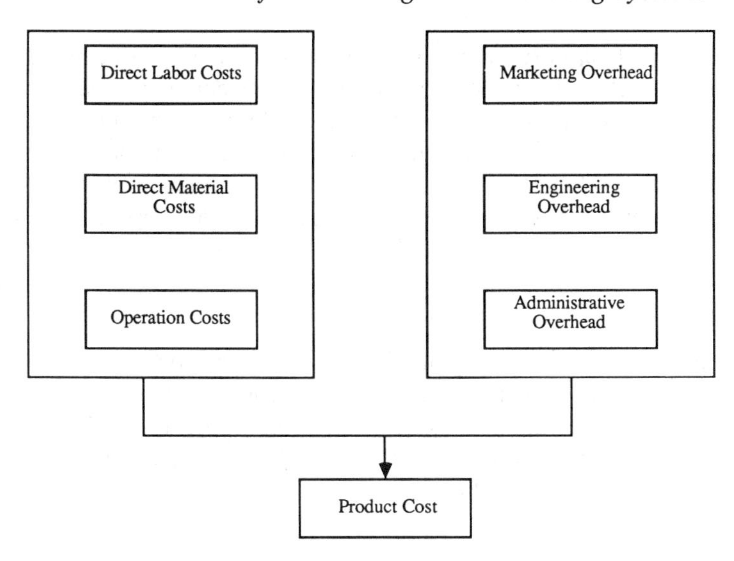

Fig. 13.3 *The required inputs to a cost-estimating system.*

The MICAPP system

MICAPP is a cost-estimating system distributed by the Society of Manufacturing Engineers. The system provides various functions for cost estimating. However, a considerable amount of manual work is required to prepare the data for input. The extensive database available in the COSTIMATOR is not available in this system. A printer is required at all times and hard copies are supplied regardless of the estimator's desire. 'What if' analysis is allowed when modifying lot sizes, standard time, machine times or shop rates. A learning curve is incorporated in the system to improve the quality of estimates over time.

The CACE system

The Computer-Aided Cost-Estimating system was developed at the Iowa State University. It was a research project in cost estimating. The program is essentially a translation of manual cost-estimation calculations. It was written in FORTRAN 77 and consists of a main program, two subroutines, and four utility functions. The data are entered following program commands and prompts which require the cost estimator to prepare data and perform calculations before running the program. The system output is a detailed cost estimate.

The AM estimator

This system was originated by the *American Machinist and Automated Manufacturing*. It was developed by Ostwald (1985) and is based on a research project on cost estimating. The system is able to generate very accurate estimates. It supplies material and machine databases which accommodate changes and new input. This system requires more training than other systems. The system outputs include individual part estimates, machine routings, feeds and speeds. Despite the accuracy of the system, it is not very practical since it takes a long time to develop estimates and it is not very flexible (Bacher, 1983; Goldberg, 1987).

The E-Z Quote system.

This system is available from the E-Z Systems Division, CIM, Inc. E-Z Quote is described as an extremely user-friendly system. It is a close translation of cost estimating using manual methods. The only database available is a materials database. It was originally designed to estimate stamping and tapping operations.

13.4.2 CACE systems for construction and process industries

The objective of cost estimating in construction is different from that in manufacturing. The construction cost estimator includes only material and labour, but the manufacturing cost estimator must also estimate operations costs. Process industries used computer-aided cost estimation before it was introduced into manufacturing.

The Economist System

This system is a fully interactive program developed by the Computer Aided Design Centre Ltd, Cambridge, UK, for estimating the capital and operations costs of chemical plants and carrying out profitability and sensitivity analysis (Williams and Gerrard, 1984). Process-design information and costs are correlated using a capital cost-estimation module. This module consists of a data bank of equipment costs and a set of methods to convert these costs to total plant cost.

The CO$T System

The CO$T system (Blecker and Smithson, 1985) is a detailed cost-estimating program which was developed in the late 1960s. The system develops cost estimates of process plant. It generates cost estimates and reports. Process specifications, plant designs and flow information are required for system input. Equipment data are also required. The user can input as many as 40 parameters for a single item description.

Figures 13.4 and 13.5 present a summary of software and hardware characteristics of the various computer-aided cost-estimating systems.

SYSTEM REQUIREMENTS

	COSTIMATOR	MICAPP	AM ESTIMATOR	EZ-QUOTE
PROCESSOR HARDWARE	IBM PC-XT	IBM PC-XT	IBM PC-XT	IBM PC-XT
PRIMARY MEMORY	256K	128K	256K	256K
SECONDARY MEMORY	Hard/Floppy Disks	Floppy Disks	Floppy Disk	Hard/Floppy Disks
DISPLAY	Monocrome Monitor (MM)	MM	MM	MM
OPERATING SYSTEM	DOS 2.0 or higher	DOS 2.0 or higher	DOS 2.0 or higher	DOS 2.0 or higher
HARDWARE COST	$1000	$1000	$1000	$1000
SOFTWARE COSTS	$18735	$1195	$795	$2495

Fig. 13.4 *A summary of requirements for various CACE systems (Casey, 1987).*

SOFTWARE CAPABILITIES				
CAPABILITY	COSTIMATOR	MICAPP	AM ESTIMATOR	EZ-QUOTE
SOFTWARE FEATURES				
DIGITIZE PRINTS	X	--	--	--
MATERIALS DATABASE	X	X	X	X
AREA, VOLUME OR PART	X	--	--	X
MACHINE CAPABILITY OVERRIDE	X	X	--	--
MACHINE DATABASE	X	X	X	--
HANDLING DATABASE	X	X	X	--
CAPABILITY TO EDIT DATABASES	X	X	X	X
PROCESS PRINTOUT	X	X	X	X
ROUTING PRINTOUT	X	X	X	X
ESTIMATE PRINTOUT	X	X	X	X
LETTER PRINTOUT	--	--	--	X
ESTIMATE BASED ON STANDARD DATA	X	X	X	--
ABILITY TO STORE ESTIMATES	X	X	X	X
BILL OF MATERIALS	X	--	--	--
INCLUSION OF OUTSIDE (SUBCONTRACTED) OPERATIONS	X	X	X	X
SPEEDS AND FEEDS CALCULATION	X	X	X	--
ACTUAL COSTS FEEDBACK CAPABILITY	X	X	--	--

Fig. 13.5 *Capabilities of various cost-estimating software systems.*

13.5 Totally Integrated Manufacturing Cost Estimating System (TIMCES)

This paper presents the design for a Totally Integrated Manufacturing Cost Estimating System (TIMCES). TIMCES may be considered as a utility program which is a part of the CIM system and is used to develop cost estimates. TIMCES's structure allows the integration of CAD, and process planning technique into a unified system. Inputs to the TIMCES are provided through CAD, process plan and several other databases. Dimensions, sizes and shapes for the parts may be furnished using an available computer aided-drafting system (e.g. AUTOCAD) or selected from a CAD database file. The system described here is capable of developing cost estimates only for products. The product cost-estimating process often consists of the estimation of the costs associated with various elements.

13.5.1 Materials cost estimating

Material is defined as substance being transformed or used in a manufacturing transformation. Materials are classified as:

1. Raw materials;
2. Commercial products;
3. Subcontract products; and
4. Inter-department transfer products.

Materials cost estimating usually includes both direct and indirect materials. The information required is obtained from bills of material, product design and inventories of direct and indirect materials. The cost estimator should identify the nature of the material, through material analysis, and determine the quantity required and then pick a cost policy.

The procedure for estimating the material cost is as follows:

1. Measure the shape and the volume of the material:
   (a) Design drawing,
   (b) Bills of material;
2. Identify the price of material:
   (a) Accounting records,
   (b) Vendors,
   (c) Surveys;
3. Find the value of any salvage material:
   (a) Accounting records,
   (b) Vendors;
4. Choose a material cost policy:

   (a) Last in first out,
   (b) First in first out,
   (c) Others;
5. Tabulate the total cost of material.

13.5.2 Labour cost estimating

Labour cost estimation constitutes the second part of the direct cost estimation. Its importance is obvious due to the extensive attention it gets from management, government and researchers. The first step in determining the labour cost estimate is to estimate the labour time. Then, a cost figure can be developed using labour cost rates (Chaibi, 1987).

The procedure for estimating the labour cost is as follows:

1. Identify operation:
   (a) Production plan,
   (b) Machine selection,
   (c) Process sequence,
   (d) Material requirements;
2. Determine labour time:
   (a) Motion and time studies,
   (b) Predetermined time standards,
   (c) Work sampling,
   (d) Man-hour reports;
3. Identify hourly rates:
   (a) Accounting records,
   (b) Personnel files,
   (c) Predetermined hourly rates;
4. Get fringe benefits:
   (a) Accounting records,
   (b) Personnel files,
   (c) Predetermined rates;
5. Tabulate the total cost of labour.

13.5.3 Cost of machinery and tools

Estimating the cost of machinery and tools used to manufacture a certain product is an integral part of the cost-estimating function. Tools can be classified as hard or soft tools. Hard tools are those that are designed and manufactured specifically for a certain manufacturing operation. On the other hand, soft tools are conventionally used in common manufacturing operations. Tooling costs are estimated for the following reasons:

1. To determine the investment necessary for tools within a time frame during the planning phase; and
2. To evaluate alternative tooling combinations and select the combination incurring the least cost.

The costs of tools or new equipment occur only once. As a result, difficulty is experienced when allocating these costs to individual operations.

13.5.4 Cost of operation

An operation involves material, labour and equipment. The estimator must have the necessary cost-estimating data in the form of trade books, handbooks and various data sources about the operations involved in the design. The necessary information includes:

1. Part design;
2. Production plans;
3. Material specifications;
4. Tooling specification; and
5. Standard time sheets.

Operation cost estimating starts by breaking down the operation elements. For each element the labour cost, material cost and equipment tooling cost are estimated. Each operation consists of three phases:

1. *Set-up.* Preparing all the conditions required for the operation.
2. *Cycles.* Performing the operation for a number of cycles.
3. *Maintenance.* Maintaining all the conditions required for the operation.

13.5.5 Overhead cost

Overhead cost in present cost-accounting practice is the portion of total cost that cannot be traced directly to particular operations, products or projects (Ostwald, 1984). Indirect expenses should be allocated, utilized and added to the unit-cost estimate. The problem with allocating overhead charges is that these costs often exist even if the product is not produced.

Companies, based on their activities, may adopt different techniques in measuring the overhead cost. Overhead charges may be determined in different ways including:

1. Overhead as a ratio of direct labour dollars;
2. Overhead as a ratio of direct labour hours; and
3. Overhead as a ratio of prime cost.

13.5.6 Cost of product

A product is the key determinant of an organization's success. The research, engineering, manufacturing and marketing departments work together with management to provide a successful product. Costs beyond the manufacturing stage include the costs of engineering, marketing and administration-related activities.

To estimate the cost of a product, the cost estimator needs to be supplied with:

1. Material cost;
2. Labour rate;
3. Labour time estimate;
4. Labour overhead rate;
5. Tooling costs:
   (a) Tool changing,
   (b) Tool cost per unit;
6. Number of units;
7. Overhead costs:
   (a) Engineering,
   (b) Marketing,
   (c) Administration.

The estimator should also decide on the techniques to be used in calculating the product cost estimate. TIMCES utilizes the operation method which is an analytical approach.

The operation product cost is the summation of the material cost, direct labour plus overhead cost, and the total tooling cost. The cost of the product is obtained by charging the overheads to the operation product cost.

An overview of the TIMCES logic is as follows:

INPUT: Product design
COMPUTER: CAD software
DESIGN
MODIFY
ANALYZE
STORE results.
OUTPUT: Product's layout

INPUT: Product's layout
COMPUTER: CAPP software
ANALYZE layout for Group Technology
VERIFY
REJECT or ACCEPT
STORE results

OUTPUT:	Route sheet
INPUT:	Route sheet
INPUT:	Material cost estimating file
COMPUTER:	COMPUTE material cost estimate
	VERIFY
	REJECT or ACCEPT
	STORE results
OUTPUT:	Direct material cost estimate
INPUT:	Labour cost estimating file, including overhead
COMPUTER:	COMPUTE labour cost estimate
	VERIFY
	REJECT or ACCEPT
	STORE results
OUTPUT:	Direct labour cost estimate and overhead
INPUT:	Operation cost estimating file

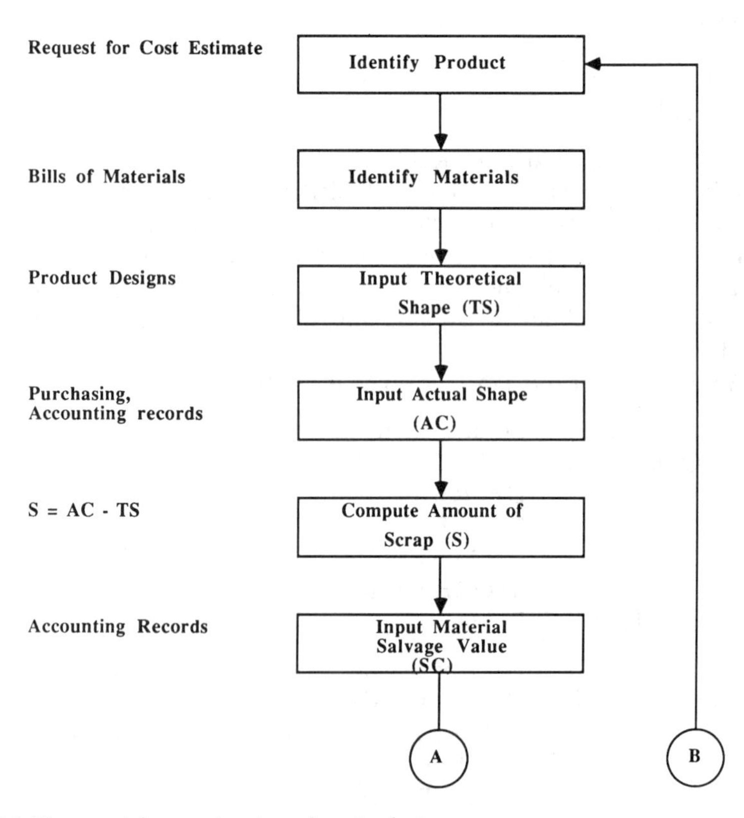

Fig. 13.6 *The material cost-estimating subroutine logic.*

COMPUTER: COMPUTE machining and tools cost estimate
VERIFY
REJECT or ACCEPT
STORE results
OUTPUT: Tool and machining cost estimate

INPUT: Overhead cost estimating file, excluding overhead
COMPUTER: COMPUTE overhead cost estimate
VERIFY
REJECT or ACCEPT
STORE results
OUTPUT: Overhead cost estimate

INPUT: Direct material cost estimates
Direct labour cost and overhead estimates
Tool and machining cost estimates
Overhead estimates, excluding labour overhead
COMPUTER: COMPUTE total product cost estimate
VERIFY
ANALYZE
VALIDATE
REJECT or ACCEPT
STORE results
OUTPUT: Final product cost estimate

13.6 A proposed design structure for various modules of the TIMCES

The logic presented above demonstrates the flow of information within the TIMCES shell. In this section the flow charts, the file structure, and the equations required for the operation of the various modules of TIMCES are discussed.

13.6.1 Automating the material cost estimating

Figure 13.6 illustrates the flow chart of the material cost estimating logic. The variables appearing in this diagram are:

TS = Volume of theoretical shape
AC = Volume of actual shape = TS + S
S = Volume of scrap = AC − TS
CMS = Material cost per volume unit.
SC = Salvage value of scrap
CDM = Cost of direct material per unit
= (AC)(CMS) − SC

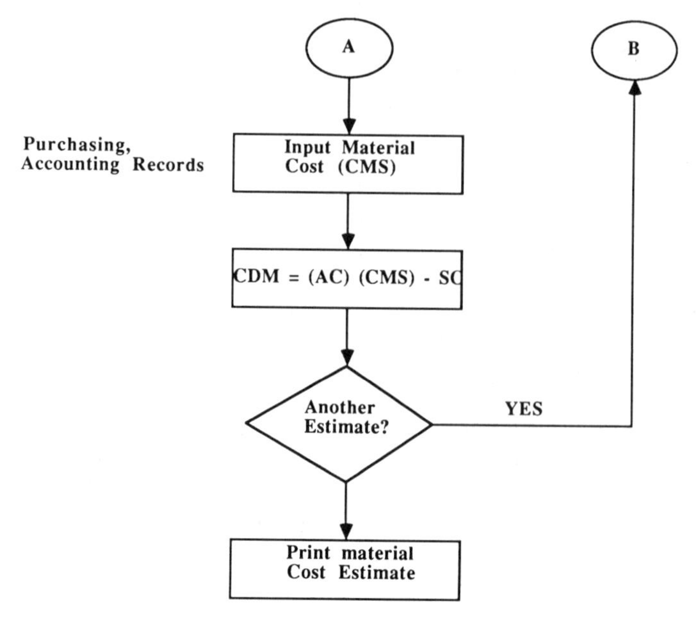

Fig. 13.6 *(Continued)*

The required material cost-estimating file structure for the system should accommodate the storage and the access needs. The information in this file may include:

MATERIAL NAME AND CODE RECORD
- Material name
- Material code

MATERIAL SHAPE RECORD
- Material name
- Material code
- Material shape

MATERIAL MEASUREMENT RECORD
- Material name
- Material code
- Material measurement

MATERIAL COST RECORD
- Material name
- Material code
- Material cost

These records should be linked by a material code which will serve as a database key to facilitate storage and retrieval of records.

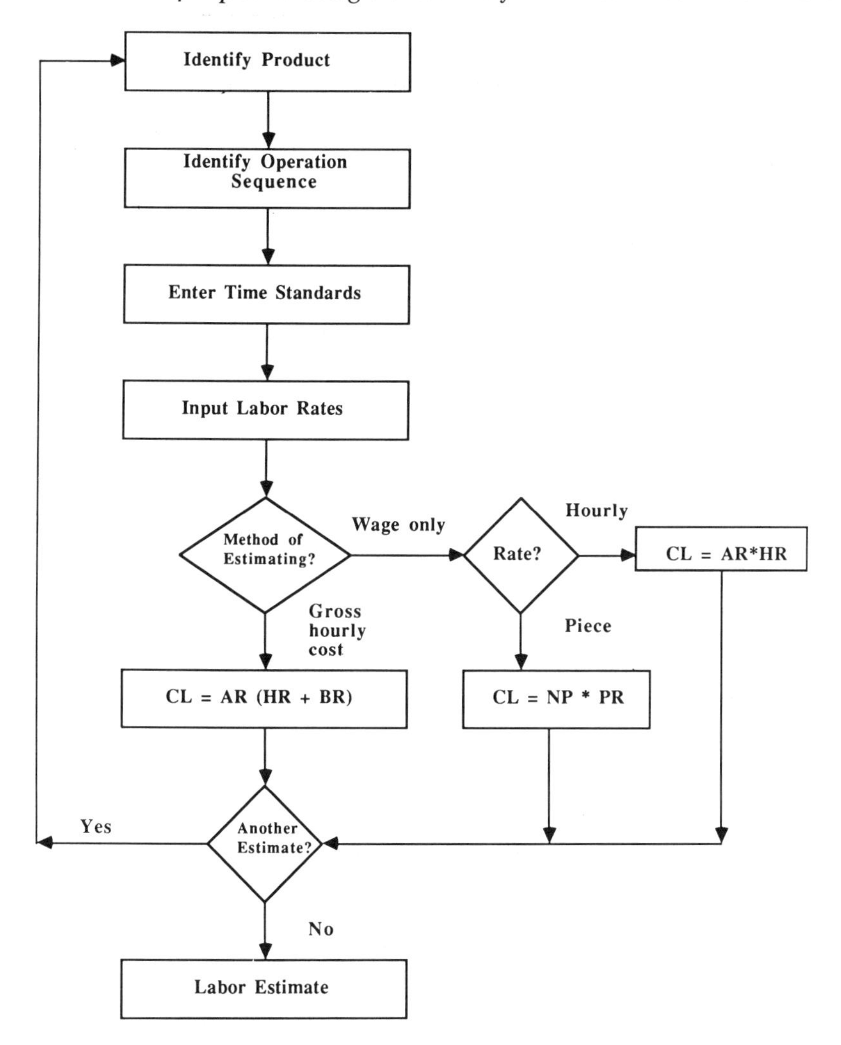

Fig. 13.7 *Labour cost-estimating subroutine logic.*

13.6.2 Automating the labour cost estimating

Automating the labour cost estimating phase may be done using different approaches including MTM, 4M, etc. Figure 13.7 illustrates flow diagram for the algorithm used to determine the labour cost estimate. The variables appearing in Fig. 13.7 are explained below:

CL = Cost of labour
AR = Actual number of hours worked
HR = Hourly labour rate

NP = Number of pieces produced
PR = Rates of labour hours per piece produced
BR = Labour overhead

Records and fields required in the labour cost-estimating file may include:

OPERATION IDENTIFICATION RECORD
Operation ID
Operation description

OPERATION ELEMENTAL RECORD
Operation ID
Number of elements
Total operation time

ELEMENTAL RECORD
Element name
Element labour time
Element labour rate

OPERATION TIME RECORD
Operation ID
Normal operation time
Standard operation time
Operation cost

13.6.3 Automating the operation cost estimating

Figure 13.8 represents the flow chart for the algorithm required for estimating the cost of operation. The variables used in this algorithm are detailed as follows:

TO = Total time required to perform the operation of one unit
TH = Handling time
TM = Time of machining
TCT = Tool changing time
T = Tool life
tu = Unit time estimate
ST = Setup time for a certain operation
N = Number of unit produced in a lot
CU = Unit cost of operation
RD1 = Direct labour rate
ROH = Overhead rate as a percentage of direct labour

An operation estimating cost file needs to include the following records and fields:

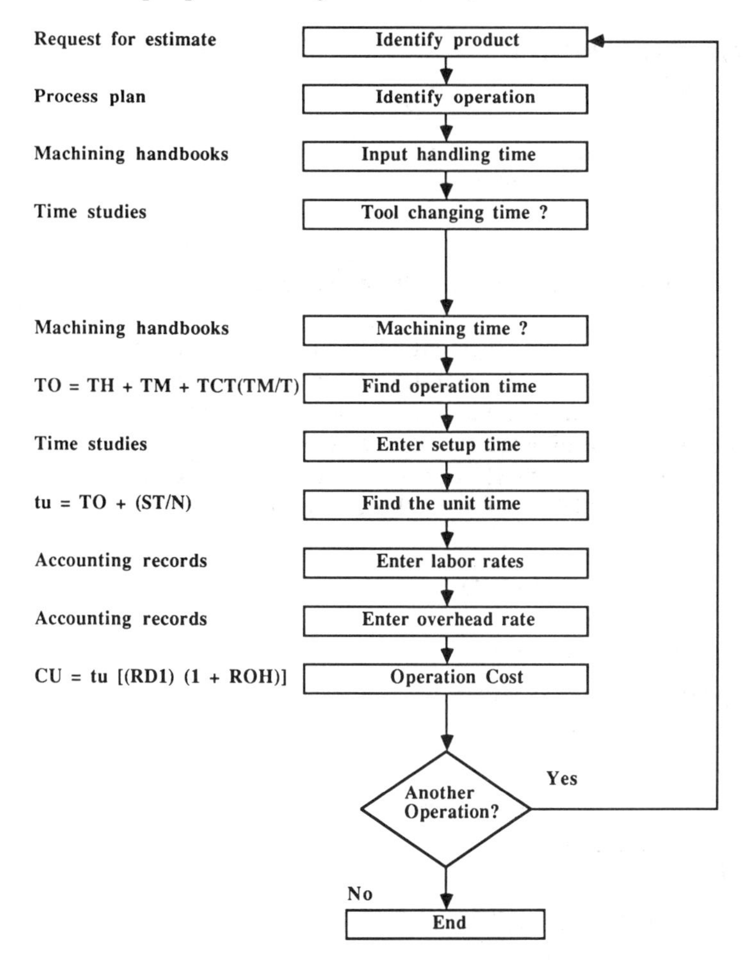

Fig. 13.8 *The operation cost-estimating subroutine.*

OPERATION IDENTIFICATION RECORD
  Operation ID
  Operation description

SET-UP TIME RECORD
  Operation ID
  Set-up time
  Set-up labour rate

MACHINING RECORD
  Machine ID
  Machining time
  Machining labour rate

TOOLING RECORD
Tool ID
Tool cost
Changing time
Changing labour rate

13.6.4 Automating the overhead cost estimating

The overhead cost is the segment of the product cost which cannot be directly traced to particular operations or products (Ostwald, 1984). The overhead rate in this system is generated by summing up marketing, engineering and administrative costs associated with the processes and operations performed at each department (activity centre).

Records and fields required in the overhead cost estimating file may include:

OPERATION RECORD
Operation code
Operation estimate

MARKETING RECORD
Marketing code
Marketing rate

ENGINEERING RECORD
Engineering code
Engineering rate

ADMINISTRATIVE RECORD
Administrative code
Administrative rate

13.6.5 Automating the product cost estimating

The final step in cost estimating is to develop the product cost estimate. This would be accomplished by adding marketing costs, administrative costs, and engineering costs to the already determined material, labour and operation costs.

The product cost-estimating file will consist of several records outlined as follows:

PRODUCT IDENTIFICATION RECORD
Product code
Product name

MATERIAL RECORD
Material code
Material estimate

LABOUR RECORD
Labour code
Labour estimate

OPERATION RECORD
Operation code
Operation estimate

OVERHEAD RECORD
Overhead code (marketing, administrative, and engineering)
Overhead charge

Figure 13.9 presents a flow diagram of the product cost-estimating logic.

Using input from described files and following each file's specific algorithms, which may include the calculation routines, optimization routines and other functions (e.g. SEARCH), the product-estimating package will tabulate an operation unit cost to which all overhead will be added. The driving module of the system will be written in C. It will function as the driving system of this software. The database files of the program can be developed using dBase IV. The updating routine for these files is done interactively by the proper C code, and the database utility functions. Figure 13.10 summarizes the TIMCES modules and the required databases.

The targeted hardware system to be used has the following components:

1. Personal computer with an 80386 or 80486 CPU equipped running DOS 3.0 or higher with:
   (a) Colour display (VGA),
   (b) Hard drive (80 MB hard drive or higher),
   (c) Communication ports (for printer, plotter, and scanner),
   (d) Math co-processor for 80386;
2. Printer;
3. Plotter;
4. Digitizer pad; and
5. Scanner.

13.7 Conclusion

One of the functions of cost estimating in manufacturing is to estimate the cost of the product prior to manufacture. This allows the designer to investigate the viability of a well-designed product in a real-world manu-

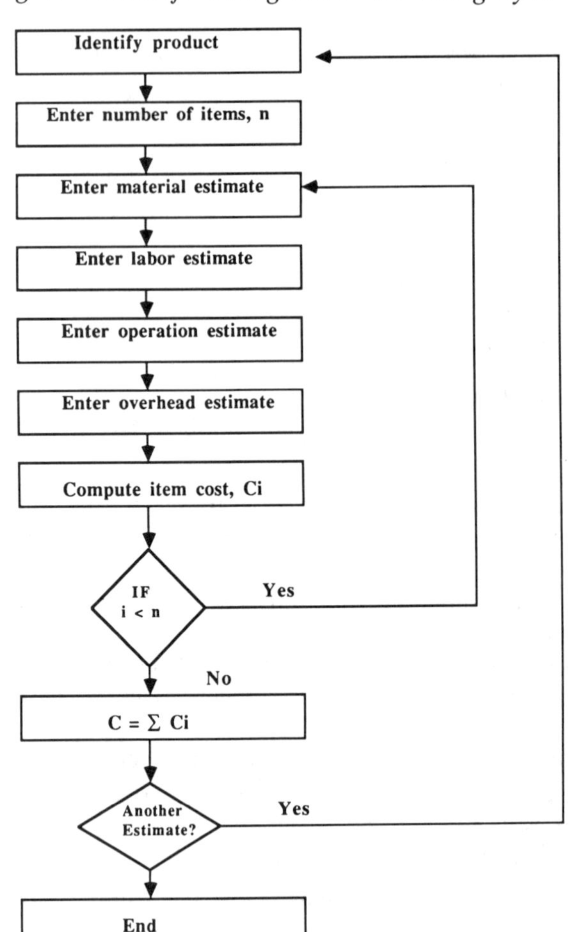

Fig. 13.9 *The product cost-estimating logic.*

facturing environment. The majority of computer-aided cost-estimating systems which are currently available in the market were reviewed. Many of these systems are basically an automated translation of the conventional manual methods and cannot be utilized in an integrated manufacturing environment.

This paper presented the design for a Totally Integrated Manufacturing Cost Estimating System (TIMCES), which can be easily integrated with the other engineering software systems. The input to the proposed TIMCES system will be provided by several databases including computer-aided design, process planning, materials cost, labour cost, etc. Since the trend in today's manufacturing world is towards complete integration of engineering, production and business functions through the computer, the

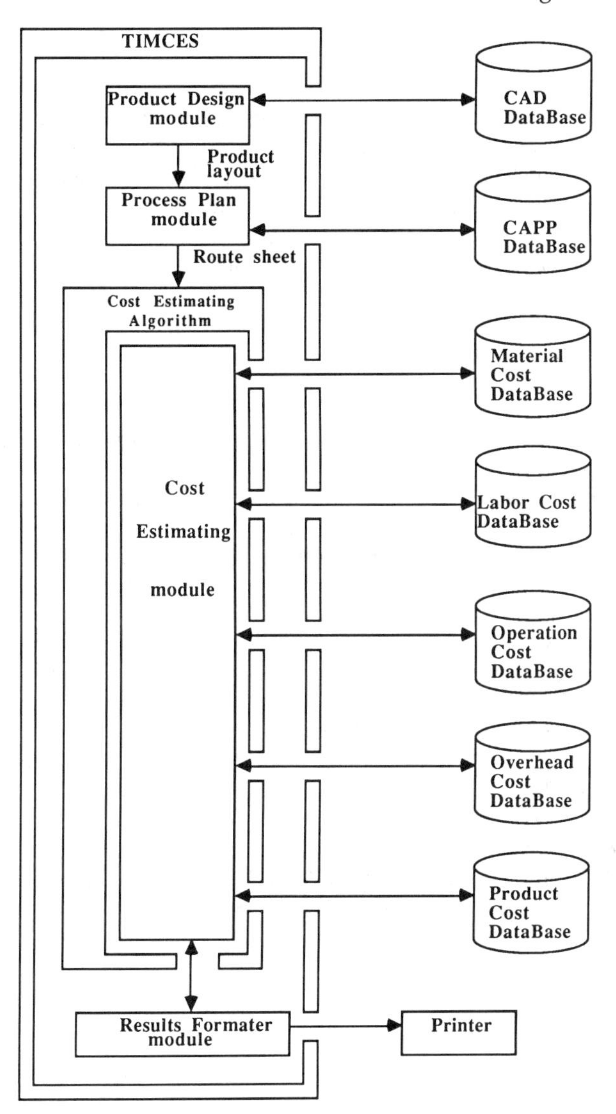

Fig. 13.10 *A flow diagram of the TIMCES modules and the required databases.*

TIMCES will be a powerful tool during design, manufacturing and cost justification phases.

13.8 Acknowledgement

The authors wish to express their sincere appreciation to the Graduate Programs and Research Office, University of Louisville for partial support of this work.

13.9 References

Bacher, L.C. (1983) Computer assisted estimating: analyzing the options. *Cost Engineering*, **25**, 65–71.

Blecker, H.G. and Smithson, D. (1985) Detailed cost estimating during the process engineering phase of a project. *Process Economics International*, **5**, 17–28.

Casey, J. (1987) Digitizing speeds cost estimating. *American Machinist and Automated Manufacturing*, 71–9.

Chaibi, A. (1987) *Design of a Computer Aided Manufacturing Cost Estimating System*, Master of Engineering Thesis, University of Louisville, Kentucky.

Clark, F.D. and Lorenzoni, A.B. (1987) *Applied Cost Engineering*, Marcel Dekker, New York.

Goldberg, J. (1987) Rating cost estimating software. *Manufacturing Engineering*, Feb. 31–4.

Jelen, F.C. (ed.) (1970) *Cost and Optimization Engineering*, McGraw-Hill, New York.

Malstrom, E.M. (1984) *Cost Estimating Handbook*, Marcel Dekker, New York.

Ostwald, P. (1984) *Cost Estimating*, 2nd edn, Prentice Hall, Englewood Cliffs, New Jersey.

Rembold, U., Blume, C. and Dillman, R. (1985) *Computer Integrated Manufacturing Technology and System*, Marcel Dekker, New York.

Torres, C.U. (1980) Statistical techniques in cost engineering. *Transactions of the American Association of Cost Engineers*, B 7.1–5

Tyran, M.P. (1982) *Product Cost Estimating and Pricing: A Computerized Approach*, Prentice Hall, Englewood Cliffs, New Jersey.

Whealon, I. (1985) Sound basis for computer aided cost estimating. *Machine and Tool Blue Book*, 54–8.

Williams, L.F. and Gerrard, A.M. (1984) Computer aided cost estimation, a survey. *Process Economics International*, **5**, 28–9.

14 Risk Evaluation of Investment in Advanced Manufacturing Technology

B.B. Hundy and D.J. Hamblin

14.1 Introduction

Many industries, for instance the consumer electronics and automotive industries, invest enormous sums in new model development, product feature research, process research and capital investment in equipment and facilities. All these items have one thing in common: the investment is being made in the expectation of some benefit and there is some risk that the benefit may not be realized. Of course, the benefit/investment ratio may turn out to be better than expected, the research programme may successfully be completed sooner, the sales of the new model may be higher, or the savings from an equipment investment may be greater than originally expected. Generally, however, this would be seen as an unexpected bonus and concern is mainly with the downside risk that the company may make a loss on the project. We would consider it a loss if the return was less than the cost of money used to finance the project.

Risk analysis of major projects is a major activity in industries such as defence and oil extraction, in terms of marketing returns and features such as political, geological and technical risk and timing (Cooper & Chapman, 1987). Risk assessment such as used in these industries, however, could involve the assembly of a large analysis team working for months, if not years, and only very major projects (e.g. construction of a new pipeline) can justify such an expenditure. In the consumer industries the introduction of a new model with its product-development costs and the associated large capital investment might fall into this category.

There are, however, many smaller projects in research and development and in capital investment where presently no risk analysis is carried out even on projects involving some millions of dollars. Investment in product and process development is justified on expert opinion of the likelihood of success and its benefit. Capital investment is normally justified by the project return, calculated using 'most likely' figures for cost and benefit, having to clear some sort of hurdle, e.g. minimum payback period or, more usually, a minimum internal rate of return (IRR). Risk is minimized by raising the hurdle, for example, by specifying a high minimum IRR. This risk allowance can be quite high as financial analysts tend to be conservative and are aware of a long history of over-optimistic engineers and marketing men. Critical appraisal of the project is limited to assessing the sensitivity of return to changes in the inputs. This is insufficient to determine the risk profile. We believe that simple methods of analysing the risks associated with a specific project provide additional useful information for management decision taking, in particular the probability of the project making a loss. They would also minimize the need for general risk allowances. Ideally, any such methods should not involve large amounts of manpower to collect the information necessary to carry out the calculations and should be completed rapidly.

In this paper we intend to consider only capital investment projects though the methods used can easily be carried over to R&D or product development projects.

14.2 Capital investment decision

Risk analysis can be carried out using any of the usual project-evaluation criteria such as payback period, NPV or IRR. We prefer to use the internal rate of return as it gives a single figure representing the return generated by the investment inside the firm which can be compared with the cost of money to the firm. This advantage substantially outweighs theoretical disadvantages. It is common in many firms considering a cost-saving project to apply a general risk factor uniformly to all projects by specifying a general hurdle rate for the IRR which has to be met by any investment project. This is often very high in relation to the cost of money to ensure a high level of security when optimistic cost assumptions are made in risky environments. Thus a hurdle rate which is twice the cost of money is not unusual. It can be argued that this is unnecessarily high for many projects which have lower risks but which could contribute effectively to the profitability of the firm. Occasionally one meets situations when a lower hurdle rate is specified for flexible investments, in which case some volume or product risks are reduced. However, this is not usual.

We argue that each investment proposal should be considered individually and the associated risks assessed so that its likely return, taking account of the risks inherent in that proposal, can be compared directly with the cost of money (with a small percentage added, if desired, to cover the absolute minimum return sought by the firm), e.g. 15% or 16% after tax, for the UK in the present economic environment.

A spreadsheet program has been developed to model the determination of cash flows and profitability in an investment project. This includes the effect of the tax regime and depreciation allowances particular to the country concerned and gives the IRR and payback for the project. Inputs include annual changes in running expenses, such as direct labour, materials, maintenance and stock holdings; one-off items in any one year, such as start-up costs and expenses, and the capital expenditure items of facilities, tooling, working capital, etc. Spreadsheets offer considerable advantages over conventional investment analysis computer programs. They allow functions rather than single values to be input where appropriate, and they have the potential to be customized readily to individual user requirements. First-time users are, however, advised to use a fully tested model rather than design their own. Using such a model it is easy to determine quickly the effect of inputting a range of variables to represent the possible risk scenarios. These results can then be weighted to reflect their probability of occurrence.

Normally the inputs to this model will be the identified tangible benefits and it has been argued (Meredith and Suresh, 1986) that ignoring some of the intangible benefits can lead to a pessimistic assessment of an investment opportunity. We show that some of the so-called intangibles, such as quality and flexibility, can in fact be quantified to some degree. Factors such as increases in market share, due to improved delivery or quality, are more difficult to assess but estimates can be made and input to the model with a range of risks to determine possible effects.

In an earlier paper (Hundy and Hamblin, 1988) it has been shown that the effects of potential changes in the future to factors affecting the project can be simulated on the spreadsheet calculator to show their quantitative effects. It is the purpose of this paper to show how various potential future changes can be combined to give an overall view of the project taking such risks into account. Three typical examples of possible investment in the automotive industry are chosen, and the two prime methods of assessing risk. Table 14.1 shows the use of the spreadsheet model for risk assessment.

14.2.1 New metal-cutting machines in a high-volume engine shop

This is a fairly straightforward case where new metal-cutting machines, automatic assembly, etc., are proposed to reduce labour manning and improve quality levels. The potential savings derive from a reduction

TABLE 14.1 *Use of spreadsheet for risk analysis*

Volume	600 000 p.a.	←
Rework cost	−£1.07 per piece	←
Warranty cost	−£1.40 per piece	
Direct labour	−8.14 minutes/piece	←
Maintenance labour	−50 hours/week	←
Maintenance matls	−£70 '000 p.a.	
Project launch	19 000 hours	
Severance	22 employees	
Facility investment	£7069 '000	←
Tooling investment	£1796 '000	←
Years of project	10	
Inflation	4.1 %	
Labour rate	0.0587 £/minute	
Fringes, etc.	82 %	
Launch rate	9 £/hour	
Severance pay	7600 £/person	
Taxation rate	27 %	
Tax paid year accrued	70 %	
Internal rate of return	21.4%	
Payback period	4.0 years	

Cumulative offset method
uses macro to:
  alter each probabilistic variable
  store offsets to IRR
  calculate mean and SD of offsets

@ RISK method
Inserts distribution formula in
place of deterministic value and
samples by choice of processes

in direct operating labour, improvements in quality (showing up as a reduction in scrap, rework of the machine parts, reduced test and repair and warranty), reduced maintenance labour and maintenance material and reduced inspection. Putting in the most likely figures (from Table 14.1) to the calculation gave an IRR of 21.4% which, in a number of companies, would be just acceptable.

Risk factors which could affect this are the anticipated sales level of the engine, the level of the anticipated improvement in quality, the level of the anticipated reduction in maintenance costs and the actual facilities and tooling expenditure. Assessments can be made as to some of the

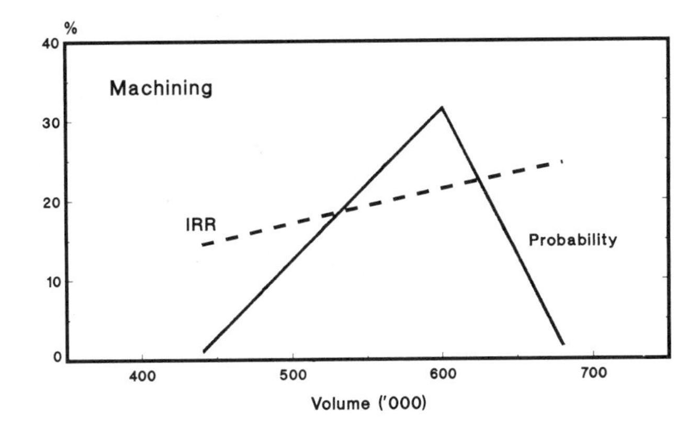

Fig. 14.1 *Probability of volume variation and resulting project return.*

TABLE 14.2 *Range of variables for the machining project*

Variable	Low	Most Likely	High
Annual volume (000)	440	600	680[1]
Scrap saving (%)[2]	0.82	1.32	1.52
Direct labour saving (min/unit)	4.21	4.61	5.01
Maintenance materials[3] saving (£000)	40	70	110
Facilities (£000)	6669	7069	7669
Tooling (£000)	1596	1796	1996

[1] Limited by line capacity.
[2] Rework (3.5% improvement), engine test and repairs (0.9 min/unit improvement) and warranty £1.4/unit. These were prorated to the scrap level for these calculations
[3] Maintenance labour (28 h/day) prorated to the maintenance materials for these calculations

possible variations in these and for this purpose a most likely low level and high level were estimated for each variable. These are shown in Table 14.2.

The actual high and low figures have been chosen to enable the range to be broken down into a small number of equal-sized bands. It will be noted that not all these distributions are symmetrical. Given the most likely low and high figures a probability can be ascribed to each, plus intermediate figures, on the basis that the total probability must equal 1.0. Figure 14.1 shows the range of probabilities for the case of volume changes and also the IRR calculated for each level.

Thus a mean IRR and variance can be calculated for the above case by multiplying the IRR for each specific volume band chosen (e.g. 460 ± 20) by its probability and obtaining a weighted answer. In the above case this would be 20.3%. Obviously other forms of distribution apart from the simple triangular one used here could be input, provided that an alternative realistic distribution can be derived.

We have observed in practical project evaluation that, within the working range of most variables, the offset/changes (in IRR) caused by a change to one variable is relatively independent of variations in other variables. For example, if a 10% increase in facilities cost gives a reduction in IRR from 25% to 20% (offset 5%) then, if the project return changes to 35% through improved labour savings, then the effect of the 10% capital change will still be about 5%. That is, it would reduce the return from 35% to 30%. This suggests that the effect of each variable can be determined independently with all the other variables set at the 'most likely' figures, and they can then be combined.

One method of combining the results is to work out the mean and variance for each potential variable, treated independently (volume, scrap, direct labour, etc.) and then add these together to give a result for the project as a whole. This is described as the **cumulative offsets method**. On this basis the mean for the project as a whole becomes 18.4% and the standard deviation (s.d.) is 3.2. Thus the mean return for the project has been reduced, by taking these risks into account, from 21.4% to 18.4%. The central limit theorem of statistics enables us to assume that the summation of a number of different distributions will (given a sufficient number), take the form of a normal distribution. From the mean and standard deviation given by the summation, the shape of this curve can be calculated and plotted (Fig. 14.2), and the risk of the project returning an IRR lower than the cost of money, plus a modest additional figure, say 16%, can be assessed. In this case there is a 22% risk of the project returning a figure less than 16% and this would be considered risky and undesirable.

There are a number of assumptions inherent in the above method and, although simple to use, it may be argued that the assumptions are too broad. An alternative is to use a computer risk-analysis package for random sampling of each variable distribution. We have used @ Risk (copyright, 1988. Palisade Corporation, USA), an add-on to LOTUS 123 (copyright Lotus Corporation), which enables any of a wide variety of distributions to be specified for each variable. The results are then sampled using either a Monte Carlo method or a Latin Hypercube. Using the same triangular distribution as before (for strict comparison) this gives a result as shown in Fig. 14.3 for 1000 iterations with Latin Hypercube sampling. The mean is 18.5 and the standard deviation 2.5, with a 29% risk of the project returning less than 16%. There is a certain amount of skewness to the distribution.

Both methods can be seen to give very similar results. The downside risks used in this case are greater than the upside opportunities, and it can be argued that this reflects actual experience in many projects. Projects

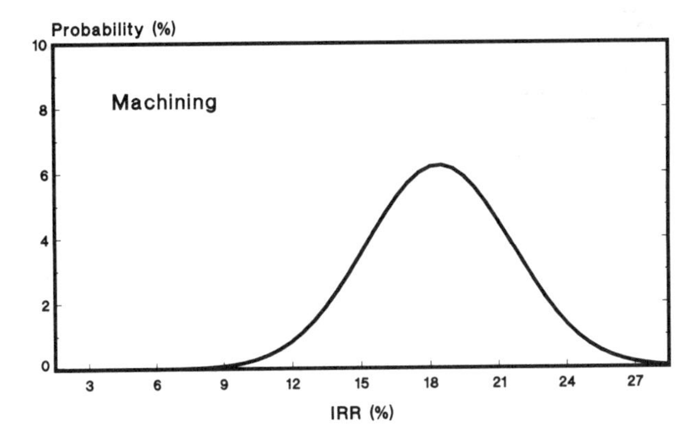

Fig. 14.2 *Distribution of returns for machining project after considering all variances.*

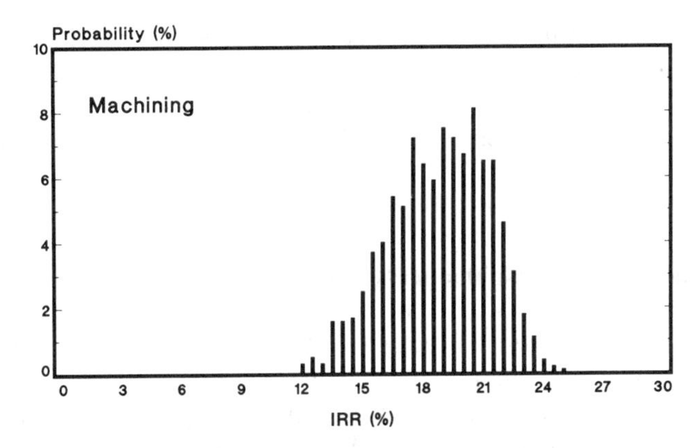

Fig. 14.3 *@ Risk © output for distribution of returns for machining project.*

can come in faster, at lower cost, showing greater savings, but on balance the opposite is generally true. The result of this rather pessimistic view is to lower the IRR for the project by about 3 points and take it to a level where the risk of a return less than the cost of money would be seen as too high.

The probable management action in this case would be to reassess the project to see if the capital spend could be reduced significantly, without losing the quality improvement, possibly by reducing some of the automation level.

Depending on the degree of certainty about the most likely value for a variable, different distributions can be used. For example, using a normal distribution, instead of a triangular distribution, will give more emphasis to the 'most likely' figure. Asymmetry in the low and high figures can be

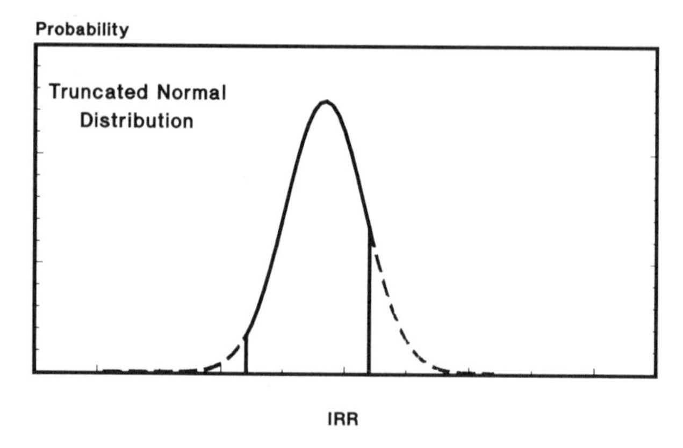

Fig. 14.4 *Form of the truncated normal distribution.*

covered by truncating the distribution at the appropriate points (Fig. 14.4). This procedure will, of course, give a mean IRR closer to the first calculated 'most likely' figure. The standard deviation selected will, of course, affect how different the final mean figure is to the 'most likely' case.

14.2.2 Transfer line press in a press shop

This is a modern concept (Brautigam, 1989) where large panels can be produced at a very high rate on a transfer press with a low level of manning and very rapid die changes. Some savings in maintenance and utilities, and an improvement in repair content are anticipated, compared with the existing lines of manned presses.

The rate of production, even in conventional press lines, is much greater than the rate of build of the vehicle and thus presses are operated in batch mode, building for a few days' usage of several different parts, before returning to the original part for the next cycle. The number of days of build covered in each cycle depends on the balance between the cost of holding stock and changing dies. Cycles of 5, 10 or 20 days are normal depending on the level of demand. In the case we are considering, a high level of usage is normal and so a five-day cycle is used.

It is assumed that the strategy will be to load the new press to its maximum capacity. Thus the number of different panels that will be put on the press will depend on the rate of output of the press (number of strokes per hour) with a higher speed leading to a greater number of panels being loaded, and thus to higher manpower savings compared with the present manually operated presses. The annual demand will also affect the number of different panels produced on the new press. Thus if

TABLE 14.3 *Risk factors considered in transfer press project*

Variable	Most Likely Value	Function
Press speed (parts/h)	700	Truncated normal (700, 60, 650, 850)
Annual vehicle demand (000)	205	Triangle (180, 205, 250)
Direct labour saving (min/panel)	1.44	Normal (1.44, 0.04)
Rework saving (min/panel)	0.14	Triangle (0.12, 0.14, 0.16)
Tool change saving (h/change)	7	Triangle (6.0, 7.0, 8.0)
Maintenance saving (h/day)	25.8	Triangle (20.8, 25.8, 35.8)
Utilities saving (£000)	78.9	Normal (78.9, 4.0)
Start-up cost (£000)	200	Normal (200, 20)
Facilities (£000)	4550	Triangle (4450, 4550, 4650)

the annual demand was for, say, 200 000 roof panels then the five-day demand would be 4400, taking nearly seven hours of press time, including tool changing, at 700 strokes/hour. If the demand was lower at 120 000, then the time taken would be a little over four hours. Thus there would be press capacity available to enable more jobs to be run in the week so as to keep the press fully loaded. With the short tool-changing time, the total number of stampings produced on the press in a five-day cycle will not change very significantly with changes in vehicle demand. This particular variable will thus not have a large effect on the profitability of the project compared with other factors, such as the press speed, manning levels, etc.

Other factors that should be considered for risk analysis include direct labour saving per job, labour saving in tool changing, maintenance costs, cost of utilities, project start-up costs and the capital expenditure. The results using the 'most likely' figures gave an IRR of 19.1% which would generally be seen as low. The @ Risk package was used for analysis.

Table 14.3 shows the 'most likely' figures and the range and distributions chosen for each of these. These covered triangular (minimum, most likely, maximum), normal (mean s.d.) and truncated normal (mean, s.d. lower cut-off, higher cut-off) distributions were based on judgment in each case. It will be seen that there was some optimism that higher press speeds might be achieved and this led to a slight increase in the mean IRR for the project of 19.8%. (Fig. 14.5). More important is the fact that

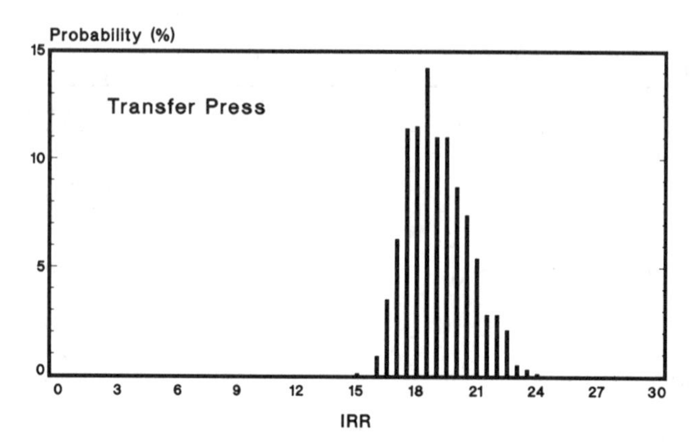

Fig. 14.5 *Distribution of returns for the transfer line press.*

the standard deviation at 1.5 is relatively low and the chances of a return below the cost of money (16%) is zero, and even at 18% is only 12%. This is a relatively safe project and investment could sensibly proceed.

14.2.3 Flexible manufacturing system for engine components

This analysis has been discussed in detail elsewhere (Hundy and Hamblin, 1988) but, in brief, it involves a comparison between two alternative methods of producing a range of parts at a demand of 50 000 per year:

1. An FMS with machining centres, AGVs, robot loaders, etc.; or
2. A conventional shop with small transfer lines and conventional drills, mills, etc.

The FMS is more expensive (even though running on three shifts reduces the capital difference somewhat), but shows a saving in direct labour content. It is one where the so-called intangible benefit of flexibility is important and, as will be shown, can be quantified using some risk analysis.

The most likely scenario shows both projects to be acceptable but, with the conventional alternative at 29.8% being slightly better than the FMS at 29.1%. Risks have to be considered and these are shown in Table 14.4.

In this case it is difficult to apply the @ Risk package approach. The problem is that very significant variations have to be considered which would demand the use of a lot of complicated logical IF statements in the spreadsheet calculator. If, for instance, the analyst is considering the impact of annual demand reducing significantly, say from 50k to 35k, then he must decide how the operations manager would react to minimize such a deleterious effect. On the conventional facilities he would prob-

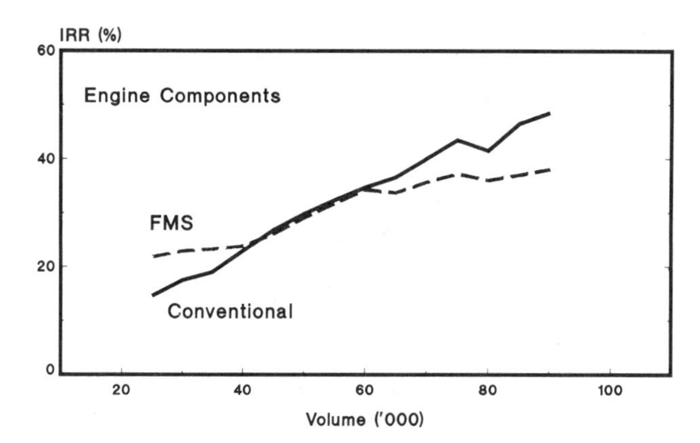

Fig. 14.6 *Effect of volume on remains for engine component manufacturing systems.*

TABLE 14.4 *Range of variables for the engine components project*

	Low	Most likely	High
Volume (000 per annum)	25	50	90
Sales price (£/part)	145	161	177
Materials cost (£/part)	84	92	100
New design of parts introduced (no. of years ahead)	4	7	11
Project life (years)	10	20	24

ably choose to operate on two shifts only and to reduce his labour accordingly. In the case of the FMS, an alternative approach would be to remain on three shifts but sell the remaining capacity of the flexible machining centres for other work, e.g. spares or prototypes. A modest increase in demand would probably be met by weekend working in both cases, but any increase beyond this would demand investment in additional facilities, which would be relatively most costly in the FMS alternative.

These different scenarios can be modelled using the calculator and the effect of each on the IRR calculated over a range of possibilities. Figure 14.6 shows the results of such calculations, assuming the optimum decisions by the operations manager, across a range of demands. The necessary IF statements to cover all these alternatives automatically in a spreadsheet analysis would be extremely difficult and consequently prone to error. It is, however, relatively easy to set out a series of volume scenarios recording the resulting IRR for each change.

A similar difficulty exists if one considers the probability of new products which require different design of parts being introduced during the

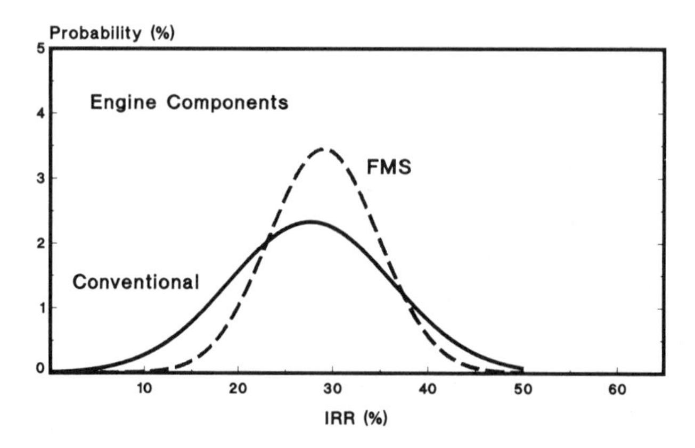

Fig. 14.7 *Distribution of returns for engine component manufacturing systems.*

project life. The FMS can be modified more cheaply and easily to provide these, compared with the conventional system. The difficulty again is in deciding when this will take place and writing the instructions in the spreadsheet accordingly, to allow the @ Risk package to be used.

Thus, in this example, we would view each of the variables as independent and would use the cumulative offsets method. On the assumption that we have a normal distribution we obtain the results shown in Fig. 14.7. The mean IRR of both projects has, of course, reduced, due to the introduction of new models, but the inherent flexibility of the FMS is now showing up to significant advantage with a mean of 29.0%, compared with 27.6% for the conventional method and a risk of only 1.2% for the project being below 16% for the FMS. The FMS solution then appears to be the best, particularly bearing in mind the high downside risk for the conventional shop (Fig. 14.7).

14.3 Conclusions

Collection and preparation of the raw data for a potential investment is normally quite time consuming. The proposed system has to be designed, alternatives considered, savings and running costs estimated and quotations obtained for the cost of equipment and tooling, etc. Once this has been done, however, the input of the data into the spreadsheet calculator or a PC takes only a few minutes, and the actual calculation of a project IRR on a personal computer takes only a few seconds. This is for an analysis over 10 years, including factors such as predicted inflation, currency levels, taxation and depreciation. It is sensible, therefore, to use the calculation frequently during the project development phase to evaluate possible alternative systems or modifications to the desired system, to determine the sensitivity of the project to changes in the various inputs, or indeed to decide whether to proceed at all with the project.

In the present paper we have shown how the PC-based IRR calculation can be used to obtain a feel for the risks inherent in any project. Again the determination of the likely levels of variation in any one of the variables is the time-consuming factor rather than the actual evaluation on the computer, which can usually be completed within an hour or so by either method described here.

The main problem is to obtain measures of the possible ranges of outcomes, which are necessary for this analysis. As an example, we can take the case of sales volume over the 10 years of the project. In two out of the three projects described here, this is an important factor influencing the IRR of the project. The marketing staff will have spent some considerable time in making a prediction of the most likely volume over each of the next 10 years for planning purposes. If asked, however, to put a probability on this outcome, it is highly likely that they would be unable to quantify this, or even to understand what is required. This problem has been dealt with at some length by Spetzler and Stael von Holstein (1975).

The analysis requires a probability function for various possible volumes, and yet the expert in the area almost certainly has little idea of what a probability function is, or how to achieve it. Our experience is that some progress can usually be made by reminding the forecaster that perhaps the only certainty is that his forecast volume will not be correct. He can then be asked what is the lowest it could possibly be, or the highest. By postulating various levels of sales and noting the reactions, e.g. 'Oh no, this is a regular runner, we have a lot of experience which suggests that even with a dip in the market we would never do as badly as that', or 'Well this is rather a new departure for us and we are pretty uncertain as to its volume and it's not impossible, though unlikely, that the sales would be as low as that', it is usually possible to establish the most likely extremes of the distribution. Depending on the firmness with which the most likely figure is defended, the analyst can then decide whether to use a triangular or a normal distribution with the standard deviation chosen to reflect the implicit degree of certainty. If the high and low figures are not symmetrical around the most likely one then truncated normal or triangular distributions can be used.

Past history can also be useful in deciding on the probability distribution and often when the analyst is an experienced engineer or manufacturing manager, he can input reasonable limits of his own choice. If nothing else, inputs of various distributions will show the sensitivity of the project to particular variables. If the project was shown to be very sensitive to a particular variable it could be researched in more detail.

The use of sensitivity and risk analysis also enables some quantification of the effects of intangible benefits. We have shown earlier that the tangible effects of, say, quality on scrap, rework, warranty, etc. can be analysed, using the calculator with probability distributions. The effects of market share would be assessed by a possible increase in contribution, arising from an estimated increase in sales. In this case a fairly pessimistic view of the possible increase in market share would be taken and if desirable the reaction of competition in the marketplace could be modelled as well.

We believe that the ease of use of the PC to enable risk analysis to be undertaken is such that is should be more widespread as an additional support to management decision making.

14.4 References

Brautigam, M.M. (1989) 20th ISATA Conference, Florence. Report 89156.

Cooper, D. and Chapman, C. (1987) *Risk Analysis for Large Projects*. Wiley, Chichester.

Hundy, B.B. and Hamblin, D.J. (1988) *Int. J. Prod. Res.*, **26** (1), 1799.

Meredith, J.R. and Suresh, N.C. (1986) *Int. J. Prod. Res.*, **24** (5), 1043.

Spetzler, C.S. and Stael von Holstein, C.A. (1975) *Management Science*, **22** (3), 340.

15 Analysis and Evaluation of Flexible Capital Investment

Carl-Henric Nilsson, Håkan Nordahl and Ingvar Persson

15.1 Introduction

Modern production systems are technologically advanced. The change in technology has made it increasingly possible to modify the flow of products and to handle different product variants. The ability to offer customized products has become an important competitive factor. The companies at the cutting edge of the advanced technology are forced to adjust to these changes and, hence, operate under a higher degree of uncertainty. Flexibility in manufacturing is, therefore, of substantial significance to all companies in the manufacturing industry.

In the industry, managers are expected to do economic analysis with respect to capital budgeting. The production system should be viewed as a tool for business development (Hill, 1989). It is thus important that the evaluation of an investment be in harmony with what the business development demands. It is also important that the tools used for economic analysis deepen the insight and understanding of the strengths and weaknesses of the installation (Persson, 1990).

Certain types of machinery are better suited than others to meet the demands for flexibility such as, for example, FMS (flexible manufacturing system) and industrial robots. Flexibility is a broad concept and is dependent upon the circumstances under which it is used. Different perspectives will generate different conceptualizations of flexibility. In this article, flexibility related to the product scope and flexibility related to the components of an investment are discussed.

The advantages of flexibility must be taken into account in the capital budgeting process. Until now, there has been a lack of CBTs (capital budgeting techniques) that evaluate flexibility. One such CBT (Nilsson and Nordahl, 1988; Nordahl *et al.*, 1988; Persson, 1988; Persson *et al.*, 1989) called capital-back method is presented in this article. The capital-back method takes into account the flexibility of the components that constitute an installation.

The capital-back method is then compared with the pay-back method. A deeper study of the capital-back method is carried out with respect to:

1. The share of flexible parts in the installation;
2. Sensitivity to the discount rate level;
3. The profitability of the total investment; and
4. The lifespan of the components.

Finally, conclusions are drawn concerning capital-back method and its implications for the management and the decision makers of companies.

15.2 Economic evaluation of capital investments – an overview

The theoretical foundations of capital budgeting techniques are: modern microeconomic theory, statistical theory for decisions under uncertainty and operations research. The discounted cash flow techniques are fundamental (Fisher 1930). With the passage of time, these techniques have become very sophisticated. The use of advanced statistical methods (Hertz, 1964; Wagle, 1968), decision-tree techniques (Magee, 1964a; Magee, 1964b), linear programming (Weingartner, 1962; Näslund, 1966) and option theory (Brealey and Myers, 1988; Cox and Rubinstein, 1985) are now possible in capital budgeting practice.

Surveys have shown that there is an increasing adoption of discounted cash-flow techniques in firms (Gitman and Forrester, 1977; Kim and Farragher, 1981; Hendricks, 1983), and the most advanced methods are still seldom used. The surveys indicate that managers still trust highly the results obtained by using capital budgeting techniques.

In a Swedish study (Yard, 1987), it was found that 40% of the companies in Sweden used the pay-back method as the primary method. Since the educational level in Swedish industry was high, the simplicity of the pay-back method cannot by itself explain the frequent use of this method. This raises an interesting question as to what important piece of information the pay-back period gives.

Weingartner (1969) argues that pay-back method is not an appropriate method for profitability evaluation according to economic theory. Yard's (1987) interpretation is that the pay-back method is to be considered as a measure of flexibility.

In order to explore the information offered by the capital budgeting process, a large capacity-expansion project was analysed in a case study (Persson, 1989b). No traditional calculations of profitability were made, even though the company regulations insisted on using traditional calculations. Instead, the investment evaluation was based on the cost-accounting analysis for the product, initial outlay, existing demand for the product and a forecast for future demand. The justification that can be given for using cost-accounting analysis is that it gives the management better perspective of the initial strength of the investment.

In the manufacturing industry, the pay-back periods considered are often short. Hayes and Garvin (1982) point out that in the early 1970s, 20% of the US companies that were studied had adopted a pay-back periods of not more than three years. A decade later, the percentage increased to 25%. Other studies confirm that pay-back periods considered by US companies are still short. The pay-back periods are especially low for rationalization investments such as, for example, robots. Our research has indicated that the pay-back periods used in the industry are as low as two years most of the times.

15.3 Flexibility

Flexibility has several dimensions. Let us elaborate on the term flexibility. Flexibility related to the design of products and the flow of goods is well analysed. Also, several authors (Browne *et al.*, 1984; Gerwin, 1983) have discussed and defined flexibility in the manufacturing process. These definitions have one major shortcoming: they are concerned with only two dimensions, the product dimension and the material flow dimension. The analyses do not consider the possibilities of reusing the equipment in different manufacturing processes. A method which takes this aspect into account at the acquisition stage ensures that the flexibility is not only evaluated but also analysed.

In the investment analysis of the manufacturing process, flexibility is given three dimensions: material flow, products and components.

To be able to analyse the flexibility part of an investment, we need to decide on the flexibility perspective to be considered. In this article, the analysis of flexibility is limited to the following perspectives:

1. Product scope (Nilsson and Nordahl, 1988; Molin and Söderlind, 1989), i.e. the width of the variety of products that can be produced in the installation; and
2. Component flexibility, i.e. the parts of the investment that can be reused in another installation.

15.3.1 Product scope

Gerwin (1983) has defined two flexibilities related to products:

1. *Mix flexibility.* The capability to handle a mixture of details that are only faintly related to each other; and

2. *Detail flexibility*. The capability to add or reduce a detail from the mix over time.

Similar thoughts have been presented by Browne *et al.* (1984) who define production flexibility as the capability to produce a wide variety of products.

The normal problem facing an installation is uncertainty about the future demands of the market. Under such condition, product scope can be defined as:

> The boundaries within which the critical parameters of the product can be varied.

One way to operationalize the product scope is to define it as the product mix or the predicted products, that can be produced without rebuilding the production system at any given time.

An interesting aspect of the product scope is the possibility to adjust it over time. Ahlmann (1987) points out the possibility to choose an installation that can handle the present situation and to add supplementary equipment when the need arises. The capability to change the product scope is achieved through an installation that is suited to be supplemented with different building blocks whenever a specific ability is needed.

15.3.2 Component flexibility

An installation can be regarded as a system composed of different subsystems or components as described in Fig. 15.1.

A component can be a standard equipment or a specially designed tool. Examples of investment components are a bed plate for a machine, a computer program and the cost of project management.

The flexibility of a component refers to the usability of the component for purposes other than those originally intended, either within or outside the manufacturing process of a company. Examples of this type of flexible equipment are robots, CNC machines and parts of FMS.

The component flexibility gets important when the demand for the installation ceases. For instance, this situation happens when the product that is being produced becomes obsolete.

Martins (1986), for example, argues that an important part of the flexibility of an industrial robot is its capability to be used as a standard component over a wide range of tasks, not related to the original task.

If we aggregate the investment costs of the flexible components, we get the total flexible part of the investment G_f.

If the same aggregation is done with the inflexible components, we get the risky part of the investment G_r. The components belonging to the

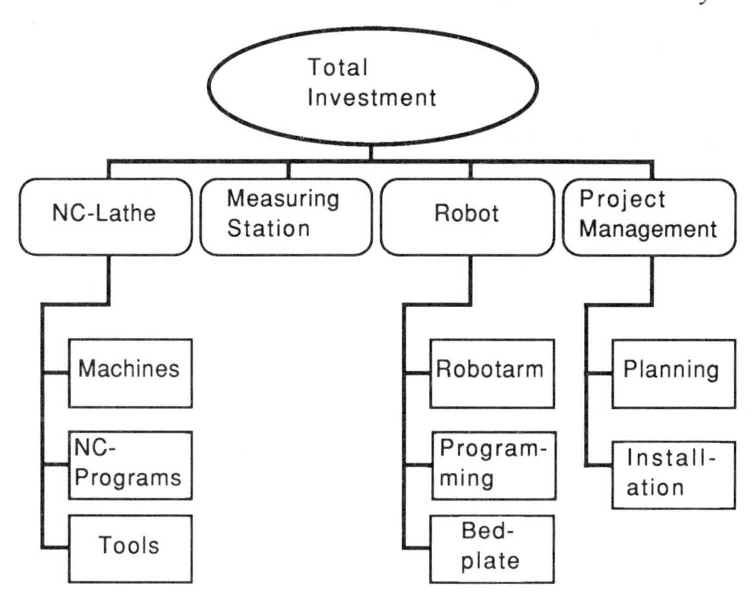

Fig. 15.1 *The sub-systems or components of an installation.*

risky investment cannot be used to produce other products in other manufacturing processes. The residual value is zero irrespective of how long the component has been used. Examples of this type of components are the fixtures and gripping appliances that are intended for a specific product. Included as part of the capital investment are also the expenditures for projection and design as well as the physical installation of these components. It is not usually possible to recover these expenditures if the project fails, and hence they form part of the risky investment.

A component can either be flexible or inflexible. If it is not possible to decide whether a component is flexible or not, the installation structure has to be further disintegrated.

The total investment G is the sum of the flexible part and the risky part of the investment.

$$G = G_f + G_r \tag{15.1}$$

This segregation of the capital investment into flexible and risky parts is by itself an important step. Component flexibility f is defined as the flexible share of an investment, as given below:

$$f = \frac{G_f}{G} \tag{15.2}$$

Once the categorization of the components that constitute an investment is done, the investment can be analysed with the capital-back method. This analysis is explained below.

15.4 Capital-back method

15.4.1 Capital-back method *v.* pay-back method

The objective of the capital-back (CB) method is to release flexible investments from the unreasonably high demands of a short pay-back (PB) period, which arises out of a high degree of uncertainty and short-term planning. The CB period should not be used as the sole determining factor of an investment as is done with other CBTs. It should be used as a complement to other CBTs such as NPV (net present value), IRR (internal rate of return) and PB. It will be seen later that the combination of CB and PB will generate interesting information that is related to NPV and IRR. The advantage of the CB method is that it takes into consideration the uncertainty of the custom-made part and the requirement for profitability for the flexible part.

The capital-back method applies the pay-back period for the risky investment, while the flexible part of the investment generates a yield that is as high as the discount rate. A low degree of uncertainty resulting from the flexibility makes a low interest rate acceptable for the capital invested in flexible equipment. This reasoning is consistent with the capital asset pricing model (CAPM) (Brealey and Myers, 1988).

When the capital-back method is used, calculation of the cost for the flexible part of the investment includes both depreciation and cost of capital. This calculation can be done in different ways, and the choice of calculation method will obviously affect the result. The most natural method used in capital budgeting is the annuity method. This provides a constant annual cost during the lifespan of the installation. According to the annuity method, if the discount rate is $i\%$ and the expected lifespan is n years, then the annual cost for the flexible part of the investment is $G_f * \text{ann}\ (n \text{ years}, i\%)$. When this amount is deducted from the annual net receipt, it gives the annual net receipt of the risky investment and it can be accumulated in the same way as in the pay-back method.

In analytical terms, the calculation of CB period is analogous to that of PB period, i.e. the investment divided with the annual net receipt, as given below:

$$CB = \frac{G_r}{a - G_f * \text{ann}\ (n \text{ years}, i\%)} \tag{15.3}$$

From equation (15.3), it may be concluded that the gradient of the lower dotted line in Fig. 15.2 decreases as the interest rate increases. The interest rate which generates a capital-back period equal to the pay-back period is of special significance. At this rate of interest, the rate of return on the risky investment is as high as the rate of return on the flexible part

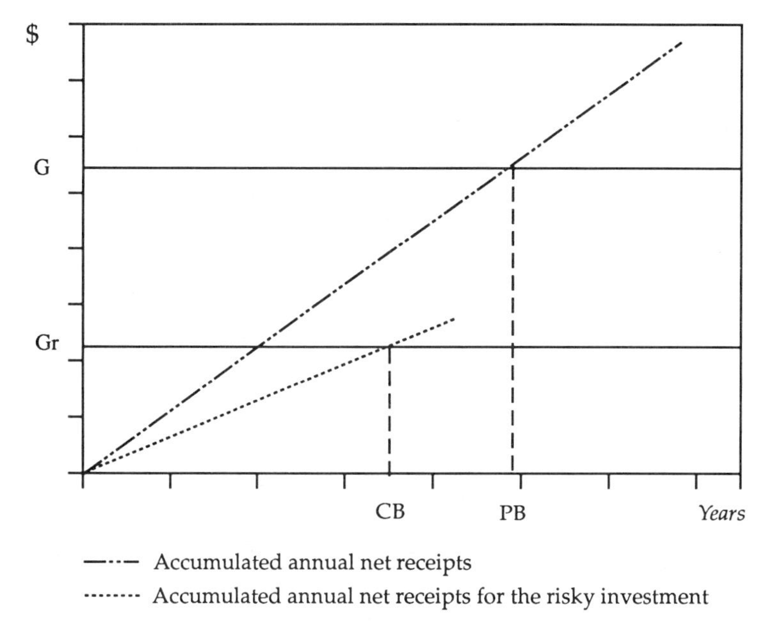

Fig. 15.2 *The relationship between capital-back method and pay-back method.*

of the investment. This calculation assumes that the lifespan of the risky investment is equal to the lifespan of the flexible part of the investments.

The capital-back method is described in Fig. 15.2. The horizontal lines show the capital investments. The lower line is the risky investment G_r and the upper line is the total investment G, which is equal to the total of G_f and G_r (equation (15.1)). The difference between the lines is the flexible part of the investment G_f.

The two lines extending from the origin show how the accumulated annual net receipts for PB and CB grow as a function of time. The lines are straight due to the fact that the annual net receipts are constants with respect to time. The steeper line is the accumulated annual net receipts a for the total investment. The interception of this line with the upper line for G gives the pay-back period.

The less inclined line shows the accumulated annual net receipts a minus the annual cost of the flexible part of the investment ($a - G_f * \text{ann}$ (n years, i%)). The interception of this line with the risky investment G_r line gives the capital-back period.

The above described calculation of the CB period is valid only if the annual net receipts are equal every year. If the annual net receipts vary, they have to be added together until the sum equals the amount invested in the risky part. This is analogous to the PB method.

The capital-back condition for the investment is then:

$$G_r = \sum_{x=1}^{CB} [a_x - G_f * \text{ann } (n \text{ years}, i\%)] \qquad (15.4)$$

a_x is the annual net receipts in year x($); n is the lifespan of the flexible part (years), and i is the discount rate of the company (%).

The PB method emphasizes on the short-term planning, uncertainty and the initial strength of the investment. But, for flexible investments, this method can lead to wrong investment decision, as the uncertainty is focused on the total investment. The advantage of the CB method is that it takes into consideration the uncertainty of the risky part and the requirement for profitability for the flexible part.

15.4.2 CB and PB *v.* IRR and NPV

The NPV rule states that an investment is justifiable if the NPV > 0. The IRR rule states that an investment is acceptable if IRR > i. Actually when the NPV = 0, the IRR will be equal to i. This means that the NPV and IRR rules will always give the same results. However, when the choice among different acceptable investments is considered, these rules can generate different results.

In order to compare the CB method with other methods, we need to rewrite the CB equation (equation 15.3) as given below:

$$CB = \frac{G_r}{a - G_f * \text{ann } (n \text{ years}, i\%)} \qquad (15.3)$$

Since the component flexibility f is equal to G_f/G (equation (15.2)), the CB equation can alternatively be written as:

$$CB = \frac{(1 - f) * G}{a - f * G * ann \ (n \text{ years}, i\%)} \qquad (15.5)$$

If we assume that the lifespan of the flexible part equals the lifespan of the risky part, then:

1. When NPV > 0 and IRR > i, it can be shown that $\partial CB/\partial f < 0$. This means the investment is acceptable and thus the CB period $\leqslant$ PB period for all values of f.
2. When NPV = 0 and IRR = i, it can be shown that $\partial CB/\partial f = 0$. This means that the CB period, as a function of the flexible share of the investment f is constant. It turns out that this constant is the PB period.
3. Lastly, when NPV < 0 and IRR < i, then $\partial CB/\partial f > 0$. This means that CB as a function of f is increasing and the investment is not justifiable.

At the breaking point, when the denominator of equation (15.5) becomes zero, i.e. when:

$$f = \frac{a}{G * \text{ann}\ (n \text{ years}, i\%)} \tag{15.6}$$

the CB function is discontinuous. For higher f, the CB period is less than zero. This, of course, does not have any practical meaning. For higher discount rates, the breaking point will have lower values.

From the above discussion, we may conclude that:

1. The CB period, together with the PB period, indicates whether or not an investment is acceptable or in agreement with the NPV and IRR rules.
2. The results we get from the CB and PB calculations are meaningful only for acceptable investments, and not for unacceptable investments.
3. When the choice among different acceptable investments is considered, the three rules, NPV, IRR, and CB and PB, can give different answers as to which investment is to be chosen.

15.5 *Analysis of the parameters that affect the capital-back period*

The CB period is influenced by four parameters:

1. The flexible share of the investment;
2. The level of the discount rate;
3. The profitability of the total investment (measured with NPV or IRR); and
4. The lifespan of the flexible part of the investment.

As there is no room to carry out a total analysis here, only an illustration of how the different parameters influence the outcome is given. The flexible part of the investment is chosen as the main parameter. The effect on the CB period is then analyzed for different levels of the discount rate, profitability and lifespan.

15.5.1 The discount rate level

Earlier research conducted is used here to select the discount rates. Research carried out on Swedish companies showed that the average discount rate used by these companies was 20%, without considering taxation. The variation of the discount rates used, however, was high. Yard (1987) found that the minimum discount rates ranged between 10% and 30% and the maximum rates were around 50%. Tell (1978) also found that the discount rates averaged slightly below 20%. But, none of them could establish if the discount rates were real or nominal. According to Gitman and Forrester (1977), the discount rates used by US companies were lower; in several cases the discount rates were between 10% and

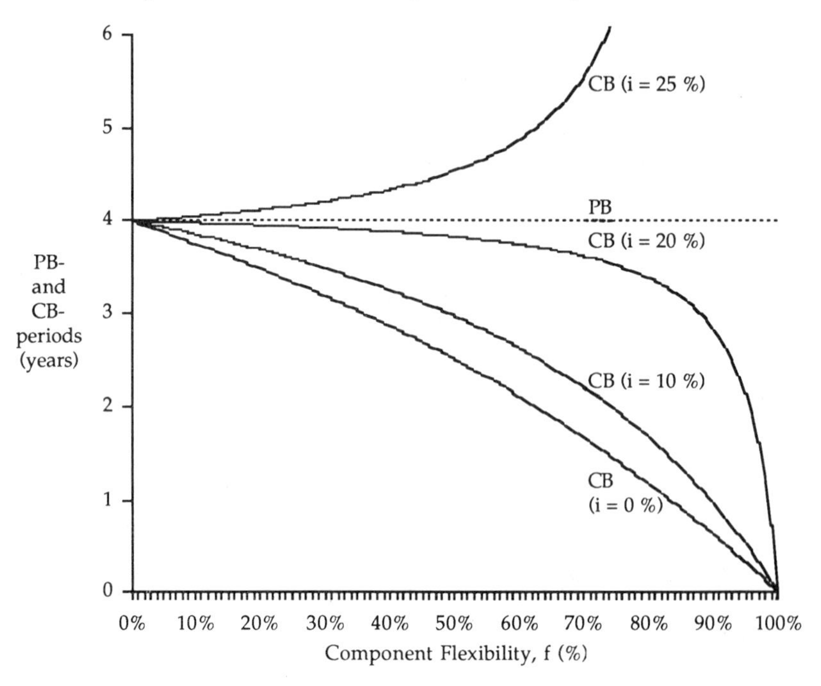

Fig. 15.3 *PB period and CB period as functions of the share of flexible part and the discount rate level.*

15%. Hayes and Garvin (1982) felt that these estimates of discount rates were considerably lower than the actual rates.

With these findings in mind, the influence of the discount rate on the CB period is analysed for discount rates between 10% and 25%. In order to save space, the total analysis is carried out for a pay-back period of only four years. This analysis is supplemented with a schematic description of what happens when the PB period is shortened to two years. The extreme value of 0% discount rate is also displayed for comparison. The analysis assumes that the annual net receipts are same every year.

Figure 15.3 describes an investment by showing the PB period and the CB period as functions of the share of the flexible part in the total investment, under following conditions:

1. G, the initial outlay, was \$1 000 000;
2. a, the annual net receipts, were \$250 000 per year;
3. n is the lifespan of 10 years;
4. i, the discount rate, ranged between 0% and 25%; and
5. f, the share ($f = G_f/G$), ranged between 0% and 100%

On the x-axis is the share of the flexible part and on the y-axis are the PB period and CB period in years.

PB is defined as G/a and is thus a constant over four years, irrespective of the variation in n, i or f.

The figure reveals that:

1. The CB curves are decreasing, which means that the CB period decreases when the share of flexibility increases. This is valid only when the internal rate of return is higher than the discount rate. When the share of flexibility approaches 100%, the CB period approaches zero.
2. The CB period increases as the discount rate increases. This has been mentioned earlier.

A closer study of the curves reveals that for a discount rate of 20%, the CB period is close to the PB period, which is four years. This is because the internal rate of return (21.4%) is very close to the discount rate used (20%) there. The share of the flexible part has to be 70% or more in order to have a significant difference between periods for PB and CB.

When the discount rate is 10%, the capital-back period decreases at almost a constant rate and the difference between PB and CB periods is more significant, even when the share of the flexible part is low. This situation is representative of the capital-back method. A low discount rate generates a low annual cost for the flexible part of the investment. Thus, the share of the annual net receipts for the risky investment increases. At higher f, the CB period becomes shorter.

15.5.2 Profitability of the total investment

In Fig. 15.3, the PB period is four years. If the PB period becomes shorter, then the CB period becomes less dependent on the level of the discount rate. This is due to the facts that a high IRR has been used and the annuity method has been chosen to calculate the annual cost.

If the IRR is high, then the annual net receipts are high and so is the share of the annual net receipts for the risky investment. For instance, an investment with a PB period of two years has a very high profitability. If the lifespan of the total investment is 10 years, IRR is close to 50%. If the demand for profitability of the flexible part of the investment is 20%, then the remaining part of the surplus that will be added to the risky investment is large. This means that if the IRR is very high, the level of the discount rate become insignificant.

The result is also influenced by the depreciation method used. When the annuity method is used, the total cost for the flexible investment is distributed as a constant annual cost for the lifespan of the flexible investment. This implies that the depreciation is lower during the earlier years. For instance, if the lifespan of the investment is 10 years, the first-year depreciation is 3.9% with a discount rate of 20%. The capital back is also compatible with other depreciation and cost of capital models. Those alternatives are, however, not discussed here.

15.5.3 The lifespan of the flexible components

The lifespan of the risky components and the flexible components need not be equal. It is, for instance, possible that the lifespan of a robot is longer than that of the gripping appliances (Björkman and Ekdahl-Svensson, 1986). This factor will influence the IRR. If the flexible components of an installation have different expected lifespans, then the CB calculations will also be affected.

The CB equation for an installation with flexible components having different lifespans is:

$$CB = \frac{G_r}{a - \sum_{j=1}^{m}[G_{fj} * \text{ann } (n_j \text{ years, } i\%)]} \tag{15.7}$$

G_{fj} is the cost of the flexible component j; m is the number of flexible components; and n_j is the lifespan of flexible component j.

In order to use this equation, an installation has to be described in terms of its components (Fig. 15.1) and these components are to be classified as flexible or risky. If it is not possible to classify a component, it has to be divided further until the classification can be made. For each component, the investment cost is specified, and for each flexible component, lifespan is also specified.

If the lifespan of a flexible component is shorter than the CB period, the component has to be replaced before the CB period is reached. Since the flexible part is treated as an annual cost according to the annuity method, the CB calculations are not affected by the replacement. Therefore, the CB equation is valid for replacement component, too.

In Fig. 15.4, the CB period is shown as a function of f, the share of the flexible part (of investment) and the lifespan n, of the flexible components.

For Fig. 15.4, the PB period is taken as four years and the discount rate as 15%. The upper curve shows the CB period when the lifespan is seven years for all components. This curve is closer to the PB line, which would mean that the investment is profitable, but the profitability itself is rather low. In this case, NPV and IRR are \$40 000 and 16.3%, respectively.

If some of the flexible components have longer lifespans, the CB period will decrease and the NPV and IRR will increase. The lower line shows the CB period when the lifespan is 12 years for all components. In this case, NPV = \$355 000 and IRR = 22.9%. The PB line and the CB

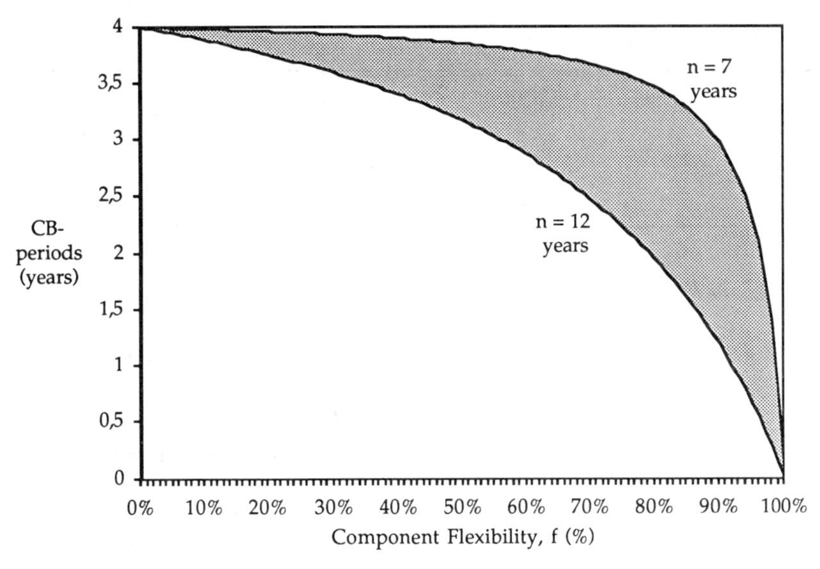

Fig. 15.4 *The influence of lifespan on CB period.*

curve are further apart and the gap widens as the component flexibility increases.

15.5.4 Capital-back and priorities

An important characteristic of an evaluation method is its ability to prioritize different investment proposals. Below, the capital-back method is compared with the pay-back method on this aspect. Table 15.1 shows two investment proposals, A and B, which are to be compared. The profitability and the flexible share of the proposals were chosen such that the priority changes when the discount rate is reduced from 20% to 10%. The proposal A has a high flexible share but has a low IRR of 21%. This makes the alternative A sensitive to the discount rate. The alternative B, on the contrary, has a low flexible share but a high IRR. This means that alternative B is not so sensitive to changes in the discount rate.

For the competing investment proposals of Table 1, the capital back method recommendations are:

1. A company with a high requirement for profitability, i.e. willing to take risks, should prefer proposal B which has a high profitability, IRR 30%, and a high risk, 70% inflexible share.
2. A company with a low requirement for profitability, i.e. averse to risks, should prefer proposal A which has a low profitability, IRR 21%, and a low risk, 70% flexible share.

TABLE 15.1 *PB, CB and IRR: a comparison*

	Flexible share	Lifespan	IRR	PB(years)	CB(years)	
					(i = 20%)	(i = 10%)
A	70%	10 years	21%	4	3.61	(2.20)
B	30%	10 years	30%	3	2.67	(2.46)

15.6 Conclusion and recommendations

The manufacturing technology is advancing rapidly and new manufacturing processes and new products are replacing the older ones at a rapid rate. This situation necessitates that increasing the flexibility of the manufacturing process ought to be a strategic goal for companies in the manufacturing industry. The CB method highlights the importance of flexibility and recommends flexible investment alternatives. Traditional capital budgeting techniques do not take flexibility into consideration, but instead, consider the flexible and the risky parts of the capital investment as equally uncertain.

The use of the capital-back method will affect the investment process as well as the crucial decisions regarding alternate investments. The CB method assumes that the components that constitute the capital investment can be categorized into two groups, i.e. a flexible part and an inflexible part. The process of categorizing the components will force the decision makers to analyse as well as evaluate the flexibility of the different investment proposals. This process by itself is, therefore, important.

A careful study of the alternative uses of the components that constitute an investment may guide the investors to a more flexible installation. An increase in flexibility does not have to result in an increase in investment cost. When the CB criterion is used, a flexible investment is treated more favourably than a less flexible one. The shorter the CB period, the better. A project with a long PB period and high component flexibility gets a fair treatment when the CB method is used. The following illustrates this above concept.

In an FMS cell, the total investment G was \$1 000 000, of which \$600 000 was allocated to flexible industrial robots and \$400 000 was meant for gripping appliances and other risky parts.

The investment led to an annual cost reduction of \$300 000. The PB period was, therefore, 3.3 years, which was longer than the company requirement of 3.0 years. The discount rate used by the company was 20%. The lifespan of the flexible part of the investment was 10 years and, therefore, the CB period would be 2.5 years. This would imply that the risky investment could be approved against the PB period requirement of 3.0 years.

Neither capital-back nor any other capital budgeting technique gives any information about the degree of uncertainty that prevails, if the inflexible investment becomes obsolete. Neither says anything about when the investment will become obsolete. This judgment is still left to the management. But what the CB method can do is to help understand investments, especially flexible investments, and, therefore, the method is well suited for the future demands of the industry.

15.7 References

Ahlmann, H. (1987) *Hög leveranssäkerhet i en dynamisk miljö*, STF Ingenjörsutbildning.

Björkman, M. and Ekdahl-Svensson, B. (1986) *Monteringsekonomi, kostnadsmodeller för kalkylering och känslighetsanalys*. Linköping Studies in Science and Technology, no. 96, Linköping.

Brealey, R.A. and Myers, S.C. (1988) *Principles of Corporate Finance*, McGraw-Hill.

Browne, J., Dubois, D., Rathmill, K., Sethy, S.P. and Stecke, K.E. (1984) Classification of flexible manufacturing systems. *The FMS Magazine*, April, 114–17.

Cox, J.C. and Rubinstein, M. (1985) *Options Markets*, Prentice-Hall, Englewood Cliffs, New Jersey.

Fisher, I. (1930) *The Theory of Interest*, Macmillan.

Gerwin, D. (1983) A framework for analyzing the flexibility of manufacturing process. School of Business Administration, University of Wisconsin.

Gitman, L.J. and Forrester, J.R. (1977) Survey of capital budgeting techniques used by major US firms. *Financial Management*, **6** (3), 66–71.

Hayes, R. and Garvin, D. (1982) Managing as if tomorrow mattered. *Harvard Business Review*, **60** (3), 71–9.

Hendricks, J.A. (1983) Capital budgeting practices including inflation adjustments: a survey. *Managerial Planning*, **31** (1), 22–8.

Hertz, D.B. (1964) Risk analysis in capital investment. *Harvard Business Review*, **42**(1), 95–106.

Hill, T. (1989) *Manufacturing strategy: The Strategic Management of the Manufacturing Function*, Macmillan.

Kim, S. and Farragher, E. (1981) Current capital budgeting practices. *Management Accounting*, **11** (June), 26–30.

Magee, J. (1964a) Decision trees for decision making. *Harvard Business Review*, **42** (4), 126–38.

Magee, J. (1964b) How to use decision trees in capital investment. *Harvard Business Review*, **42** (5), 79–96.

Martins, G. (1986) *Lönsamhet och industrirobotar*, Sveriges Mekanförbund, 12.

Molin, J. and Söderlind, M. (1989) *Investering i flexibel produktionsutrustning-något för Volvo PV?*, Inst. för Ind. Org., Lunds tekniska högskola.

Näslund, B. (1966) A model of capital budgeting under risk. *Journal of Business*, **39** (2), 257–71.

Nilsson, C-H. and Nordahl, H. (1988) *Investeringsbedömning av industrirobotar-värdering av flexibilitet och andra svårkvantifierbara aspekter*, Inst. för Ind. Org., Lunds tekniska högskola.

Nordahl, H., Nilsson, C-H. and Martins, G. (1988) Beurteilen der Investitionen von Industrierobotern. *Werkstatt und Betrieb*, **121** (12), 963–5.

Persson, I. (1988) Capital-back – method för riskanalys eller method för lönsamhetskalkylering. *Aktiv Industriautomation*, **1** (6), 56–62.

Persson, I. (1990) Analysis of capital investment: a conceptual cash flow model, *Engineering Cost and Production Economics*, **20**, 277–84.

Persson, I. (1989b) Prerequisites for use of information technology in the capital investment process. Working paper, Institute for Management of Innovation and Technology, Gothenburg.

Persson, I., Nilsson, C-H. and Nordahl H. (1989) Evaluation of flexible capital investments – with consequences for decisions and suppliers. Tenth International Conference on Production Research, Nottingham, Taylor & Francis, 392–3.

Tell, B. (1978) Investeringskalkylering i praktiken. *EFI*, 168–70.

Wagle, B. (1968) A statistical analysis of risk in capital investment projects. *Operational Research Quarterly*, **18** (1), 13–33.

Weingartner, H.M. (1962) *Mathematical Programming and the Analysis of Capital Budgeting Process*, University of Chicago.

Weingartner, H.M. (1969) Some views on the payback period and capital budgeting decisions. *Management Science*, **15** (12), 594–607.

Yard, S. (1987) *Logic and Calculation in Company Investments – Analytical Studies of Investments in Large Swedish Corporations* (Swedish with an English summary), Lund University Press.

Part Four

Peripheral Issues

16 Firm Size and Computer Integrated Enterprise Concept: CIM/CIE Related Strategic Issues for Small Businesses

Raja K. Iyer and Donald L. Liles

16.1 Introduction

The concept of computer-integrated manufacturing (CIM) has led to an unprecedented revolution in manufacturing in recent years, perhaps even comparable to the earlier industrial revolution. The CIM concept has fostered a growing and still-emerging focus on both upstream and downstream aspects of manufacturing. Upstream CIM applications include design engineering, CAD/CAM, robotics, and flexible manufacturing systems (Bernard and Guttropf, 1988; Fogarty and Hoffman, 1983; Plossl and Wight, 1967; Taraman, 1980). Downstream manufacturing considerations utilizing CIM include MRP II, KANBAN, JIT, and shop floor scheduling and control (Bernard, 1987; Kivenko, 1981; Kusiak, 1986; Lubben, 1988; Milacic, 1988; Ranky, 1983; Rolstadas, 1988). Increasingly, the CIM concept is not only being viewed as a means to solve problems which inhibit excellence in manufacturing, but also as a foundation upon which to build a computer-integrated enterprise (CIE) which can provide both intra-organizational and inter-organizational integration of information and production systems (Gaylord, 1987; Gessner, 1984; Hunt, 1989; Hamilton, 1989; Savage, 1985; Sheridan, 1989).

The CIM concept is not problem free, however. The technology to implement the CIM concept is not only expensive but also requires considerable time and effort to develop and implement. Also, there has been a lack of attention to the integration of CIM and all other functional areas with respect to management information systems (MIS) (Sartoni, 1988; Doll and Vonderembse, 1987; Vollman *et al.* 1988; Wight, 1974). Furthermore, the extension of the CIM concept to enterprise-wide and inter-organizational integration creates additional challenges that will require expensive solutions. These challenges include a lack of implementable CIE architecture, lack of standards of communication, and a lack of integrated systems built upon uniform architecture and common data.

Although these problems are common to all manufacturing firms, small firms are especially vulnerable for several reasons. These include the small firms' limited financial resources, dependence on large manufacturers in a supplier–customer relationship, lack of volume to achieve economies of scale, and limited ability to provide a variety of product design and configuration. While the latter two factors can successfully be addressed by CIM, the former two factors tend to restrain small firms from major CIM investment.

This paper provides a synthesis of the existing CIM and CIE frameworks and develops an integrating CIE framework to address small business firms' concerns in particular. The integrated framework includes an extension of SME's CIM wheel to incorporate the following specific needs for intra-organizational and inter-organizational integration:

1. Integration of CIM and MIS into all functional areas in organizations to achieve intra-organizational CIE; and
2. Definition of all tangible and intangible flows across organizational boundaries to specify an infrastructure for inter-organizational CIE.

The paper also develops strategic issues in small businesses which are particularly relevant to the implementation of CIM and CIE concepts. These CIM/CIE-related strategic issues include competitive factors, value-added chain (VAC) concepts, cost-benefit considerations, and inter-organizational sharing of CIM/CIE technology, facilities and information.

The paper concludes with an exploratory discussion of how the existing and emerging CIM and CIE technologies correlate with the strategic issues faced by small firms. The discussion also includes suggestions for implementation plans to achieve an integrated CIM and MIS architecture for intra- and inter-organizational CIE.

16.2 Firm size and the emerging CIM/CIE concepts

Fundamental economic principles suggest that efficient and effective markets result from a large number of small firms. However, recent years have witnessed the reversal of this key factor, that is, a small number of large firms has emerged due to the formation of hierarchical structures (Malone *et al.*, 1987; Williamson, 1975). Such structures have resulted from the need for the market-based coordination of the flow of materials and services and have led to mergers and conglomerates of firms and sole-sourcing of materials and services. These forces have tended to stunt the growth of small and medium-sized companies and thereby stifle competition.

Given the recent advances in computers and communications technologies, it is now feasible to achieve a new type of integration – the computer-integrated enterprise. Unlike the traditional physical integration of enterprises, it is now possible to achieve electronic integration of enterprises. This approach can result in an optimal integration of economically independent enterprises, regardless of size. In other words, the CIE concepts can achieve integration of information and business transactions, and not of the businesses themselves as practised heretofore under the guises of vertical and horizontal integration strategies. Thus, it is possible to revert back to the original economic principle of efficient markets through the preservation of a large number of small firms.

One of the major impediments to the embracing of the CIE concepts by most firms is the prohibitively large investments needed to implement them. In the case of small firms, this problem is especially pronounced due to several significant limitations imposed by the firm size. Some of these are discussed below.

16.2.1 Limitations on financial resources

Small firms typically do not possess vast financial resources nor do they have the necessary financial clout to raise capital for major investments. This points out the need for the **shared CIE** concept, an idea becoming increasingly popular as evidenced by the endorsement of the Shared Flexible Computer-Integrated Manufacturing (FCIM) Systems program (produced by the Office of Technology Commercialization) by the US Department of Commerce. The need for CIE architectures to support small firms is further strengthened by the shared CIE concept.

16.2.2 Dependence on large manufacturers

Small firms depend heavily on large companies in typical supplier–customer relationships. It is true that large firms are becoming increasingly aware of the competitive advantages in preserving and strengthening such relationships (Cash and Konsynski, 1985; Doll and Vonderembse,

1987; Johnston and Carrico, 1988; Johnson and Vitale, 1988; Porter, 1980; Porter, 1985; Porter and Millar, 1985). However, the fact remains that small firms' output is directly tied to their customers' demand. Such reliance makes small businesses vulnerable not only in terms of long-term survival, but also short-run uncertainties in production runs and resource utilization. CIE concepts and architectures, as discussed later in this paper, can assist small firms in smoothing production runs and levelling resource utilization.

16.2.3 Inability to achieve economies of scale

Small firms suffer from the lack of economies of scale, resulting in higher per-unit production costs compared to firms who achieve economies of scale. While this issue brings up the debate between 'smallness of firm and economies of scale', it is worthwhile to examine if it may now be feasible to retain small firms and, yet, achieve economies of scale as well. That is, can we integrate small firms electronically and achieve economies of scale in the industry and/or the economy at large?

16.2.4 Limited ability to provide variety in product design

Owing to resource limitations, small firms are limited in their ability to provide variety in product design and configuration. Recent advances in technologies such as CAD, CAM and group technology, are aimed at supporting flexibilities in design and manufacturing. It is, therefore, possible now for the small firms to remain small and efficient, yet meet the demands for product variety. Furthermore, with CIE concepts applied across organizational boundaries, firms can share information about design changes quickly and without any ambiguities in communications, thereby assuring that even small firms can react to the need for product design changes in a timely fashion. Improved connectivity through the CIE concepts can lead to increased competitiveness for firms, regardless of size.

16.3 Bringing CIM/CIE to small firms: the Is have it!

It seems clear that CIM/CIE concepts can lead to the preservation of small firms in the economy by removing the financial and economic efficiency aspects of the arguments for vertical and horizontal integration. It would indeed be difficult to overlook potentials for significant competitive advantages from CIM and CIE, as pointed out in the previous section. The problem, however, is that investment in CIM/CIE could be so high that small firms may be reluctant to adopt these new technologies.

A recent industry survey (Sheridan, 1989) points out that, although over 88% of the presidents and CEOs regard CIM as an essential or very important competitive weapon in a world-class manufacturing environment, about 30% of these respondents state 'unavailability of funds' as one of the primary obstacles to more rapid adoption of the CIM technology by US manufacturing companies. The same respondents also indicated that lower manufacturing costs (53.9%), product quality (50.4%), improved production control (44.6%), responsiveness to the market (41.0%), flexibility (37.4%), reduced inventories (37.4%), and small-lot manufacturing (25.9%) are some of the major benefits from CIM. While these benefits are beyond any dispute even for the small firms, the issue of up-front investment still remains. Does CIM/CIE have to remain only with the FORTUNE (fortunate?) 10 – i.e. only the first 10 of Fortune 100?

It is suggested that some of the answers may lie in the interpretations of what the letter I means in CIM/CIE. 'Integration' is obviously one of these interpretations. We believe that 'information', 'inter-company', 'intra-company', and 'interfaces' are other interpretations that could, together, lead us to an acceptable framework to bring CIM/CIE to small firms. Specifically, we propose that the benefits of CIM/CIE can be maximized within and across industries if information interchanges at the inter-company as well as intra-company interfaces can be integrated with active participation by all interested parties: vendors and customers of CIM/CIE users (the so-called value-added chains – VACs) (Porter, 1985) and vendors and customers of CIM/CIE providers. Presented below is a framework which extends the SME's CIM wheel to include the information integration and interchange aspects of a CIM/CIE platform for firms of all sizes, and especially for the small firms.

16.4 A framework for the computer-integrated enterprise

The framework for the CIM-based CIE is a synthesis of the existing CIM and CIE frameworks in general and SME's CIM wheel in particular. The framework consists of a series of propositions which are utilized to develop an integrating framework. It is an extension of SME's CIM wheel and addresses two major needs:

1. To integrate 'islands of automation' (for example, CAD, CAM, group technology and FMS) and 'islands of management-information systems' to achieve an intra-organizational CIE (Fig. 16.1); and
2. To define all tangible and intangible information flows across organ-

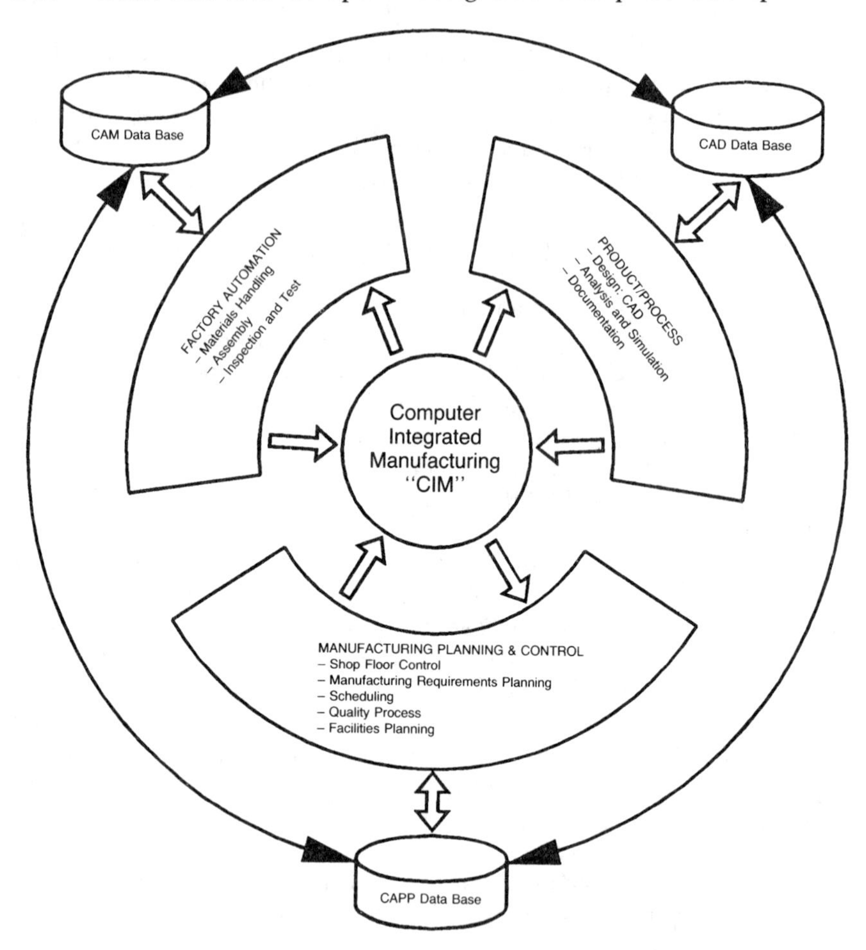

Key:

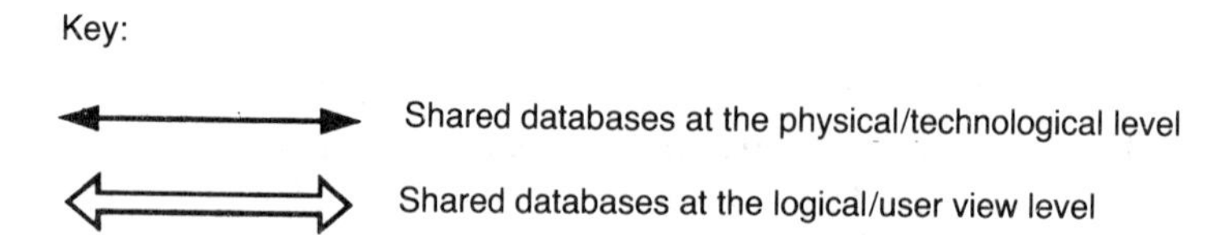

Fig. 16.1 *Computer-integrated manufacturing integration among islands of automation (based on SME's CIM wheel).*

izational boundaries to specify an appropriate infrastructure for an inter-organizational CIE (Fig. 16.2).

These propositions also discuss strategic issues, such as competitive factors, value-added chain (VAC) concepts, and cost-benefit consider-

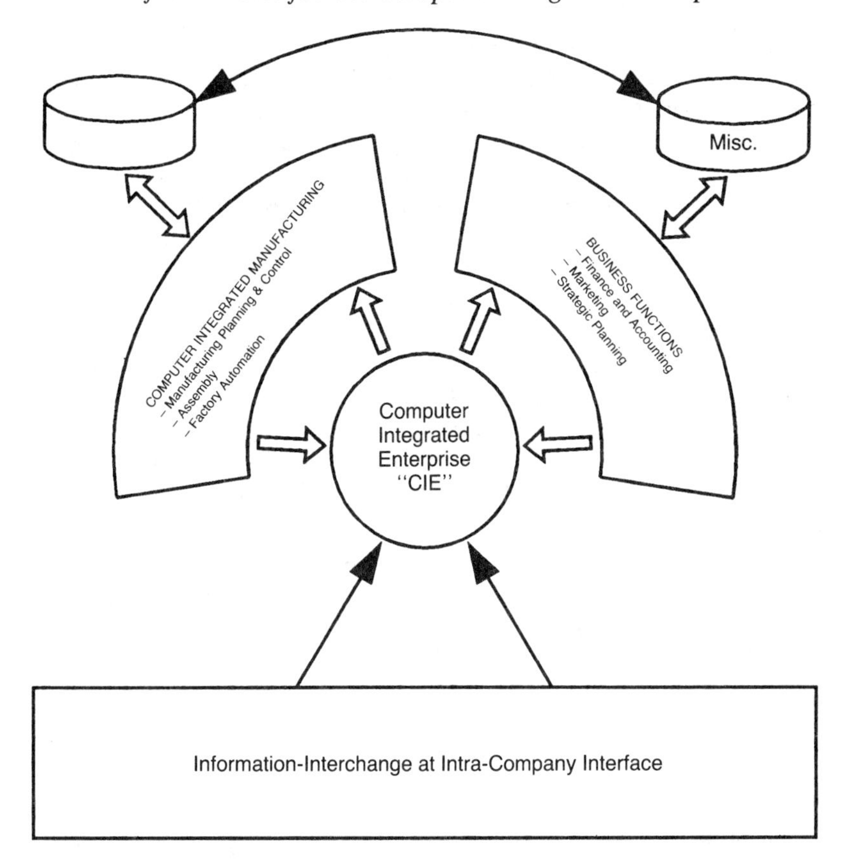

Fig. 16.2 *Information interchanges at the inter-company and intra-company interface.*

ations, which are particularly relevant in implementing CIM/CIE concepts for small businesses.

16.4.1 Proposition 1

> Interfaces between islands of automation should be viewed as consisting of several layers of information systems: for instance, transactions processing systems, management information systems, decision support systems, and knowledge-based and expert systems.

Islands of automation have proliferated over the recent years in explosive proportions. While these attempts at automating individual functions have resulted in significant flexibility and efficiency improvements, the jury is still out on the issue of such contributions at the organizational level. Consider, for example, the debate between market-based coordination versus hierarchy-based coordination (Malone *et al.*, 1987). While CAD enables flexibility and ease in handling requests for complex products, such product requests require more information exchanges, thus increasing the coordination costs. This means that hierarchical coordination through subsidiaries would enjoy coordination cost advantages over market-based coordination among independent firms.

For small companies, higher costs of coordination between islands of automation could lead to significant competitive disadvantages and eventual business failures. As shown in Fig. 16.1, interfaces between islands of automation must ensure that the total system is optimized with respect to factors such as flexibility, effectiveness, efficiency, costs and productivity. One should also recognize that layers of information systems exist not only within the islands of automation, but also at the interfaces between the islands of automation. That is, information interchange at the interfaces must provide all types of information systems: transaction processing, management information, executive information, decision support and knowledge-based expert systems.

16.4.2 Proposition 2

> Integration of information from various islands of automation should be independent of specific physical views or schema of information and be based only on logical and user views that are relevant for the various layers listed under proposition 1.

Information interchange at the interfaces in CIM/CIE can occur freely only if freedom from physical representation of data is assured. Standards, both at the industry and corporate levels, are necessary to make partial upgrades and replacements of integrated systems when new technologies appear. While several standards, such as MAP, IGES, EDIF and BDI, seek to address the standards issue within islands of automation, interfaces still suffer from lack of acceptable views of data.

Consider, for example, the interface between CAD and CAM. The language of CAD is primarily geometry-based with the results from designers being stored as points and equations. These allow geometrical shapes, such as lines, arcs and circles, to be combined and displayed to represent a finished part. Downstream in CAM, however, product features are essential, with attributes like 'hole', 'notch', 'slot' and 'groove' having manufacturing meanings to prepare programs for numerically controlled (NC) machines.

One of the major challenges of CIM/CIE is to make the languages of CAD, CAM and other islands of automation a common one by representing parts and business features in a data structure that can be accessed and interpreted by any application. Recent advances in relational database technology and database management systems allow us to specify data definitions which are independent of the physical representation of data. By storing data as well as data about relationships between data, relational DBMS provide the much-needed capabilities to integrate information from various islands of automation independent of physical views of data and to assure information interchange based only on logical and user views.

16.4.3 Proposition 3

> Intra-company and inter-company information interchanges can be achieved through the shared databases concepts which are not only commonplace today but also are not prohibitively expensive even for small manufacturers. While separate databases may be maintained within the islands of automation, sharing of these databases, consistent with propositions 1 and 2 above, is essential and economically feasible with the currently available database management technologies.

Progressing beyond computer-integrated manufacturing into computer-integrated enterprise creates additional interfaces and the need for information interchanges at these interfaces. Similar to our earlier argument, efficient performance at the firm level means that coordination at the interfaces between CIM and other business functions be efficient as well. As pointed out by Doll and Vonderemse (1987):

> achieving full strategic benefits from CIM will require a broadened partnership of top management as well as engineering, manufacturing, marketing, and management information systems (MIS) executives who share a common vision of how CIM makes possible new approaches to designing business systems. Developing this common vision requires a broad conceptual framework which integrates CIM, environmental analysis and business strategy.

A framework for integrating CIM and other business functions into a computer-integrated enterprise (CIE) is shown in Fig. 16.2. Sharing of databases maintained at all islands of automation and islands of MIS is imperative to realize full strategic and competitive benefits from CIM. Integration across business functions will require the development of cross-functional decision-support systems which use an integrated engineering, manufacturing and business database to build a smaller and more adaptive technological core (Doll and Vonderemse, 1987). Shared database concepts along with relational DBMS enable the utilization of information technology to manage the complexity of multi-product and multi-configuration demands on firms of all sizes.

With CIM integrated with other business functions into a CIE, small firms can gain competitive advantages through **economies of scope** instead of economies of scale. In this context, the CIE concepts lead to implementation of key strategy-oriented operational strategies. These include more frequent product entry/exit decisions, integrating product design and process selection decisions, and coordinating the revision of complex production plans and schedules in a frequently changing manufacturing mix. Without such enhanced information processing capabilities, the 'factory of the future' cannot be implemented (Doll and Vonderemse, 1987).

16.4.4 Proposition 4

> Inter-company information interchanges can be achieved through linked database concepts, such as electronic data interchange (EDI), and be consistent with propositions 1 and 2 above, i.e. layers of information systems and logical, and not physical, views of data. Linked databases are essential to lower transaction costs and improve coordination between firms at all layers of the value-added chain (VAC), thereby leading to financial and economic benefits. Since these benefits accrue across the VAC, it should be expected that the investments in this type of integration also be shared by all firms in the VAC.

The aspects of information and integration are even more important in inter-company interchanges to achieve enterprise integration. Information must be more timely than ever – online and realtime. Integration must create interfaces between firms to operate as seamlessly as possible with all vendors' products (Hamilton, 1989).

Linked databases are essential to lower transaction costs and improve coordination between firms at all layers of the value-added chain (VAC), thereby leading to financial and economic benefits. Making the data readily and easily accessible to all end users in all firms in the value-added chain is an absolute requirement for the success of CIM/CIE integration attempts. What vendors and users need to focus on next are report writers and query tools that put information users in a position to react immediately to the information they are getting. Linked databases with easy and ready access based on user views will avoid the buildup of backlogs of requests for custom programming from traditional data processing sources.

Benefits of integrated information interchanges accrue across the VAC in several ways. Firms may utilize information, computers and communication resources that exist at one layer of the VAC to improve performance in firms at other layers of VAC. This means lower transaction costs and/or improved coordination between firms at various layers in the VAC. With these benefits, CIE concepts are in a position to ensure that the economy can derive benefits of (almost-perfect) markets and competition while also achieving efficiency and effectiveness of hierarchical coordination through the CIE concepts.

Small and medium-sized companies have the potential to gain more from CIE compared to larger firms. These firms typically form the lower layers of the VAC and depend on the flow of reliable and timely information from the upper layers of the VAC. Therefore, inter-organizational sharing of information is extremely important in order to foster competition among small and medium-sized companies. For instance, when a major Department of Defense (DOD) contractor shares non-proprietary information on his proposed bid with all firms in the lower levels of the VAC, these firms will be in a better-informed position to generate subcontract bids effectively and efficiently. Similarly, if the infrastructure exists for the small firms to transmit the subcontract bids electronically upward in the VAC, significant reductions in transaction costs can be achieved. Furthermore, with a seamless interface for information interchange at the organizational boundaries, firms at the higher levels of the VAC are in a well-informed position to select from competitive subcontract bids.

Since all firms in the VAC can and should gain from such electronic information interchange and integration, it should be expected that the investments in this type of integration be shared by all firms in the VAC as well. Sharing of investments in CIM/CIE can make it less painful on the small and medium-sized firms. However, the logistics of a shared investment can be difficult to develop, and there are the possibilities of anti-trust and other legal problems. We believe that a combined user–vendor body can iron out these difficulties and work toward a **shared computer-integrated enterprise** concept, which can lead to significant cost savings, competitive benefits and strategic advantages for firms of all sizes in general and small firms in particular, as discussed in the next section.

16.5 Shared CIE concepts and strategies: benefits for the small firms

One of the major benefits from a shared CIE is that major initial investments can be avoided by small and medium-sized companies. Also, any risks of such investments can be avoided by small firms or, at least, be spread over several firms. Large firms, especially those who are prime contractors on major contracts, can look forward to significant improvements in the quality and timeliness of information interchanges and other transactions among firms, thereby leading to improved productivity and reduced costs. Other specific benefits from a shared CIE include the following:

1. CIEs can substantially increase productivity which lead to major improvements in quality control and inspection. This can reduce major costs of rework and rejects.
2. CIEs can facilitate the introduction of new products, including inexpensive manufacturing of just enough units for market testing.
3. CIEs can make feasible rapid changes in volume of output and shifts in product mix, thereby increasing global competitiveness of firms of all sizes in any industry.
4. CIEs allow continuous incremental adaptation to changing requirements for products and systems that otherwise would require major retooling and down time.
5. CIEs allow just-in-time manufacture and delivery, and substantially reduce the costs of inventory.
6. CIEs can lower investment break-even points substantially and lengthen the time before a plant becomes obsolete, providing more favourable returns on investments.
7. CIEs, when implemented in the defence industry, can improve defence readiness and constrain procurement costs by producing high-quality components, reducing procurement lead times and inventory costs, stemming the trend to offshore sourcing, increasing surge capacity during a military emergency, and providing geographic dispersion of the strategic manufacturing base.

A recent attempt at the shared CIE concept is the shared Flexible Computer-Integrated Manufacturing (FCIM) centre program offered by the Office of Technology Administration of the US Department of Commerce (1989). These shared FCIM centres would consist of a manufacturing service operation which has off-the-shelf flexible manufacturing systems that would lease manufacturing time to small firms. This is the same technique used by computer mainframe manufacturers to introduce computers to the business community in the 1950s.

Education and training for both management and the workforce are crucial to the effective adoption and use of FCIM, whether the FCIM is leased or used in-house. Also, FCIM must be understood in a broader, total business context. Thus, the shared FCIM centre becomes a production site as well as a teaching factory. While serving as a manufacturing service centre, the shared FCIM can also be a component for educating client companies' management and workforce. According to the Office of Technology Administration, the resulting benefits of the shared FCIM include:

1. Immediate competitive manufacturing capability with no up-front investments and other costs, especially for the small and medium-sized firms which cannot afford off-the-shelf FCIM systems nor do they have the in-house expertise to develop them;
2. Strategic planning for the appropriate degree of automation;
3. Training of management and workforce before an investment;
4. Testing of systems to assure they meet individual firm needs;
5. Faster utilization of capacity when firms install their own equipment on an incremental and modular basis – which also provides for an expanded life cycle for a constantly updatable production system.

The office of technology administration enumerates several other benefits as well. In rural areas, the shared FCIM centres can be established as part of a state's efforts to diversify from agriculture, oil or mining. Urban states may be interested in the centres as a way to modernize their existing industrial base and support small companies that are under increasing pressure from overseas competitors. The Department of Defense is examining the use of this concept as a tool to upgrade the manufacturing capabilities of third- and fourth-tier subcontractors. This posture reinforces our propositions earlier in which we advocated the use of CIE infrastructure to ensure two-way information interchange between defence contractors and small and medium-sized companies which are typical in the third and fourth tiers of the VAC in defence-contract oriented manufacturing.

The shared FCIM centres are still a relatively new data. The first two shared centres are in pilot operation: one in Meadville, Pennsylvania (National Institute of Flexible Manufacturing), and another in Huntington, West Virginia (Marshall University). Two more centres were planned to be in operation by the end of 1989, with several additional centres being expected to start up in 1990. About 10 others are in initial active planning stages and 15 additional groups are beginning consideration. One group's plans call for a central teaching factory and additional branch micro-production centres in a single state. These shared FCIM centres are being financed primarily by the private sector with some initial support from state governments.

16.6 Concluding remarks

In this paper, we have provided a framework for the 'I' in the CIM/CIE concepts, assuring ourselves that the 'C' for computer, the 'M' for manufacturing and the 'E' for enterprise are already familiar to everyone concerned with the CIM/CIE concepts. Our concern in this paper has been to provide computer-based integration of information interchanges at the intra-firm as well as inter-firm levels to lead toward the computer-integrated enterprise concept.

Our emphasis has been on the concepts and strategies for the small firms which are typically at the lower tiers of the value added chains in most industries. Since these small firms suffer from lack of capital and know-how for CIM/CIE concepts, unique CIM/CIE related strategies, such as the shared CIE concepts, are essential. We have discussed an example of these shared CIEs – the shared FCIM centres sponsored by the Office of Technology Commercialization of the US Department of Commerce (1989).

16.7 References

Bernard, T. (1987) *Arificial Intelligence in Manufacturing: Key to Integration?* North-Holland: Elsevier.

Bernard, T. and Guttropf, W. (1988) *Computer Integrated Manufacturing: Communication/Standardization/Interfaces*, North-Holland: Elsevier.

Cash, J.I. and Konsynski, B.R. (1985) IS redraws competitive boundaries. *Harvard Business Review*, March–April, 134–42.

Doll, W.J. and Vonderembse, M.A. (1987) Forging a partnership to achieve competitive advantage: the CIM challenge. *MIS Quarterly*, June, 205–220.

Fogarty, D.W. and Hoffman, T.R. (1983) *Production and Inventory Management*, South-Western, Cincinnati, Ohio.

Gaylord, J. (1987) *Factory Information Systems: Design and Implementation for CIM Management and Control*, Marcel Dekker, New York.

Gessner, R.A. (1984) *Manufacturing Information Systems: Implementation Planning*, Wiley, New York.

Hamilton, D. (1989) *CIM Primer*, EDGE, Sep.–Oct., 1–20.

Hunt, V.D. (1989) *Computer Integrated Manufacturing Handbook*, Chapman and Hall, New York.

Johnston, H.R. and Carrico, S.R. (1988) Developing capabilities to use information strategically. *MIS Quarterly*, March, 37–48.

Johnston, H.R. and Vitale, M.R. (1988) Creating competitive advantage with interorganizational information systems. *MIS Quarterly*, June, 153–65.

Kivenko, K. (1981) *Managing Work-in-Process Inventory*, Marcel Dekker, New York.

Kusiak, A., (ed.) (1986) *Flexible Manufacturing Systems: Methods and Studies*, North-Holland, Amsterdam.

Lubben, R.T. (1988) *Just-In-Time Manufacturing* McGraw-Hill, New York.

Malone, T.W.J. Yates, and Benjamin, R.I. (1987) Electronic markets and electronic hierarchies. *Communications of the ACM*, June, 484–97.

Milacic, V.R. (1988) *Intelligent Manufacturing Systems II*, Elsevier, Amsterdam.

Plossl, G.W. and Wight, O.W. (1967) *Production and Inventory Control*, Prentice-Hall, Englewood Cliffs, New Jersey.

Porter, M.E. (1980) *Competitive Strategy*, The Free Press, New York.

Porter, M.E. (1985) *Competitive Advantage*, The Free Press, New York.

Porter, M.E. and Millar, V.E. (1985) How information gives you competitive advantage. *Harvard Business Review*, July–August, 189–203.

Ranky, P. (1983) *The Design and Operation of Flexible Manufacturing Systems*, IFS, North-Holland, UK.

Rolstadas, A. (1988) *Computer-Aided Production Management*, Springer-Verlag, New York.

Sartoni, L.G. (1988) *Manufacturing Information Systems*, Addison-Wesley, New York.

Savage, C.M. (ed.) (1985) *A Program Guide for CIM Implementation: A Project of the CASA/SME Technical Council*, The Computer and Automated Systems Association of SME, Dearborn, Michigan.

Sheridan, J.H. (1989) Toward the CIM solution. *Industry Week*, October, 35–51.

Taraman, K. (1980) *CAD/CAM: Meeting Today's Productivity Challenge*, The Computer and Automated Systems Association of SME, Dearborn, Michigan.

US Department of Commerce, Office of Technology Administration (1989) *Flexible Automated Manufacturing Systems*, Report published by and available from: Office of Technology Commercialization, US Department of Commerce, Washington, DC.

Vollman, T.E., Berry, W.L. and Whybark, D.C. (1988) *Manufacturing Planning and Control Systems*, 2nd edn, Dow Jones-Irwin, Homewood, Illinois.

Wight, O.W. (1974) *Production and Inventory Management in the Computer Age*, CBI Publishing Company, Boston, Massachusetts.

Wight, O.W. (1984) *Manufacturing Resource Planning: MRP II*, Oliver Wight, Essex Junction, Vermont.

Williamson, O.E. (1975) *Markets and Hierarchies: Analysis and Antitrust Implications*, The Free Press, New York.

17 Application of Robotics and Automation Technologies to Industrialized Housing Manufacture

Admad K. Elshennawy, William W. Start
and Subrato Chandra

17.1 Introduction

Current technologies in the construction and industrialized housing manufacturing industry are capable of producing a home in a factory in much the same way a home is built on-site. The efficiency of this industry is affected by many factors that include the increased demand for such homes and the type of existing traditional building practices that are not capable of meeting such increased demand. The productivity of this industry, however, can be improved by employing existing as well as emerging technologies in its operations. The repetitive nature of the different processes involved in the manufacture of homes and their components in a factory lends itself to automation, similar to other manufacturing operations. Other countries such as Sweden and Japan have already implemented automation technologies in their production of factory-built homes.

Research and development efforts in the industrialized housing manufacturing industry have realized the need to set standards for improving the way we build our homes today. Automation of the industry is an obvious solution for the 21st-century building facility.

The 21st-century industrialized housing manufacturing facility can be visualized as a fully integrated facility relying on automation that is capable of producing homes of higher quality and affordability. The transfer of automation and robotics technologies in discrete-part manufacturing into the construction environment is technically feasible as it appears in the following discussion in this paper.

The concept for the 21st-century manufacturing facility for industrialized housing employs advanced and emerging technologies. The facility employs different aspects of advanced information and manufacturing technologies including CAD/CAM (computer-aided design and manufacturing), MRP (material requirement planning), automated material handling and advanced control systems.

17.2 Automation in industrialized housing

Application of automation into industrialized housing manufacturing requires building blocks for full integration and automation as follows:

1. Development of an effective information system;
2. Material requirement planning system;
3. Computer-aided design;
4. Computer-aided manufacturing;
5. Automated material handling;
6. Control system at factory, workstation, and machine levels;
7. Safety requirements; and
8. Flexibility.

Each of these modules (or blocks) is briefly explained in the following sections.

17.2.1 Information system

An important element in any integrated and automated operating environment is an effective information and communication system. Such a system should perform important tasks such as:

1. Translating customer requirements into design and production control data needed for the design and manufacture of the home;
2. Determining machine, process, tool and material requirements;
3. Developing process plans that detail how the home is to be made. In the factory, these plans include: routing-slips which are used to schedule the movement of materials into different workstations, and operation sheets which detail the sequence of events for each workstation; and
4. Generating programs to instruct shop-floor equipment, i.e. machines and robots to perform different operations.

17.2.2 Material requirement planning (MRP)

Inventory control has two major objectives: minimizing inventory cost and maximizing the level of service to customers and different departments within the company. There are four different types of inventory of concern:

1. Raw materials and purchased components;
2. In-process inventory;
3. Finished homes; and
4. Maintenance equipment and tooling.

Inventory management within the factory should consider both inventory accounting, which is concerned with inventory transactions and records, and inventory planning and control which can be accomplished by a computerized MRP system development. Functions of the MRP system within the factory include: minimizing inventory cost within the factory, and improving production scheduling as well as the purchasing of materials.

The MRP system constitutes an important element for plant operations. It receives inputs from the bill of materials file, master production schedule which is generated based on sales forecast and customer demands, service parts file, and inventory record module. The MRP system converts such inputs into useful output reports, and then orders are issued to suppliers for delivery.

It is important that a reliable relationship between the factory and suppliers exists. Employment of the JIT (just-in-time) concept is useful in this situation. JIT is an inventory system which insures that materials will be delivered with the required quality at the required time in the required location.

17.2.3 Computer-aided design (CAD)

This function is the module which translates customer demands into production data. The CAD system generates hard-copy architectual and structural drawings for the home to be built, directly from its database. The computer-aided design system also generates a bill of material which can then be handled by the MRP system. An expert advisory system for home design can assist designers making decisions at different phases of the design process. The expert system applies artificial intelligence methodology into the industrialized housing manufacturing industry.

17.2.4 Computer-aided manufacturing (CAM)

Among the major advances in computer-aided manufacturing is the development of modern numerical-control (NC) systems and industrial robots. The modern factory for manufacturing homes and their components will employ a number of NC machines and robots for different operations. The machine controller receives instructions in a computer-readable format from the information-preparation system through the process planning module. Such instructions contain a detailed step-by-step set of directions which tell the machine or the robot what to do. Machines and robots will perform operations and tasks such as:

1. Cutting or sawing;
2. Assembly;
3. Welding and inserting joining strips; and
4. Loading and unloading.

17.2.5 Automated material handling

An effective and reliable material-handling system is essential for increasing the efficiency of operation and smoothing the flow of material through the plant. It is envisioned that an automated overhead conveyor system that is computer controlled is necessary. Employment of other belt conveyors is also obviously necessary for handling small parts and materials. The material-handling system will receive instructions through the factory and workstation control system levels as necessary. It is important to design the material-handling system in such a way that its movement does not interrupt different tasks at different work stations.

17.2.6 Control system

The concept behind any fully integrated and automated factory is that it is possible for both the hardware and software to be compatible and easily integrated with new products as they become available. The control system will operate at the factory, workstation and machine/robot levels. It receives orders from a computer terminal, defines the necessary resources, and then schedules, coordinates and monitors the activities at each level involved.

17.2.7 Safety

Several safety issues are of concern and require careful consideration when designing the automated housing facility. Such issues include the movement of large structures such as roof and floor systems and wall panels. Since automation is to be employed, the risk involved in having workers in the work area is significantly reduced.

17.2.8 Flexibility

The standardization of operations, which is one objective of automating the industrialized housing industry, will allow the factory readily to accommodate manufacturing and fabrication jobs with variability in size, finishing and customized features. The conceptual design of the factory uses flow-line layout where the concept of group technology (GT) can be easily applied. GT is a concept whereby parts or products with similar design and production features can be manufactured in a cell or a flow line. This facilitates the movement of parts within machines and increases the productivity of operations.

17.3 GE House 1000 manufacturing facility

This section is a summary of the work performed which was partially funded by GE Plastics in Pittsfield, Massachusetts in 1989. The General Electric Company's Plastics Division in Pittsfield, Massachusetts has developed prototype engineered plastic-based foundation, floor, wall and roof panels for a 21st-century home. Conceptually, these large multi-function insulated panels made of recyclable plastics and other materials would be produced at a few centralized plants using different molding and forming technologies and shipped to panelizing plants throughout the country. These panelizing plants would cut the large panels, install doors and windows, and ship them to the building site. The conceptual design of the manufacturing facility for producing such panels was developed by the Department of Industrial Engineering and Management Systems at the University of Central Florida. The facility provides a basic array of manufacturing systems and equipments. It includes several types of modern automated construction machines, such as numerically controlled cutting machines, automated material-handling equipment and a variety of industrial robots to tend the machines and to perform different operations. The entire facility is operated under computer control using advanced control and sensory systems.

The goals of building such a facility are:

1. To provide the US building industry with a new way of manufacturing homes with high quality and improved durability;
2. To encourage the modernization of the US building industry by

providing the necessary technology and information based on a new generation of automation and manufactured systems; and
3. To introduce new materials for the manufacture of industrialized housing.

The proposed factory layout satisfies GE's requirements within the conditions imposed by:

1. The nature of material to be processed:
   (a) Physical configurations and chemical properties, and
   (b) Size and corresponding weight;
2. The manufacturing processes required:
   (a) Sawing/cutting/drilling,
   (b) Welding/joining/inserting, and
   (c) Assembly;
3. Production rates;
4. Material handling;
5. Capital equipment and cost;
6. Capabilities of the raw material suppliers;
7. Customer needs; and
8. Existing as well as future advances in manufacturing technology.

Production rates were the first requirement assessed in the design and development process for the model factory. Given GE's specifications of 2600 houses per year as the required output, along with other specifications for panel types, quantities and sizes for each of four different houses, it was determined that the factory would need to be capable of processing 1000 panels per day. This was a worst-case estimate for the exclusive production of 2500 sq. ft houses.

The desire for an efficient and effective factory layout then triggered the assumption that the facility would operate for three shifts per day. Assuming two production lines (one line is used for the production of exterior and interior wall panels while the second line is used for the production of roof, floor, and foundation panels) it was apparent that each line would require a capacity of 20 panels/line/hour.

17.3.1 Functional tasks

Functional tasks associated with the operation of the facility are: information preparation and handling, computer-aided manufacturing, and shop-floor operations. As soon as customer requirements and specifications have been defined for a home, the tasks of information preparation and handling start. Computer-aided design activities generate architectural and structural drawings as well as a bill of materials for the home. Such output is then translated into information that can be used by computer-aided manufacturing equipment including NC (numerically controlled) machines and robots. Such information defines the event sequence for the different manufacturing operations, assigns tasks to machines and robots, and directs the motion of tools for different operations on the shop floor.

The flow of information described above is depicted in Fig. 17.1 which also shows that such tasks are executed in three different areas in the factory: sales and marketing, design and product engineering, and factory operations.

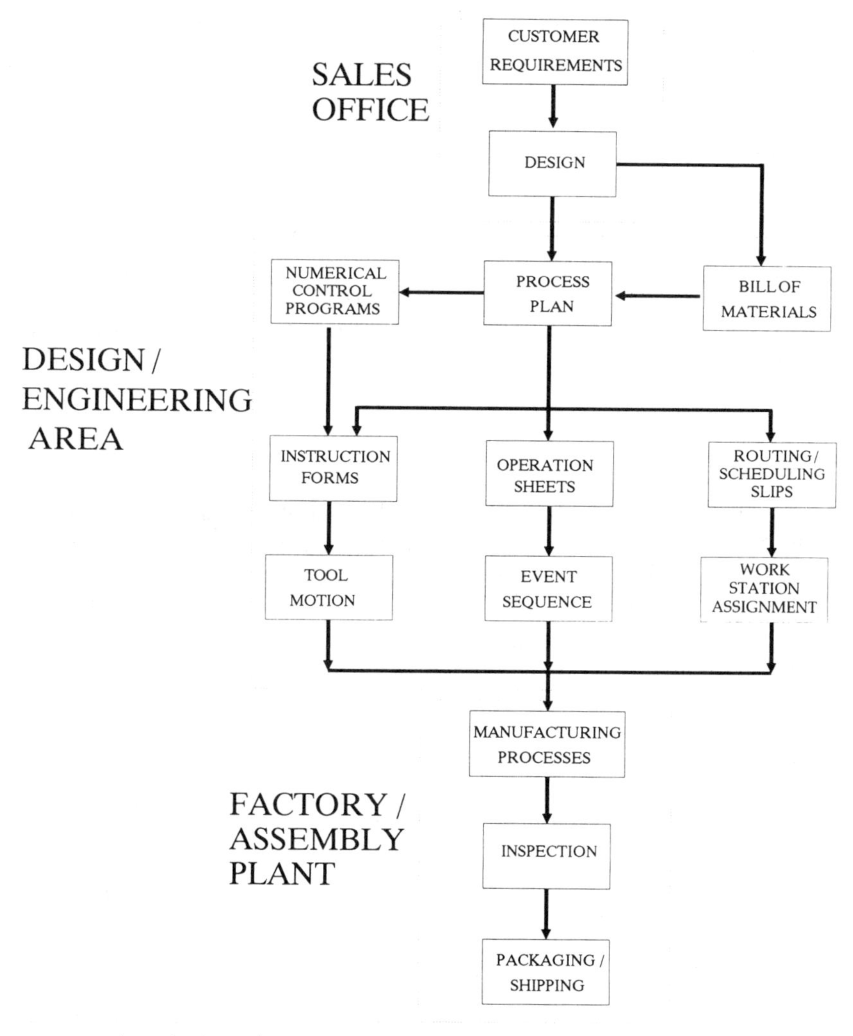

Fig. 17.1 *Flow of information.*

17.3.2 Information preparation and handling

Building an automated facility like this involves equally important tasks of information preparation and transforming the customer's requirements for a particular house into the production control data needed to manufacture that house.

Information and preparation tasks include designing the house, determining which materials and tools are required, developing the process plans that detail how the house is to be made, and generating the numerical control programs that instruct the robots and machines actually to make the components for the house.

Of equal importance is verification of the plans to insure that design meets manufacturability and safety requirements.

Computer-aided design

This is the actual start of the process. Once a house is designed, all of the information necessary to manufacture its components should be made in a standard, computer-readable format so that it can be used by other elements of the factory's control systems. Such information includes architectural and structural drawings, material specifications and process/machine/tool selection.

Material requirement planning

Based on the output information from the computer-aided design phase, a computer-generated bill of materials is issued for suppliers that defines material specification, quantity, building material (blanks) and industrial components (windows and doors). Building material for the factory are roof, wall, floor, and foundation 'blanks', i.e. large segments which are structural components in themselves. The quantity of material required

TABLE 17.1 *Component parameters for different homes*

Home type	Floor size sq. ft	Int. Wall # L.R.*	Ext. Wall # L.R.	Roof/Floor # L.R.	Foundation # L.R.
A	1000–1200	18 2-18′	6 2-40′	10 4-16′	6 2-40′
B	1201–1600	24 2-24	12 2-40	14 4-18	12 2-40
C	1601–2000	28 2-24	16 2-40	18 4-20	12 2-40
D	2001–2500	30 2-30	18 2-40	20 4-20	12 2-40
Thickness		4″	5″	8″	7″

* L.R. = Length range

for a home is dependent on the customer specifications and requirements. The factory is capable of producing the final for four different types of homes: A, B, C, and D. Each type has different sizes and component parameters including the number and length-range of panels for interior walls, exterior walls, roof, floor, and foundation panels, as shown in Table 17.1. Blanks are received in a 40 ft. × 10 ft. size with different thicknesses as also shown in Table 17.1.

Process plans

Such plans detail the actual manufacturing steps in the proper sequence required to produce a home (or its components). In the factory, these plans are generated by computers and include: routing slips which are used to schedule the movement of materials into assigned workstations, operation sheets which detail the sequence of events for each workstation, and instruction forms which directs the tool through the motions required to perform the task/operation. The routing slips, operation sheets and instruction forms are electronically transmitted to machine and robot controllers via computer terminals.

17.3.3 Computer-aided manufacturing

The concept for factory control assumes that both the hardware and software aspects of automation are compatible and can be integrated with new and advanced products as they become available. This may not be true today, but continuous advances in technology will help to make it possible tomorrow.

The control system receives orders from a computer terminal, defines the necessary resources, and then schedules, coordinates and monitors the activities of the workstations required to manufacture the components (panels) for a home.

The factory layout shown in Fig. 17.2 has two production lines. Panels will move vertically using an overhead conveyor system to a position where machines and robots will perform the required operations at different workstations.

17.4 The production process

Process flow charts and information flow charts are shown schematically in Figs. 17.3 and 17.4 for the exterior and interior walls production line and in Figs. 17.5 and 17.6 for the roof, floor and foundation production

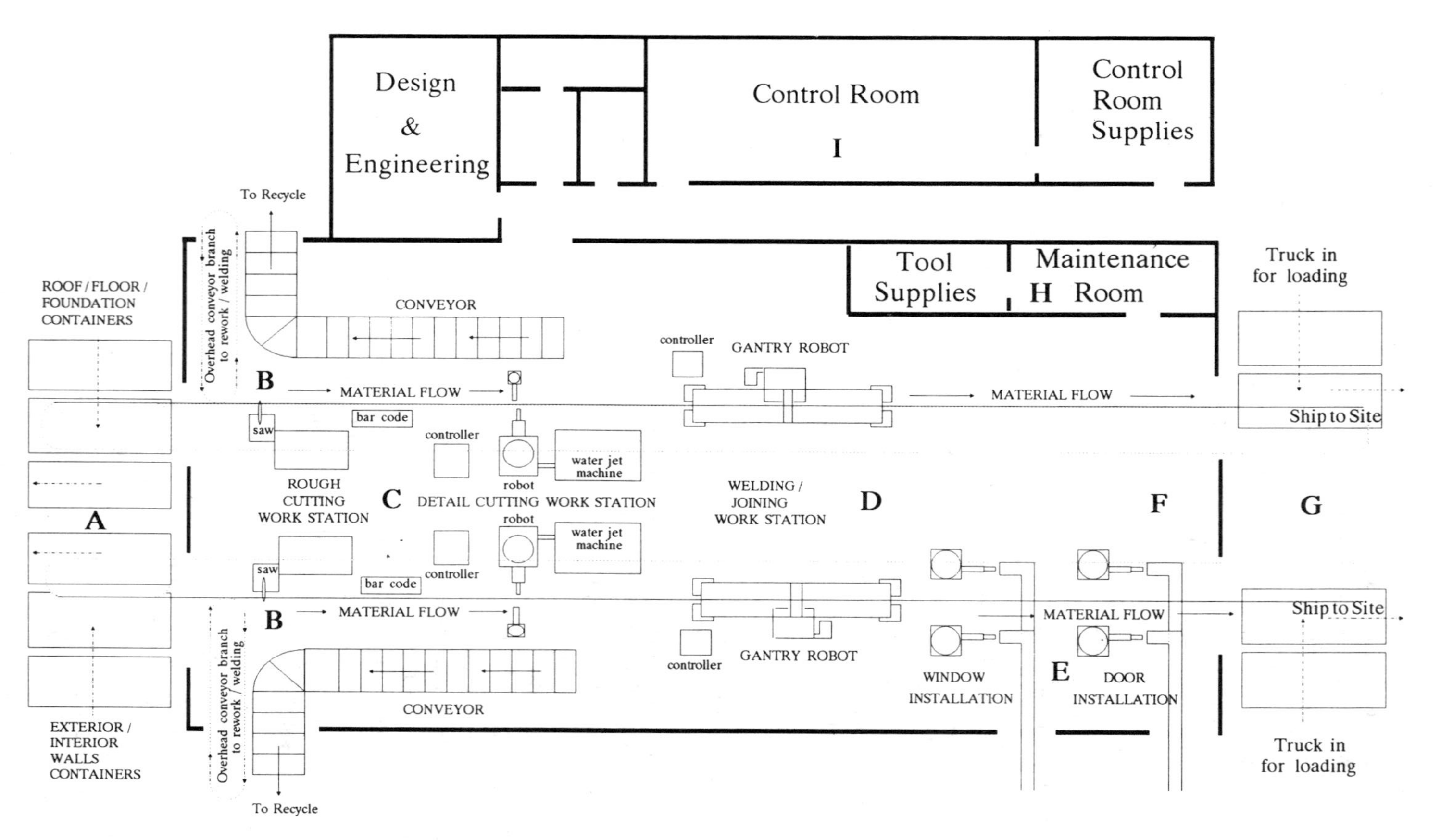

Fig. 17.2 *Conceptual factory layout.*

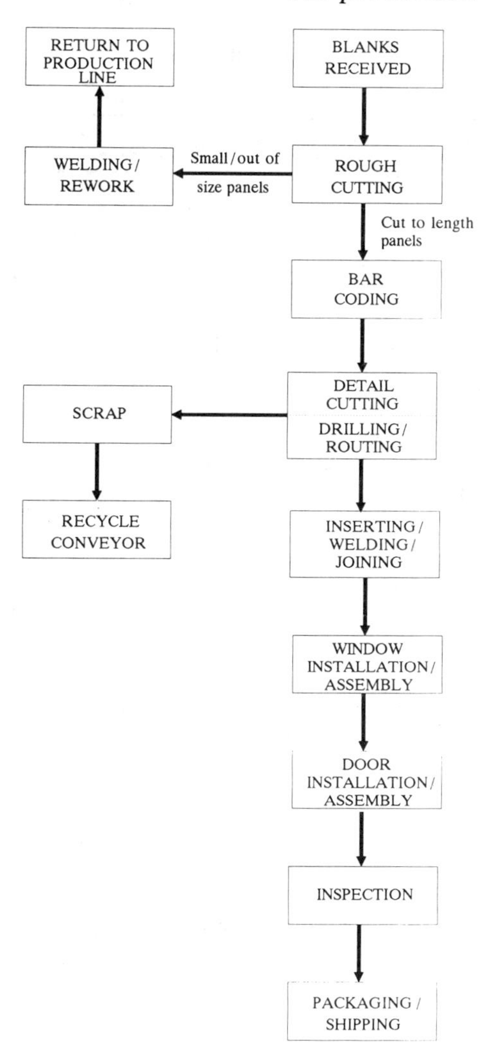

Fig. 17.3 *Manufacturing process flow for exterior and interior walls.*

line. The following is a brief description for the different operations in the factory.

17.4.1 Material arrival and unloading:

Building material (blanks) and other components (windows, doors) will be shipped to the factory in containers. Roofs, foundation and floor panels will be manufactured in one production line, while exterior and

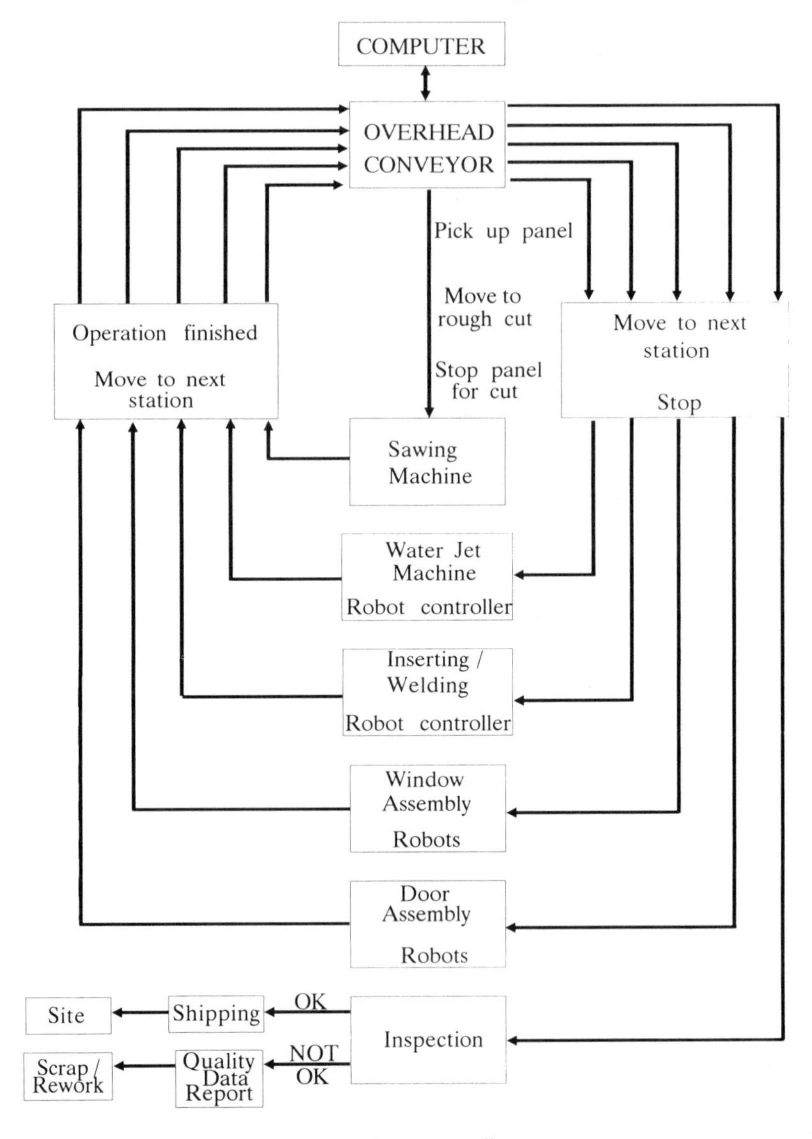

Fig. 17.4 *Information flow for exterior and interior walls.*

interior walls will be manufactured in another production line. Containers will be unloaded from the trucks and placed on a rail system that moves towards the entrance of each production line. As the container is emptied, it will move into a position where a tractor will pull it to the shipping area to be loaded with finished panels and another container moves in for unloading. This scenario is depicted schematically in Fig. 17.7.

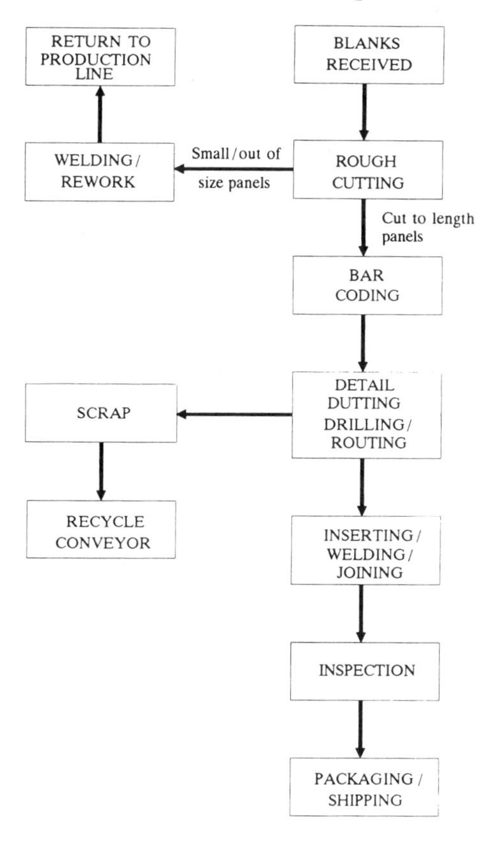

Fig. 17.5 *Manufacturing process flow for floor, roof and foundation.*

17.4.2 Material handling

Individual blanks will be moved to the production line using the overhead conveyor system. The conveyor is composed of carriers (or engagers) that will move into matching slots inserted on top of the blanks causing them to move along the different workstations as shown in Fig. 17.8(a). The overhead conveyor system consists of a structured beam track from which a powered conveyor chain is suspended and an unpowered trolley that contains the carriers (engagers) is attached (Fig. 17.8(b)).

17.4.3 Recycle material disposal

A conveyor is used for handling scrap for recycling. Two conveyor stations are used for this purpose. The conveyors are located near the

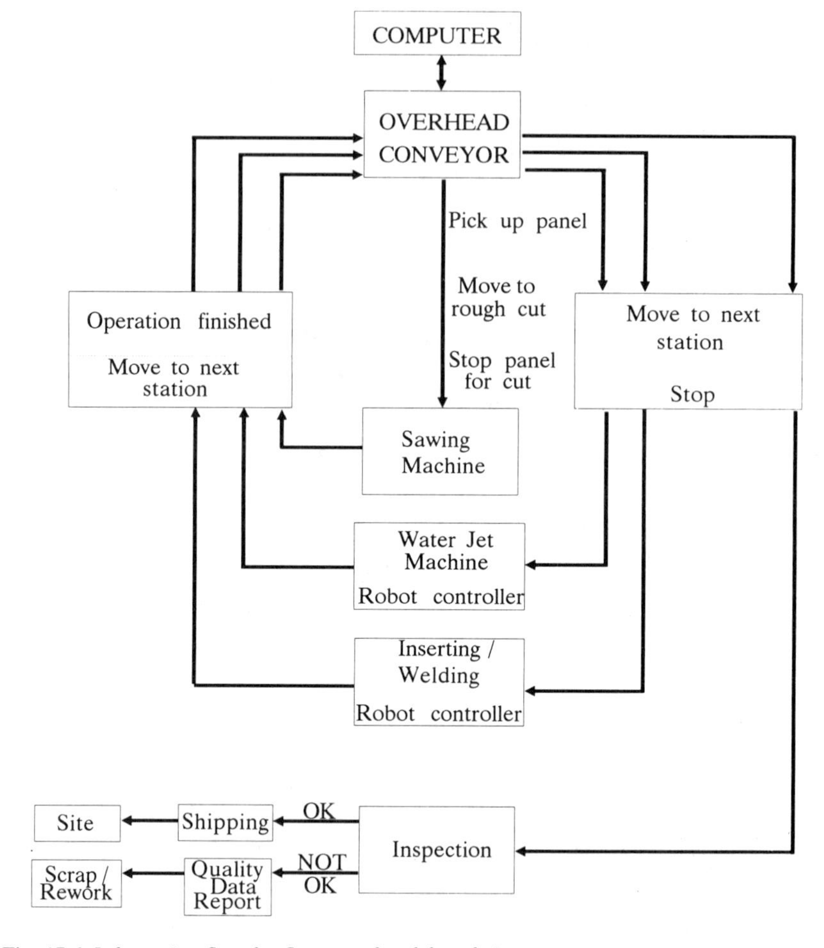

Fig. 17.6 *Information flow for floor, roof and foundation.*

detail cutting workstations (see Fig. 17.2) for both production lines. Scrap from the detail cutting process (e.g. door and window openings) is loaded on to the conveyor using a robot. The scrap material goes from the conveyor to a recycling area.

17.4.4 Workstation scenario

At each workstation, a robot or a machine controller receives information from a computer terminal that includes instruction sets, operation sheets and routing slips. A signal is passed to the machine controller from the previous workstation after its operation has been completed. The

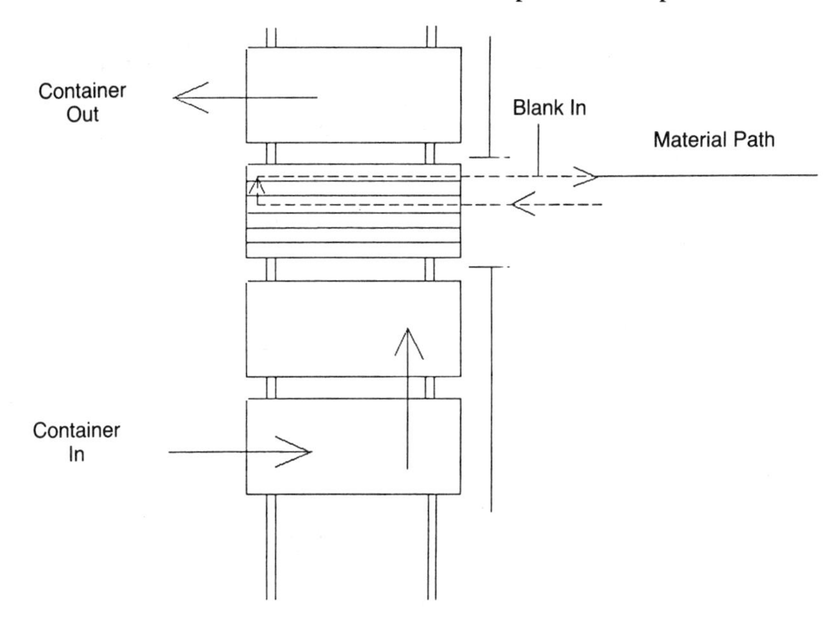

Fig. 17.7 *Unloading of incoming materials.*

previous workstation also sends a signal to the overhead conveyor to move the panel to the next workstation. A sensor at the beginning of each workstation detects the panel arrival and passes such information to the robot/machine controller. The robot/machine then performs the operation. This scenario is shown in Fig. 17.9.

17.4.5 Panel support

At each workstation, three important events take place:

1. *Laser bar-code scanners.* The bar-code scanners are located near the end of each workstation. They read labels on the panels as they move into the workstation to check that the conveyor is carrying the proper panel.
2. As the panel moves into a workstation, a sensor sends signals to the machines and robots in the workstation to ready them for operation. As soon as the panel stops, the robots and machines start their tasks according to the set of instructions sent by the control system computers. After all tasks are complete, the panel starts moving to the next workstation for other manufacturing operations.
3. *Panel support.* When the panel being processed stops at each workstation, a hydraulically activated support system moves to support the

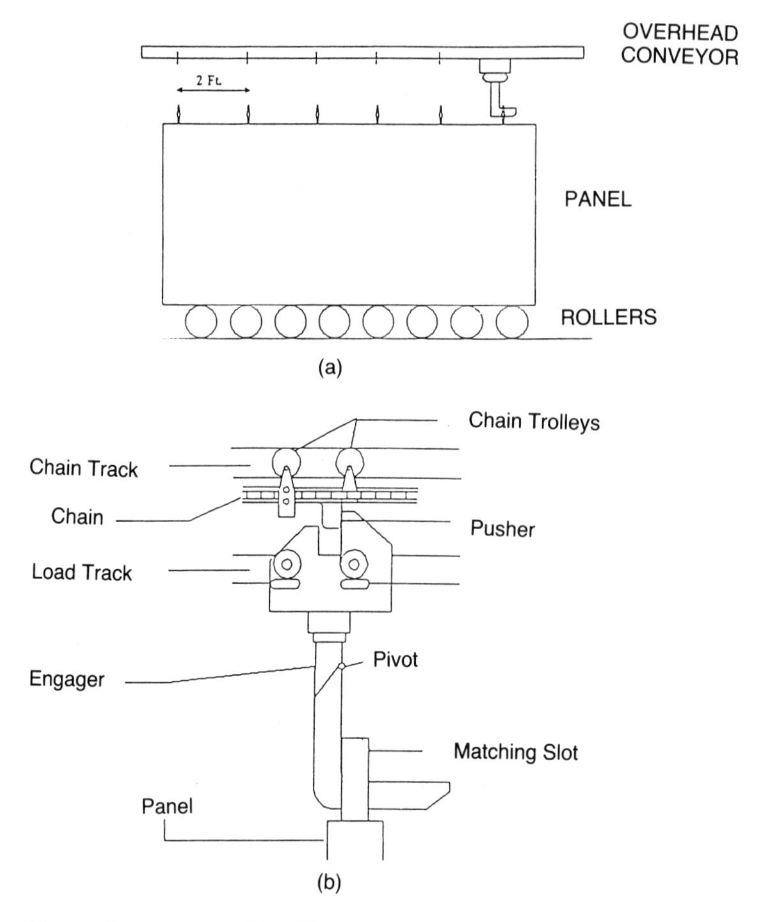

Fig. 17.8 *Panel movement and handling.*

panel during operations. The system consists of several hydraulic cylinders located at both sides of each production line.

17.4.6 Workstations description

The following is a brief description for each of the workstations in the factory.

Rough cutting workstation

Rough cutting is performed by a numerically controlled radial sawing machine. One machine would be employed for each production line

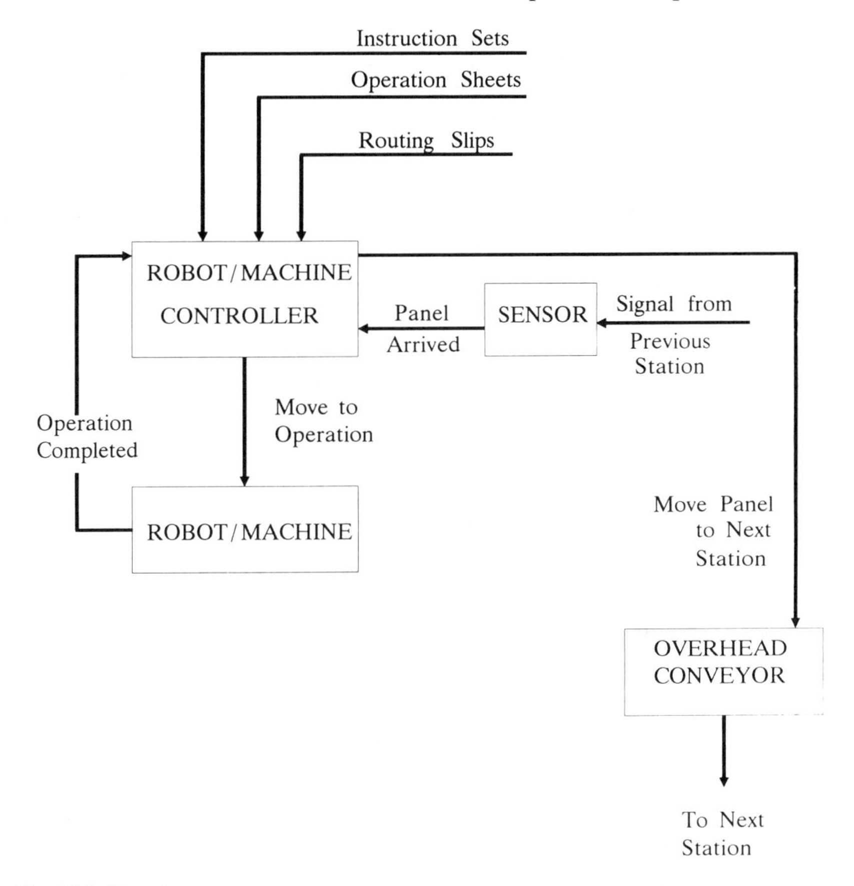

Fig. 17.9 *Transformation of information for workstations.*

where straight cuts as well as angular cuts may be performed as necessary. As shown in Fig. 17.10, when the panel stops at the rough cutting workstation, the sawing machine moves to cut the panel as instructed by its controller.

Detail cutting workstation

Water-jet machining was selected as the means of achieving precise cuts (including drilling) in the panels. Although cutting rates for this process are relatively low, it is predicted that future advances in this area will make higher rates possible. The detail cutting is performed by a robot carrying a nozzle connected to the water-jet machine with a hose. The robot is of Cartesian configuration. Once the panel stops, the robot nozzle will start operation at a position defined in the instruction sheet

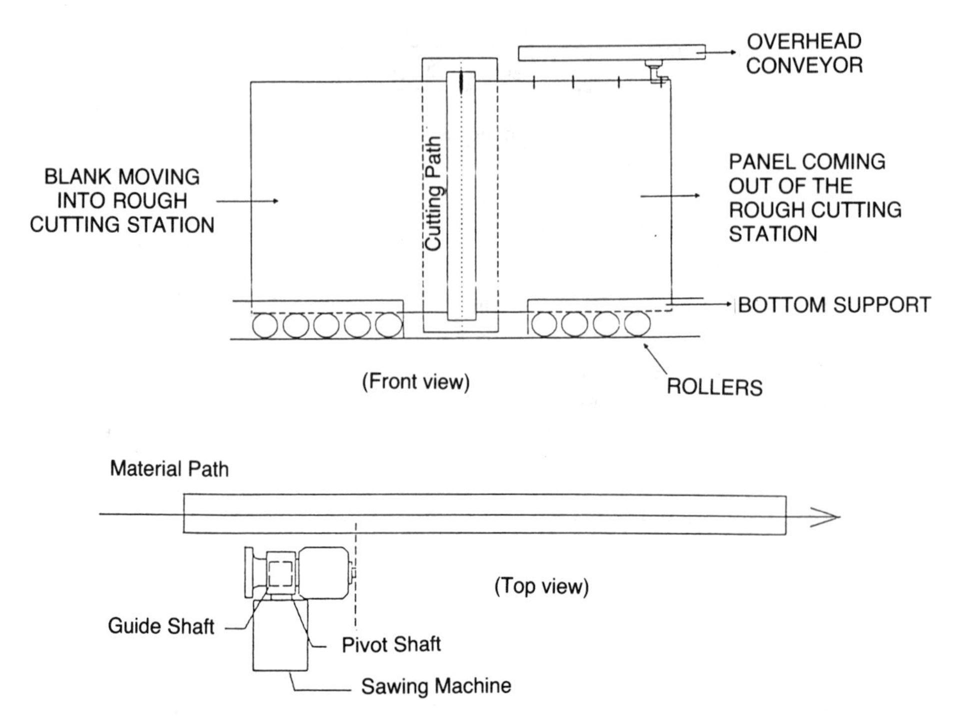

Fig. 17.10 *Rough cutting workstation.*

sent to the robot controller. After the cut is complete, the panel advances to the next station. The robot moves on a track that is long enough to perform more than one cut in each panel, if required. Figure 17.11 shows a schematic top view of the workstation layout.

Welding workstation

As the panel comes from the detailed cutting station, a joining strip is installed at its side to allow different panels to be joined when necessary. The strip will be inserted into and clamped to surrounding material with the use of a gantry (overhead) robot. The strip will be welded into the panel with an ultrasonic welding device that will accomplish either spot welds or seam welds as necessary. After the weld is complete, both the robot and machine release the panel to move on. Figure 17.12 shows the strip being welded to the panel after its insertion.

Window/door installation workstation

Windows and doors are provided in halves and must be snapped in from both sides of the panel. The windows and doors are brought into the

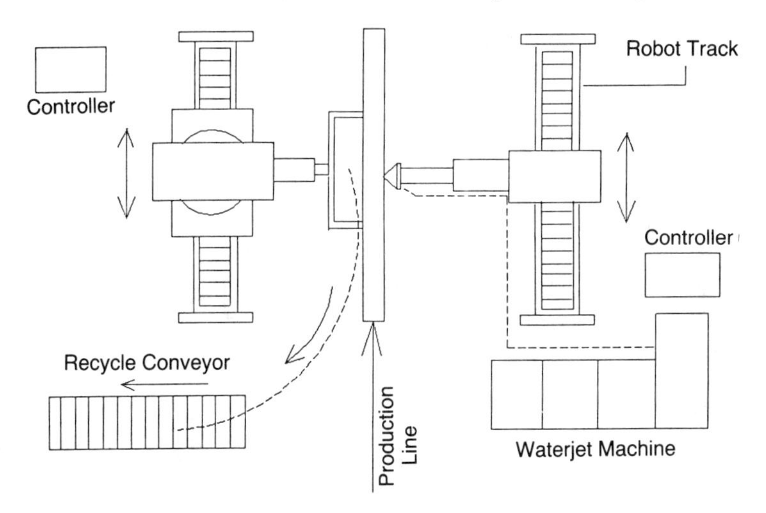

Fig. 17.11 *Detail cutting workstation.*

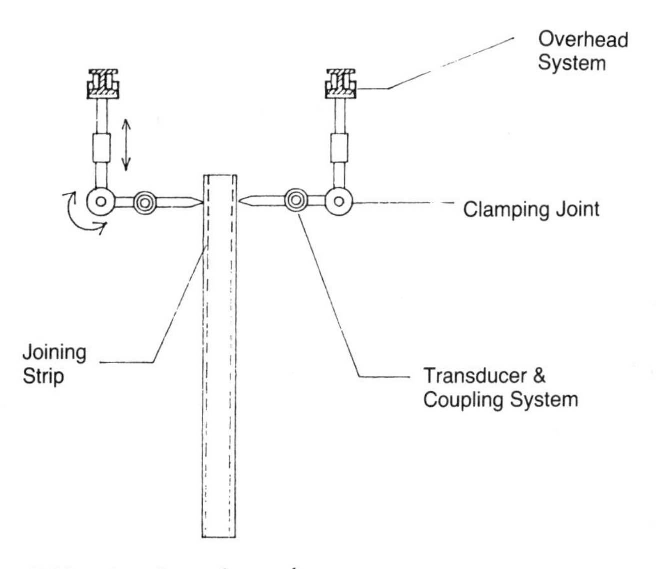

Fig. 17.12 *Welding the strips to the panel.*

assembly station loaded in magazines. These magazines are lowered at the point where a robot needs to take a window or door out of the magazine. The robot then turns to the panel and inserts the window or door into the opening made at the previous detail cutting station. This is schematically shown in Fig. 17.13. These windows or doors are snapped

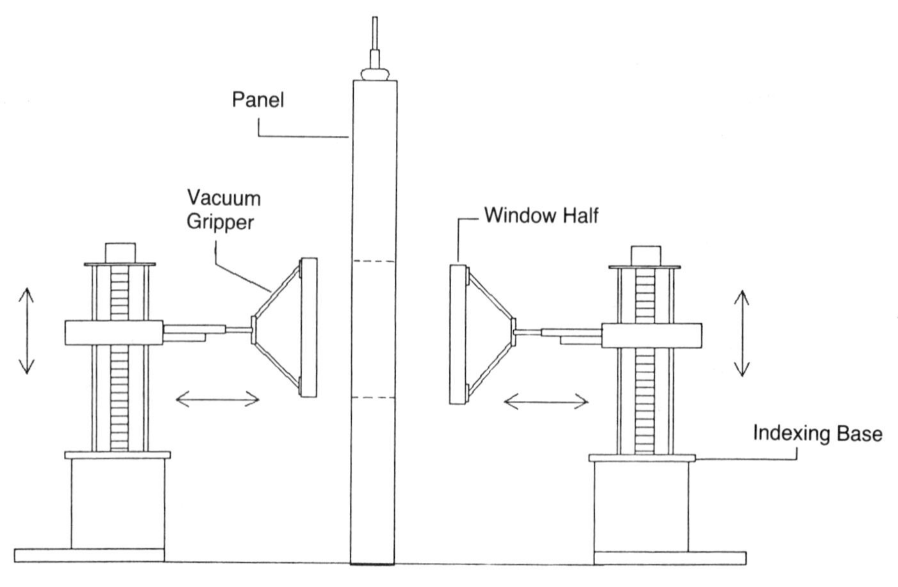

Fig. 17.13 *Window installation.*

into the holes in a clamshell type configuration. Both halves of either are installed simultaneously.

17.5 Staffing analysis

Operation of the facility requires 10 workers whose responsibilities are mainly supervision, inspection, maintenance and control. The primary location for each of these individuals is shown in the plant layout in Fig. 17.2. Table 17.2 shows the different tasks or responsibilities and the number of workers or supervisors required.

17.6 Conclusions

This paper provides a brief description of the development effort of a conceptual design for an automated facility for building the 21st-century homes using new materials which are to be introduced by the General Electric Company in Pittsfield, Massachusetts.

The facility design provides a new application of robots and automation in the construction and industrialized housing manufacturing industry. It is fully integrated and automated and the machines and robots as well as the material-handling systems are computer-controlled using advanced

TABLE 17.2 *Manpower assignment*

Task(s)	Location/layout	Number
Supervise unloading of incoming material for each production line	A	1
Visually inspect for defects in material		
Supervise recycle conveyor operation	B	2
Supervise overhead conveyor branch for out-of-size panels		(one/line)
Record any defects observed in rejected panels (if any)		
Supervise rough cutting workstations	C	1
Supervise detail cutting workstations		
Observe any failures		
Supervise welding/inserting workstations	D	1
Supervise window/door installation workstation	E	1
Final inspection before shipping	F	1
Supervise shipping area	G	1
Maintenance/service	H	1
Control room operations	I	1
Observe any computer failures		
Observe any error messages		
Total manpower required		10

control-system concepts. Employment of robots and machines for such operations as cutting, drilling and welding is not new. What is new is the application of such technologies in an environment where the material size and characteristics as well as the nature of operations and working conditions are basically different from those in other manufacturing systems.

In summary, the paper presented a description of a conceptual design for an integrated and automated facility for the manufacture of home components for the industrialized housing manufacturing industry.

18 Training and Cost Effectiveness in the Development of Training Devices

Lucy C. Morse and Brian Goldiez

18.1 Introduction

The advance of military technology in recent years has resulted in an increased reliance on the use of training devices to achieve and maintain a state of combat readiness. In many cases, a training device is used to prepare personnel to operate a parent system which is too expensive, time consuming or dangerous to be used for initial training. The success of training devices in reducing costs and training time for major weapon, aircraft and other systems has led to the development of training devices for tasks that are not specific to a system. Training devices are now accepted and vital elements in the training programmes of all branches of the service. Manufacturing quantities for these military devices range from single items to lots of 100 devices. Lot sizes of less than 10 units are typical. The same is true in the civilian sector which uses, for example, training devices for airline pilot training and student driver training.

Although there is a considerable amount of data about specific training devices, there is no organized body of information on what is necessary to manufacture cost effective and training effective systems, especially in the military sector. As a result, the design of effective training devices is an effort that includes imperfect data, opinion-based design rules, and an increasingly large number of choices in the large array of technologies that can be used to address any single training problem.

One problem in the area of training is that research on methods to estimate training effectiveness and cost effectiveness lags behind the pressing need of the user to acquire and field effective systems. The goal of training-device formulation is to develop and manufacture a training device that meets the training requirements at minimum cost, or provides the maximum training benefit for a given cost.

Standardized methods to estimate the cost of designing and manufacturing training devices have eluded the agencies which procure such devices. Methods to estimate software design and production have been particularly elusive. In addition, methods to ascertain and link cost with a measure of training effectiveness is cumbersome, at best. Each organization which procures training devices has a preferred method of estimating cost and, in some cases, training effectiveness. In the case of the Department of Defense, no standards exist. Format is often articulated using the work breakdown structure (WBS) requirements in contracts. The WBS represents a method to communicate cost information. The underlying cost-estimating methodology is rarely verifiable.

18.2 Statement of problem

Recently, several methods have been developed for assessing the training effectiveness of various training strategies. These methods assign a numerical rating which an analyst can use to compare alternative training approaches. Two current software packages under evaluation within the Department of Defense are Optimization of Simulation-Based Training Systems (OSBATS) which was developed and is described by Singer and Sticha (1987) and the Automated Simulator Test and Assessment Routine (ASTAR) (American Institutes for Research, 1988).

The idea behind OSBATS is to organize the large amount of training technology and learning theory currently available and develop a model for aiding training developers in evaluating training-device alternatives. This project is currently in the final prototype stage.

Originally it was assumed that some of the cost data required by the OSBATS model had been collected by agencies within the Federal Government. Organized databases were not found to exist. Various estimating methods were used to provide costs for the cost elements (Singer and Sticha, 1989). Partially due to lack of databases, OSBATS is currently device specific, rather than generic.

ASTAR is a series of computer programs designed only to measure training effectiveness. These programs guide an analyst through an evaluation of the effectiveness of various methods which can be used to train for a given task or set of tasks. The programs are multi-dimensional with respect to the amount of information required and the types of analyses provided. ASTAR can deal with either general information on the type of tasks to be trained or extremely detailed task lists. Intermediate amounts of information can also be utilized. This dimension of information content becomes directly related to the quality of results to be expected. General information will yield general results, while detailed task and other input data can be expected to provide the ASTAR software programs with sufficient data to yield more precise results. This approach is advantageous because ASTAR is not required to carry or create large databases. Also, one is able to use ASTAR to investigate very preliminary training issues in a general way and in an interactive format.

ASTAR provides four analyses as outputs: an analysis of the training problem, an analysis of the acquisition efficiency, an assessment of the transfer problem and an analysis of the transfer efficiency. The training problem defines the skill and knowledge deficiencies that the trainees would have relative to certain performance criteria. Acquisition efficiency examines the instructional features and training principles that help the trainees overcome their deficits with that particular training device.

The transfer problem determines the deficiency-to-performance criterion that remains after training on the device. Finally, transfer efficiency indicates how well use of the training device will promote transfer of the learning that has occurred to the parent or actual equipment (Rose and Martin, 1988).

Multi-attribute economic analysis techniques can be applied to ASTAR results to create a complete picture of the cost and training effectiveness of a training device. The economic analysis must include those costs associated with design, manufacturing and support hardware. Software costs should be partitioned across each boundary. Several potential shortcomings with ASTAR will be noted with respect to cost-estimating techniques which can be applied to the model.

18.3 Discussion of methodology

The problem with relating cost to training effectiveness is related to the fundamental origin of each process. Economic analysis is oriented to quantitative measures of analysis. Training effectiveness, on the other hand, is oriented to either qualitative evaluation or inferences made from available data. In the case of training effectiveness, an expert is necessary not only to interpret data, but also to develop experiments and to accumulate the data from subjects. Relating these two areas is not a clear process. ASTAR provides an additional complication. It appears from ASTAR documentation (American Institutes for Research, 1988) that numerical scores from ASTAR must be related against other ASTAR scores to determine which alternative is 'best'. ASTAR, in other words, is a program to evaluate relative training effectiveness, but is lacking in absolute measures.

Specific cost information on various components and the various development and manufacturing phases of a training device are also necessary to implement this analytical tool. Combining cost information and training could then provide an excellent means to analyze the effect of using different subsystems on a training device. For example, this method may be used to decide if the change in a visual system on a training device should be implemented. The marginal increase in training effectiveness, as reflected in the ASTAR scores, needs to be balanced against the potential increase in the design and manufacturing cost. Detailed cost knowledge is necessary on various subsystems to perform this analysis. This information is normally difficult to obtain, but relative costs of a subsystem of a training device can be derived easier than can absolute costs. One can estimate the percentage increase in cost of a visual system of a training device, for example, when the field of view is increased by a factor of two.

To make these evaluations, the selection of what attributes of a training system or a subsystem of a training device should be considered in the evaluations is critical to the eventual outcome. The word 'attribute' is used to describe what is important in a decision problem and is often used interchangeably with objective and criterion. The attributes selected would depend on what device is being evaluated and what component of that device is being evaluated. The most important criterion in the naming of attributes is that they be independent of one another. In selecting attributes, many may be critical, but it is important to keep the total number within manageable limits and only consider those that affect the final decision. One way to select attributes is to identify one or a few attributes in the beginning, do an initial analysis, and then add more as justified by the decision. For training devices a useful way to separate attributes is to separate the attributes associated with design, manufacturing and support.

Whether looking at a total training device or a few attributes of one, a training device is most effective if it reduces total training time to a minimum. The faster or more efficiently trainees reach and retain operational proficiency, the better the device. Using ASTAR, this measure pertains to the transfer score and the acquisition score; the lower the score, the better the device. There is no direct translation of the various ratings into hours of training. Comparisons between transfer and acquisition scores of different devices can only be made when the operational performance objectives are the same. For comparison purposes these scores are assumed to be ratios. The way to tell the significant difference between two different scores on a single device is with a sensitivity analysis.

Care must be made in the selection of the ASTAR outputs used as attributes. The higher the scores on some ratings, the better the training device or training strategy. High transfer efficiency and acquisition efficiency scores reflect poorer training strategies than do lower scores.

The ASTAR ratings are used as attributes relating to training. Attributes relating to cost factors would include the device's investment cost, fixed cost per year for operating the simulator, variable cost per hour for use, life-cycle, and projected utilization.

These attributes are each then employed in a simple linear additive model, and a benefit-cost analysis is used with the results. The linear additive model is a decision tool that aggregates information from the independent attributes in a linear fashion to arrive at an overall score for each course of action being evaluated. In the past, this model has been used successfully to justify automated manufacturing technologies (Parsaei and Wilhelm, 1989). The general form of the model is (Sullivan, 1986):

$$V_j = \sum_{i=1}^{n} w_i X_{ij}$$

where V is the score for the jth alternative, w is the weight for ranking assigned to the decision attribute, and X is the ASTAR ranking.

When several attributes of cost and/or training are analysed, the weighting of attributes method or utility models described by Canada and Sullivan (1989) is used. These methods attempt to balance several attributes. Both of these methods provide measures of the relative importance of attributes to be used in economic analysis. The method of weighting of attributes is preferred because a complete set of various combinations of attributes can be analysed in an orderly fashion.

Since numerical ASTAR scores must be related against other ASTAR scores, final scores for individual alternatives are meaningless. Using the benefit-cost method, the incremental approach is required when comparing these mutually exclusive alternatives. Combining the linear additive model with benefit-cost analysis for a single alternative is described below.

$$E_j = \frac{B_j = \sum_{i=1}^{p} U_i A_{ij}}{C_j = \sum_{i=1}^{p} V_i X_{ij}} \quad \text{for } j = 1 \text{ to } q$$

where: E_j is the final score for each alternative; B_j is the score representing the ability of the jth alternative to achieve the training objectives of the tasks; C_j is the score representing the ability of the jth alternative to achieve the cost objectives of the specified tasks; U_i is an ordinal scale ranking assigned by the evaluator to the ith training attribute to reflect its importance; V_i is an ordinal scale ranking assigned by the evaluator to the *i*th cost attribute to reflect its importance; A_{ij} is the value assigned to the attribute by ASTAR; X_{ij} is the traditional cost factor attributes; p is the number of attributes common to all alternatives; and q is the number of alternative device configurations under consideration.

The value in using ASTAR in combination with traditional financial justification methods is the rapid process which can be used to obtain training and cost-effectiveness information about one or a few tasks of training devices. When one task is being examined, only those questions in ASTAR which relate to that task need to be answered.

It is desirable to relate the ASTAR results with cost in absolute terms. A melding of two approaches is suggested to bridge the gap between the subjective world of training effectiveness and the quantitative world of cost effectiveness. One approach is to use a heuristic in combination with weighted evaluations drawn from subjective and objective evaluations of alternatives. Weighting factors could be based upon what level of ASTAR information is available combined with a judgment of the heuristic values used in the analysis.

Another approach to relate ASTAR to cost is to use utility theory and assess the utility of the ability of a training device to meet training goals. The value of utility theory is the ability to assess the value of an attribute of a training device as compared to other attributes. This methodology is typically used in systems engineering processes, where one seeks an optimal design to a given set of requirements. The drawback of utility theory is the subjective nature of the evaluation process. One evaluator's opinion of weighting or scaling factors for a particular attribute could be very different from another evaluator's assessment.

Cost justification implies that sufficient rationale is available to support the cost-estimating process. However, the approach suggested above requires justification from two points of view: namely training effective-

TABLE 18.1 *VIGS*

Rank		Alternative 1		Alternative 2	
U_i	Attribute	A_{i1}	U_iA_{i1}	A_{i2}	U_iA_{i2}
1	Transfer efficiency	1.06	1.06	1.84	1.84
0.5	Acquisition efficiency	19.28	9.64	18.75	9.38
	$B_j =$		10.7		11.22
	$B_2 - B_1 = 0.52$				
1	Investment cost	8 000	8 000	10 000	10 000
0.5	Fixed costs	2 500	1 250	3 000	1 500
0.2	Variable cost	3 000	600	2 000	400
	C_j		9 850		11 900
	$C_2 - C_1 = 2050$				
	$E_j = B_2 - B_1 = 0.0253$				

ness and cost justification. Each of these points of view can be treated individually from a quantitative measurement point of view. In each case historical data provides the best basis for justification.

Databases for historical data must be accumulated to support cost justification. The historical cost data should be classified into groups which parallel the ASTAR costing attributes. In this way a database is created which grows with existing data. Ideally, cost reporting would be in a similar form to provide an expanding database.

The difficulty in justification methods is when factors do not match the existing data exactly or when extrapolation is necessary. In these cases, an appropriate regression technique is in order to extend current data. Justification also requires validation to assure appropriateness of data.

18.3.1 Example

This training/cost effectiveness model could be used to examine the VIGS (Video Disk Interactive Gunnery Simulator) in more detail. The objective of VIGS is to create a ballistically correct training device to teach a trainee the correct manual procedures to acquire and engage targets. Two important considerations with the current VIGS is that the targets are not interactive and the scenarios are canned, but may be preprogrammed.

If there is a need to improve a part of the visual system of VIGS, the methodology presented in this paper could be used to determine the 'best' of these mutually exclusive alternatives in terms of training and cost effectiveness during the developmental stage.

First, the tasks involved with the part of the visual system need to be identified and appropriate attributes selected. Once the attributes are selected and all changes in the training effectiveness for each alternative are specified, ASTAR can be run for each alternative. At the same time, attributes for the design and manufacturing cost factors for each of the alternatives would be calculated. Since only a partial task is being determined, in this example, some of the cost factors, such as project utilization, will be the same and can be ignored in this analysis. The attributes for the training effectiveness and cost effectiveness for the task are given in Table 18.1. The weights from the ordinal ranking scale estimated by one evaluator are also listed. (Data is hypothetical.)

In this example, the transfer efficiency for this visual task is more important than the acquisition, and is ranked higher. The initial investment cost is considered to be very important and is given a weight of one. There is an increase in the service contract for this improvement and the service attribute must be considered as a necessary function. There is an unknown factor in the service contract and each contractor has estimated possible additional costs which are included with the variable cost attribute, but are considered basically unimportant in the scale weights.

The benefits increment of the mutually exclusive alternatives $(B_2 - B_1)$ is given as 0.52. The cost increment $(C_2 - C_1)$ is 2050 and the benefit-cost ratio (E_{2-1}) is 0.0253. Since this is less than one, the increment of the additional investment is not justified in this case. Once the initial analysis is completed, a sensitivity analysis may be performed by changing the ordinal scale weights for ranking.

18.4 Conclusion

This paper has presented a methodology to assist the decision maker in evaluating the training effectiveness and cost effectiveness in developing and manufacturing training devices using the software package, ASTAR. This may be used on an entire training device, or a task of a training device or a subtask of that device. Although the approach presented provides an adequate means of adding a cost component on to the ASTAR program, several improvements are worthy of further investigation. The first, and easiest improvement to implement is to deal with cost in relative terms. It is suggested that marginal increases or decreases in cost, as appropriate, to a total training device cost be estimated for a given substitute subsystem. However, it is also desirable to develop a quantitative measure of cost per ASTAR point.

It is recommended that input data be investigated for use instead of output scores to develop a cost model. This recommendation is based upon several factors. First, the methods used to obtain ASTAR scores are not provided in ASTAR documentation. Accepting scores without knowing their source is risky. Second, it appears that the questions asked during an ASTAR database development session could lend themselves to the application of utility theory or the weighted evaluation method. Third, benefit-cost analysis is essentially the only type of economic analysis available when using ASTAR outputs. Benefit-costs analyses are useful, but absolute cost is another desirable measure when performing an economic analysis of a training device. Use of input data could provide some measure of absolute cost. Fourth, use of input data could aid in the development of a heuristic which would be useful in training-device cost estimating if a database of utility factors were maintained and made available to evaluators. As stated previously, although benefit-cost ratios are desirable, only the incremental change in the benefit-cost ratio can presently be determined using the method described by Morse (1989). This is quite satisfactory when several mutually exclusive alternatives are under consideration and the most advantageous alternative is desired. This approach has been employed in a general sense with respect to corporate decision-making processes by Parsaei and Wilhelm (1989). However, if different cost data are necessary to determine a true benefit-cost ratio, it might be better to use that cost information to determine the absolute cost of designing and manufacturing a training device.

18.5 References

American Institutes for Research, (1988) *ASTAR User Manuals*, July.

Canada, J.R. and Sullivan, W.G. (1989) *Economic and Multiattribute Evaluation of Advanced Manufacturing Systems*, Prentice Hall, Englewood Cliffs, New Jersey.

Martin, M.F., Rose, A.M. and Wheaton, G.R. (1988) *Applications for ASTAR in Training System Acquisitions*, American Institutes for Research, Washington, DC.

Morse, L.C. (1989) Determining the cost and training effectiveness of training devices. *Proceedings for Society for Integrated Manufacturing Conference*. Atlanta, Georgia, November.

Parsaei, H. and Wilhelm, M.R. (1989) A justification methodology for automated manufacturing technologies. *Proceedings of the Eleventh Annual Conference on Computers and Industrial Engineering*. Orlando, Florida, March.

Rose, A.M. and Martin, M.F. (1988) *Implementation of ASTAR: Evaluation of the Combat Talon II Maintenance Trainer*, American Institutes for Research, Washington, DC.

Singer, M. and Sticha, P. (1987) Designing training devices; the optimization of simulation-based training systems. *Proceedings of the 9th Interservice/Industry Training Equipment Conference*. Washington, Florida, November.

Sullivan, W.G. (1986) Models IEs can use to include strategic, non-monetary factors in automation decisions. *Industrial Engineering*, **18** (3), 169–76.

19 On Safety of Workers in FMS Environments

Suebsak Nanthavanij

19.1 Introduction

In today's manufacturing systems, a high degree of automation, high reliability in operations, and low maintainability contribute significantly to an increase in productivity. Trends in manufacturing are now moving toward computer-integrated systems which use computers to plan the production schedules and control the operations of machines. Among those, several manufacturing industries have found flexible manufacturing systems (FMSs) to be most promising. This is due to the fact that an FMS is able to automate most of the manufacturing processes, including the handling, storing, and retrieving of materials between workstations and storage areas. Several functions which used to be performed by human workers are now handled by machines as one result of these trends. However, while jobs are preferably assigned to machines where applicable, a number of workers are still found on the factory floor and they perform certain important functions such as installing and programming the machines, monitoring their operations, maintaining them, and repairing the malfunctioning equipment. Such preference in job assignments is also believed to reduce human errors which sometimes lead to injury-causing accidents. Here a question is raised whether the safety of workers in the automated systems is more than that in the traditional ones.

In a traditional manufacturing system where workers and the system's other components interact on a continuous basis or at regular intervals, workers are usually exposed to and aware of hazardous situations. As time progresses, they learn how to avoid and handle these situations. In other words, they may be able to develop subconsciously what is known as **hazard awareness**. For an FMS, such high automation level leads to what is believed to be a lower exposure of workers to hazards. That in turn reduces their safety consciousness and makes the hazardous situations less recognizable to them (Nanthavanij and Abdel-Malek, 1990).

An FMS introduces new types of machines, different work environments, and different job functions to workers. It is not clear whether the protective and warning devices commonly used in the traditional manufacturing systems can provide similar protection to workers in these new automated systems. This paper investigates several worker–machine interactions in FMS work environments which may be hazardous and capable of producing injury-causing accidents. The paper is organized as follows; it begins by addressing an economic influence of workers' safety on manufacturing industries; then, it discusses some protective and warning devices currently used in FMS components' work areas; and finally, recommendations are presented regarding an improvement of those devices and appropriate precautions which should be taken to secure a minimum level of hazards.

19.2 Safety and its economic influence

In order to emphasize the economic influence of workers' safety it is necessary to discuss the related safety regulations, penalties and costs involved when accidents occur (e.g. cost of investigations, compensation, insurance premiums, etc). This section describes the history of the United States' safety regulations (OSHAct and workers' compensation law) and their enactments. The practitioners should be aware that each country has its own safety regulations and, as a result, the levels and costs of penalty can differ considerably from what is shown in this section. Nevertheless, it is to our belief that the economic impact of safety is still significant and cannot be neglected when justifying the FMS economically.

Historically, an injured worker whose injury resulted from employment could obtain indemnity only by proving the employer's negligence. The process of proving such negligence was generally difficult and it could take years before the court reached a final decision. Prior to enactment of workers' compensation laws, workers had been protected under common law which obligated their employers to provide them with safe places to work, safe tools, knowledge of any hazards which were not immediately apparent, competent supervisors (and fellow workers), and safety rules (Hammer, 1989). Nevertheless, under the same law, the employers had three powerful and successful defences which usually disqualified the injured workers from being indemnified. They were: 1. the defence of contributory negligence; 2. the defence of assumption of risk; and 3. the fellow servant rule (Grimaldi and Simonds, 1989). In 1902, the US Congress passed the first workers' compensation law which required the employer to compensate the injured worker whether or not negligence could be proved. Currently there are 57 workers' compensation laws, one for each of the individual 50 states, the District of Columbia, Guam, Puerto Rico, the Virgin Islands, and three for the federal government.

Each of the laws, however, is different in terms of its interpretation, legal provisions and coverages, penalties and benefits (Hammer, 1989).

The costs of workers' compensation are paid for by employers. Depending on the requirements of individual states, employers are required either to obtain insurance from state funds, to use a private insurance company, or to have self-insurance (if they are qualified). These costs could impose significant financial burdens to employers who have a high record of work-related injuries and may eventually force them to go out of business. When workers' compensation law was initially enacted, it was believed that such compensation costs would motivate employers to institute safety programmes and provide safe working conditions for their employees in order to reduce work injuries to an irreducible minimum. This enactment was found to be insufficient since some employers assumed that their purchase of insurance would absolve them from their safety responsibilities. This obstacle was later overcome in part by an incentive plan which allowed employers to use an experience rating, the procedure most commonly used today, to determine the compensation rate. With this plan, employers paid only the cost of the injuries they experienced during some past few years, plus a service charge to the insurance carrier for administering the indemnification process (Grimaldi and Simonds, 1989).

In 1970, the US Congress enacted the Williams–Steiger Occupational Safety and Health Act (OSHAct) which became effective in 1971 (Hammer, 1989). The act requires employers to provide a safe and healthful work environment. Under this act, one of the employer's duties is to furnish to each employee a place of employment that is free from recognized hazards. Failure to comply with this act would result in receiving citations and subject them to pay for penalties depending on the level of violations. Four types of citations are possible: imminent danger, serious violation, nonserious violation, and *De minimis*. The monetary charge varies with the type of penalties, whether civil or criminal, and can range between $1000.00 per day or violation and $20 000.00 per conviction or violation (Grimaldi and Simonds, 1989).

Without doubt, both workers' compensation laws and the OSHAct put financial burdens on manufacturing industries whose work environments are found to be unsafe and have recognized hazards. Instituting a good safety programme, providing safe places to work, and safe tools to work with can help to reduce this burden and motivate workers to work more efficiently; thus, increasing overall productivity.

When a work environment is hazardous, an accident (either injury or no-injury type) tends to occur. Even though that accident does not cause any injury to the workers, its occurrence can be costly. This may be due to those costs arising from property damages, repairs, loss of working time, and production-line interruption. For an injury-causing accident, additional costs such as medical expenses, investigation costs, workers' compensation, and rehabilitation expenses (if necessary) may be found to accompany the above mentioned.

Basically, there are two major classes of costs resulting from injuries and accidents. They are the insurance cost and the uninsured costs. The insurance cost is the cost of workers' compensation insurance which the company has to pay to its insurance agency. On average, this insurance premium will be large enough to cover the money paid for medical expenses on compensatable cases and compensation to the injured employees. The uninsured costs, previously referred to as the indirect costs, are those not included in the insurance cost. Grimaldi and Simonds (1989) listed several uninsured cost elements as follows:

1. Cost of wages paid for working time lost by nearby uninjured workers;
2. Property and equipment damages and repair costs;
3. Cost of wages paid for working time lost by injured workers;
4. Extra cost due to overtime work necessitated by an accident;
5. Cost of wages paid supervisors while their time is required for activities necessitated by the injury;
6. Wage cost due to decreased output of injured worker after return to work;
7. Cost-of-learning period of new worker;
8. Uninsured medical cost borne by the company;
9. Cost of investigations and compensation application forms processing; and
10. Miscellaneous unusual costs.

These uninsured costs vary with the nature of injuries and accidents. However, a formula and techniques which may be applied to approximate them can be found in Grimaldi and Simonds (1989).

As the hazard level of a work environment increases, the uninsured costs increase, and, vice versa. In a manufacturing environment, especially FMS, the number of accidents can be reduced by eliminating hazardous situations or making them recognizable, safe-guarding machines, providing warning signs and signals, installing appropriate protective measures, and educating workers to be more safety conscious.

19.3 Safety of workers in the vicinity of FMS components

In this section, we shall confine our discussion to those situations where workers and possible causes of accident are found in the area within the machines' vicinities which is defined as the 'danger zone.'

19.3.1 Computerized numerical-controlled (CNC) machines

A CNC machine is an NC which uses a microcomputer for its control. This microcomputer is usually attached to the machine or, in many cases, is a separate unit which is located next to the machine. When a worker is needed at the CNC machine site to program its operations, one is found at this control unit. During programming and program editing, most CNC machines are in their nonoperational mode and considered to be relatively safe. Some sophisticated CNC machines have monitors which have a graphics presenting capability which enable workers to preview the machines' operations as programmed and correct mistakes without actually having to run them. For those which do not have this type of monitor, workers are required to run the machines through the whole program for at least one cycle in order to check for errors in programming. During the program testing, machines are in their operational mode (which may be at slow speed). Workers should be warned not to stay within the machines' work areas which may be indicated by painted lines (for example, yellow and black zebra striped), a fence or guard rails surrounding the CNC machines. If they are required to stay within those areas, designated locations should be assigned which are far from the machines' moving parts so that workers are in a safe situation. Additionally, emergency stop buttons should be installed on the machines and located where workers can easily reach and activate them for emergency stoppage. Warning of the machines' operations may also be provided by presenting one or more flashing lights.

For a typical CNC machine, all machining processes are usually performed in an enclosure which has a transparent window for viewing purposes. During the test, close visual inspections of the machine's operations are not recommended since workers may be tempted to open this window for better viewing. Though running at slow speed, the CNC machines could still eject high-speed metal chips which might strike and injure workers. Therefore this window must be kept closed during the machine's operational mode and should have an interlock switch so that when the window is open, the machine will automatically stop.

Once the program editing is completed, the CNC machines may be run at full speed. Workers are generally released from the control functions. They may, however, periodically be required to check the incoming part feeding mechanism and/or inspect the quality of finished parts while the machines are running. Any manual adjustment of the part feeders, including realignment of the parts, must be performed only when the machines are completely stopped. Additional emergency stop buttons should be provided near both part feeder and finished part disposal sites.

At the part feeder, an additional emergency stop bar (which would emit auditory warning signals when activated) should be installed. In the event that workers attempt to realign the incoming parts manually while the machines are still running, their hands or clothes may be trapped and drawn into the machines. In this situation, with these stop bars and the warning alarm installed, they are still able to stop the machines' operations by pushing these bars with their hip or knee. The accompanying warning signal also serves to alert fellow workers of such emergency situations. However, the use of these stop bars under normal situation should be avoided.

19.3.2 Robots and robotic cells

Basically, industrial robots fall into two categories: i.e. stationary and mobile. Of particular interest are the stationary ones since they are mostly found on the factory floor. Similar in many ways to the CNC machines, these robots require humans to program their operations and repair them when they malfunction. These tasks are usually performed when the robots are in their nonoperational mode. Surprisingly, the majority of accidents have been reported to occur during this mode where worker–machine interactions are considered to be relatively safe (Pearson, 1984). These unexpected accident statistics may be due in part to a reduction in workers' hazard awareness discussed earlier. During the robots' operational mode, workers are warned not to enter the robotic cells whose boundaries are indicated by wired mesh fences, painted lines on the floor, or some other types of physical barriers (Engelberger, 1980; Potter, 1983b). Additional protection is provided by having motion sensors installed and/or pressure-sensitive mats placed within the robotic cells to detect human intruders. Those state-of-the-art devices are needed since, in many cases, workers are found to enter the robotic cells for unknown reasons or carelessly intervene in the robots' operations; thus, leading to either nonfatal or fatal accidents. Several examples of robot-related accidents were discussed in Nagamachi (1988) and Ryan (1988). The applications and effectiveness of those protective and safety devices have been extensively discussed in the literature (Potter, 1983a; Potter, 1983b; Nanthavanij and Abdel-Malek, 1987; Parsaei *et al.*, 1987). Some suggestions for improvement can also be found in the work of Nanthavanij and Abdel-Malek (1988).

Programming the robots can be done using one of the following three procedures: lead-through, teach pendant, and slave mimic. The details of these procedures have already been discussed elsewhere (Jiang, 1987). Nanthavanij and Abdel-Malek (1988) also presented several situations where hazardous worker–robot interactions could exist during the teaching mode and suggested valid recommendations regarding both protective and warning devices. On teaching devices (pendants), Cousins (1988) addressed the lack of standardization among pendants and discussed a design standard entitled 'Human Engineering Design Criteria for Hand-Held Robot Control Pendants'. It is obvious that during the teaching and maintenance modes, some safety devices such as motion sensors and pressure-sensitive mats are deliberately deactivated so that workers in the repair crew can enter and perform their functions. Therefore, their safety relies heavily on their degree of care in performing jobs and their knowledge of existing hazards. The following paragraph discusses one situation which may be hazardous, yet is not recognized as such by the workers.

Let us consider a case where the failure of the robots' power causes them to stop their operations. The workers who are monitoring these robots may assume that this stoppage is due to the malfunction of the robots' mechanism. Consequently, they may deactivate the safety devices and enter the robotic cells to check for possible causes of the malfunction. If this power failure is temporary, the robots could momentarily resume their operations while the workers are still within their work envelope and this could injure them. This mishap can be avoided if a warning light is used to indicate the power failure situation. The warning light and its label should be placed at the entrance to the robotic cell so that entering workers would be made aware of the situation. Furthermore, the robots may be programmed so that when their routine operations are interrupted either by temporary power failure or system malfunction, they can only be restarted manually. The location of those restart buttons should be outside the robotic cells, for safety reasons.

19.3.3 Automated guided vehicles (AGVs)

Automated guided vehicles are generally found on the factory floor of most FMS facilities. They serve as a means of transporting materials from one workstation to another and between storage areas and workstations. An AGV usually moves along its pathways which are defined using wires embedded in the floor or reflective paint on the floor surface. Due to its mobility, the danger zone of an AGV is not confined to one specific location; instead, it moves along with the AGV. In order to provide adequate safety to workers on the factory floor, several types of safety and warning devices must be installed on the AGVs and along its routes as well.

Virtually all of today's commercial AGVs are required to have emergency bumpers surrounding the front of the vehicles. When this bumper makes contact with an object, human, or another AGV, the resulting impact would activate the vehicle's emergency brake and simultaneously cut power to the driving motors. Typically, an AGV moves at a very slow speed to avoid loads falling off the vehicle and to minimize the severity of injury to workers who may be struck by the vehicle. However, one should be aware that the AGVs which carry heavy loads, even though they travel at slow speed, can build up considerable momentum at the time of impact – possibly enough to cause a severe injury. Such physical contact is therefore undesirable and should be avoided with all efforts. Workers should be warned of the vehicles' existence, their defined routes, and their approach. In the area of warning devices, a flashing or rotating light is usually installed on the vehicle. An additional auditory warning alarm is sometimes found on some AGVs and it emits signals to indicate the existence of the vehicle. Egawa (1988) investigated several types of warning alarms for AGVs and suggested the use of 'continuous sound with periodicity' for indicating traveling AGVs.

Consider a case where a worker stands on the AGV's route and he/she is not aware of the vehicle's approach. Though equipped with an auditory warning device, the alarm may be masked by the noise from machines on the factory floor. That worker could then possibly be hit by the vehicle. To avoid a physical contact, some AGVs are equipped with non-contact obstacle sensors. These sensors would send signals to an on-board computer which subsequently controls the AGV by ordering it to either slow down or stop. It is obvious that the stopping distance can vary from several inches to several feet, depending on the AGV's speed, its payload, floor surface, and other factors. These factors should be taken into consideration along with the design characteristics of the sensors to assure a complete stop of AGV before making contact with an obstacle.

The sensors which are used to detect obstacles are based on signals which may be ultrasonic, photoelectric or optical. Basically, a transmitter emits signals on a continuous basis which bounce back when they hit an object. Upon receiving signals, an on-board computer controls the AGV by ordering it to either slow down or stop. To improve the sensors' effectiveness and the safety of workers, we suggest the following.

When the sensors detect an obstacle, the on-board computer would compute the stopping distance D_s of the vehicle from its instantaneous speed and payload, and compare it to the distance between the vehicle and obstacle D_1 at that time. It is essential that this D_s must be somewhat greater than the actual stopping distance of the vehicle to account for the variables which may not be included in its determination. If D_s is less than or equal to D_1, the AGV should be ordered to stop immediately; otherwise, it gradually decreases its speed while the warning device increases the intensity of the alarm. These two processes (a decrease in AGV's speed and an increase in alarm's intensity) should be designed so that they take into account the AGV's speed and the distance between the vehicle and the obstacle, and continue sounding the alarm at an increasing level as the distance decreases. If the obstacle is nonhuman, the AGV, as a result of this continuous decrease in its speed, would come to a complete stop without making contact. The warning alarm would notify somebody to relocate this obstacle and manually restart the AGV so that it could proceed to its destination. In the case of workers, an increase in signal intensity would alert them of the AGV's approach and qualitatively indicate how close the vehicle is to them. Further, it serves to overcome the possible masking effect from noise on the factory floor. A decrease in AGV's speed would also allow them enough time to move out of its route. When the sensors no longer detect any obstacle, the AGV would then be ordered to resume its normal travelling speed. The flow chart describing this safety system is given in Fig. 19.1.

Since the pathways of the AGVs are usually defined and remain fixed, it is possible to indicate these travel zones by drawing lines on the floor along the AGV's routes. Workers should also be advised to avoid travelling or standing in these hazardous zones. Fencing the AGVs' pathways, though appearing to be the most effective, is not appropriate in practice since fences are costly and occupy useful floor spaces.

19.3.4 Storage/retrieval (S/R) machines

An S/R machine is a mechanism which travels back and forth between storage aisles and materials handling equipment (e.g. conveyors and AGVs) to store and retrieve materials. S/R machines are generally included as a part of an automated storage and retrieval system (AS/RS). Though various categories of AS/RS exist in the FMS environments, let us focus on the system which utilizes man-on-board S/R machines to transfer materials since they require workers to ride along on or near the carriage of these machines.

Basically, an S/R machine is capable of horizontal and vertical travels along the AS/RS aisle with a horizontal speed up to 500 ft/min and a vertical speed up to 100 ft/min (Groover, 1987). Loads are placed on the carriage of the machine and deposited into (or extracted from) the storage compartments using a shuttle mechanism.

For the man-on-board type, workers are required to travel with the S/R machines. Considering the machines' relatively high travelling speeds and the height of storage aisles, onboard safety and warning devices must be installed to assure the workers' safety. By taking both acceleration and deceleration of the machines into account, it is clear that the workers' stance would be unstable. Therefore, the platform should be fenced with guard rails to prevent workers from falling off the S/R machines. The gate of the fence should feature an interlock and be designed so that the S/R machine can only operate when the gate is properly closed and locked. A safety belt or harness fastened by a line to a secure anchorage

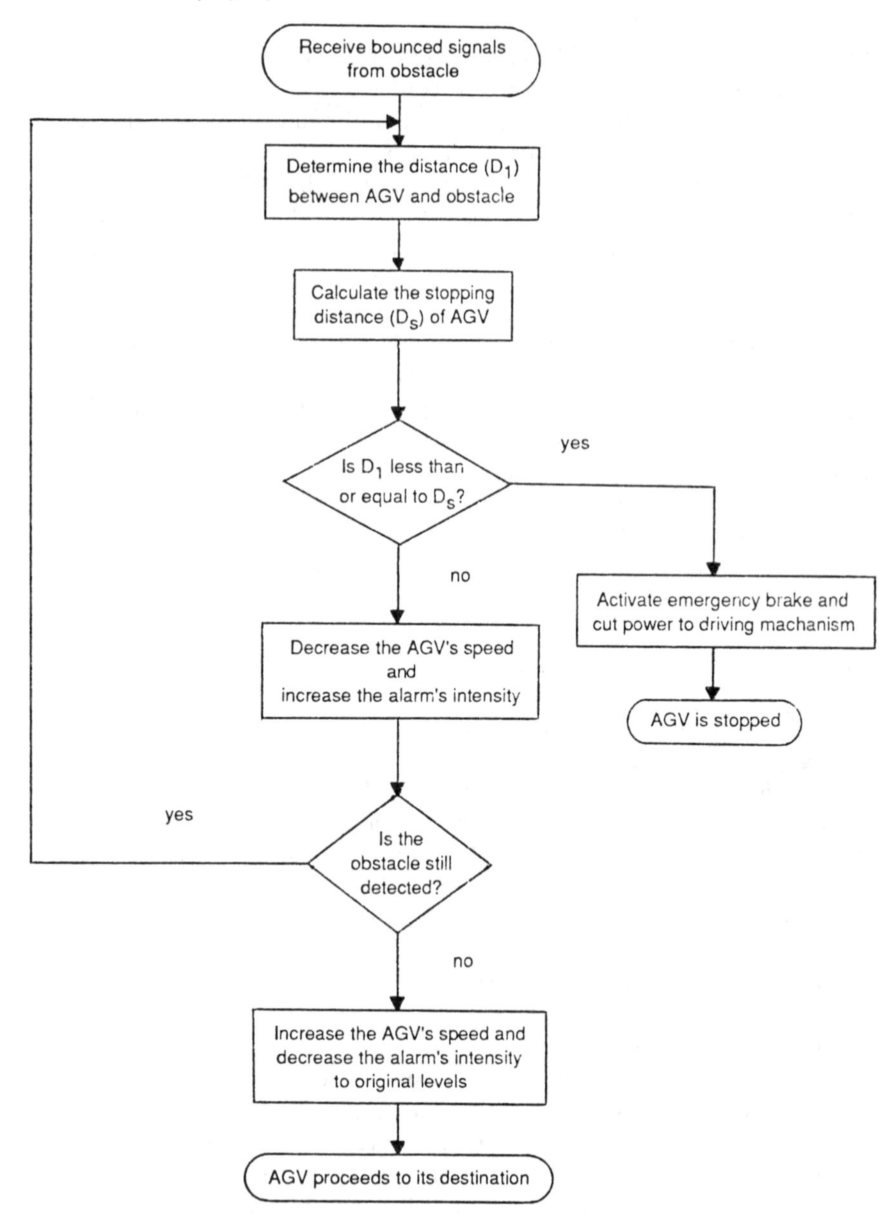

Fig. 19.1 *Flow chart of an AGV's safety system for avoiding vehicle–obstacle collisions.*

can also be used to provide additional protection. The length of this belt should be relatively short in order to secure and stabilize their positions. Emergency stop buttons should be installed at several convenient locations for emergency stoppage.

Visual warning devices should be installed both on S/R machines and the floor at the entrance to storage aisles. These devices serve to warn other workers on the factory floor of the machine's operation and its presence in certain active aisles. This indication of activities within storage aisles is necessary since other workers may not be aware of the situation and enter the aisles while S/R machines are operating.

19.4 Conclusion

This paper discusses the safety issues of workers in FMS environments and some economic influence of safety in manufacturing industries. It also addresses the possible lack of workers' hazard awareness as a result of a change from traditional manufacturing systems to those with a higher level of automation. A reduced level of safety consciousness there could lead to injury-causing accidents and may explain in part surprisingly high accident statistics in today's manufacturing systems despite the fact that fewer functions are now being performed by workers. In FMS environments, a number of protective and safety devices must be used together to secure a minimum level of hazards. The discussion and recommendations presented thus far can be summarized as follows.

During the test runs of CNC machines, close visual inspection of their operations should not be permitted or carried out with their transparent enclosures improperly closed. At locations such as part feeding and disposal sites where workers sometimes intervene in machines' operations, additional stop bars with accompanying auditory alarms should be installed for emergency situations.

For robots, an interruption in their routine operations due to temporary power failures should be distinguished by an indicating warning light. To resume operations, manual restart of the robots is highly recommended in order to assure workers' safety.

Regarding the application of AGVs, we have suggested an algorithm which describes simultaneous processes for decreasing an AGV's speed and increasing the intensity of its warning alarm when on-board sensors detect obstacles. This safety system will prevent any physical contact between the vehicles and the obstacles (whether they are objects or workers).

In order to secure or stabilize workers while traveling on man-on-board S/R machines, fence or guard rails with interlocked gates, along with safety belts should be used. An active aisle's indicator is also recommended to alert other workers and prevent them from entering.

In summary, several hazardous worker–machine interactions are still found in the workplace. Those which may not be yet recognized by workers should be revealed even though their possibilities of occurrence are remote. By realizing their existence, workers may more readily learn to handle those situations and protect themselves from being injured. Any economic evaluation of advanced manufacturing systems performed by excluding the safety issues should be considered incomplete since their impacts are too significant to be ignored. It is our objective to point it out, share this area of investigations with other practitioners in the area of safety, and contribute to an enhancement of the safety of workers in these new and promising manufacturing systems.

19.5 Acknowledgements

This work was supported by a separately budgeted research grant from the Division of Sponsored Programs, New Jersey Institute of Technology. Dr Howard Gage is specially appreciated for his time and effort in reading this paper and providing invaluable comments.

19.6 References

Cousins, S.A. (1988) Development of a human engineering design standard for robot teach pendants. In *Ergonomics of Hybrid Automated Systems I* (ed. W. Karwowski, H.R. Parsaei and M.R. Wilhelm), Elsevier, 429–36.

Egawa, Y. (1988) A study of auditory warning alarms evaluation for automated guided vehicles. In *Ergonomics of Hybrid Automated Systems I* (ed. W. Karwowski, H.R. Parsaei and M.R. Wilhelm), Elsevier, 529–36.

Engelberger, J.F. (1980) *Robotic in Practice: Management and Applications of Industrial robots*, Amacon.

Grimaldi, J.V. and Simonds, R.H. (1989) *Safety Management*, 5th edn, Irwin.

Groover, M.P. (1987) *Automation, Production Systems, and Computer Integrated Manufacturing*, Prentice-Hall.

Hammer, W. (1989) *Occupational Safety Management and Engineering*, 4th edn, Prentice-Hall.

Jiang, B.C. (1987) Safety considerations for programming industrial robots. In *Trends in Ergonomics/Human Factors IV* (ed. S. Asfour), Elsevier, 465–72.

Nagamachi, M. (1988) Ten fatal accidents due to robots in Japan. In *Ergonomics of Hybrid Automated Systems I* (ed. W. Karwowski, H.R. Parsaei and M.R. Wilhelm), Elsevier, 391–6.

Nanthavanij, S. and Abdel-Malek, L. (1987) On safe design of robotic cells, *Proceedings of the IXth ICPR Conference*, 2720–5.

Nanthavanij, S. and Abdel-Malek, L. (1988) Ergonomic evaluation of safety devices in robotic systems. In *Ergonomics of Hybrid Automated Systems I* (ed. W. Karwowski, H.R. Parsaei and M.R. Wilhelm), Elsevier, 753–60.

Nanthavanij, S. and Abdel-Malek, L. (1990) Human factors evaluation for protective measures in FMS. *Proceedings of the Manufacturing Management in the Nineties Conference*, 291–302.

Parsaei, H.R., Wilhelm, M.R. and Karwowski, W. (1987) Safety monitoring strategies in the design and utilization of robotic work cells: a review. *Proceedings of the IXth ICPR Conference*, 2726–31.

Pearson, G.W. (1984) Robot system safety issues – best considered in design phase. *Occupational Health and Safety*, September, 38–41.

Potter, R.D. (1983a) Sensing devices extend applications of robotic cells. *Industrial Engineering*, **15** (3), 24–32.

Potter, R.D. (1983b) Requirements for developing safety in robot systems. *Industrial Engineering*, **15** (6), 21–4.

Ryan, J.P. (1988) Safety considerations in robot design. In *Ergonomics of Hybrid Automated Systems I* (ed. W. Karwowski, H.R. Parsaei and M.R. Wilhelm), Elsevier, 483–90.

Part Five

Economic Analysis and Justification Resources

20 Information Solutions to Advanced Production and Manufacturing Systems (APMS): A Discipline Impact Factor Analysis

Janardan Kulkarni, Hamid R. Parsaei and Fariborz Tayyari

20.1 Introduction

The concept of advanced production and manufacturing systems (APMS) and its allied components including computer-aided design (CAD), computer-aided process planning (CAPP), group technology (GT) and just-in-time (JIT), has created a new challenge to manufacturing firms around the world.

Advanced production and manufacturing system in brief is the integration of all activities, technologies, and information flow, required to manufacture product through the use of computer technology.

While manufacturing technology is becoming more sophisticated and its benefits numerous, as has been reported in literature, its adoption and successful implementation is yet a challenge. This is mainly due to a lack of: 1. commensurate development, and 2. relevant use of appropriate economic justification procedures (Parsaei and Wilhelm, 1989). It has often been reported aptly that the implementation of these new technologies requires a significant amount of investment, top management commitment and employee involvement.

Some of the common objectives in implementing advanced production and manufacturing system (APMS) include productivity improvement, reduction of work in process inventory, better and safer working conditions, better quality products, and quick response to changing consumer needs.

Traditional discounted cash flow for many years has been used as the only financial decision technique to evaluate the desirability of advanced manufacturing technology projects. These conventional methods including net present value (NPV), internal rate of return (IRR), payback period, and benefit-cost ratio are single objective and often concentrate on easily quantifiable factors such as labour and material (Parsaei and Wilhelm, 1989). Thus, the main benefits such as high flexibility, better product quality, and safer working conditions are often ignored by these methods.

20.2 Methodology

In order to assess the retrieval effectiveness of different kinds of indexing language and their behaviours in online bibliographic information systems, we selected the topic of advanced production and manufacturing systems. Three kinds of search vocabularies were selected for use in the test (Feng, 1989).

1. *Natural Language (N)*. This is a language composed of simple concepts when the search request is formulated without the use of thesauri, dictionaries and similar tools.
2. *Controlled vocabulary (C)*. All words or phrases are selected from the thesauri.
3. *Natural/controlled language (NC)*. This is a mixture of both terms in 1 and 2 above, and many others.

Using the above three concepts, we formulated the thesaurus described later. Various databases were searched using the terms described in the thesaurus. Each database allows Boolean logic searching, word truncation and proximity searching. The results of retrieval were evaluated according to two well-known criteria. In this test, the relevance of a citation, with respect to a search topic, was judged on the basis of the entire context of a retrieved record, and the degree of relevance was assessed at three levels (Blair, 1986).

1. Relevant (R);
2. Partially relevant (P); and
3. Irrelevant (I).

Precision has been defined as the percentage of retrieved documents that are relevant, (Feng, 1989).

$$\text{Precision} = \frac{(\text{number of relevant citations retrieved}) \times 100}{\text{total number of citations retrieved}}$$

$$= \frac{(\text{number of } R \text{ retrieved} + \text{number of } P \text{ retrieved}/2) \times 100}{\text{total number of retrieved } (R + P + I) \text{ citations}}$$

Recall has been defined as the percentage of relevant documents retrieved, i.e.:

$$\text{Recall} = \frac{(\text{number of relevant citations retrieved}) \times 100}{\text{total number of citations retrieved}}$$

$$= \frac{(\text{number of } R + P/2) \text{ retrieved} \times 100}{\text{total } (R + P/2 \text{ retrieved}) \text{ citations}}$$

Our aim is to present a model of comprehensive sources for generating the scientific and technical information in the field of advanced production of manufacturing systems (APMSs). To this end we used computer-based information-retrieval systems or databases as they are frequently called. These computer databases have been organized according to their subject coverage on the lines of their printed counterparts, i.e. they are mostly discipline oriented, however, one will see that APMS falls into a multidisciplinary category. The coverage of a specific topic of a multidisciplinary nature, therefore, can only be improved by accessing not one but several databases. Use of this method, resulted in generating duplicate references. Therefore, the knowledge, scope, coverage and the quality of the databases are very important when you are doing a computer search.

20.3 The plan

Nine technical databases were searched on DIALOG Search System, a subsidiary of the Lockheed Corporation, using the algorithm given in the next paragraph. The names of the databases are: COMPENDEX, EI Meetings, INSPEC, NTIS, Harvard Business Review, Japan Technology, CAD/CAM, Books in Print, Dissertation Abstracts, Conference Paper Index and Supertech. They carried the best coverage of the literature on the advanced production and manufacturing systems (APMSs) subject searching. Table 20.1 shows the precision and recall values for various databases. A total of 438 citations, which are listed in the Appendix, were retrieved as described in Table 20.2.

The subject descriptors were qualified to the title field (TI), descriptor field (DE) and identifier (ID) field. Since the databases vary in size, the output also varied for the time period searched. APMS is a relatively new interdisciplinary research area so a multidisciplinary and multi-year subject search was required to yield a reasonable sample size. Each

TABLE 20.1 *Precision and recall values of databases (Feng, 1989)*

Database	Relevance level	No. of citations retrieved	No. of citations retrieved (divided by searching vocabularies used)		
			NC	C	N
ISMEC	R	3	2	0	1
	R	7	6	0	1
	P	8	8	0	0
	I	5	5	0	0
	I	1	1	0	0
	R	25	8	0	2
INSPEC	R	14	11	0	2
	P	4	7	0	1
	I	5	4	0	2
	I	1	1	0	0
	R	27	17	0	0
Harvard Bus. Rev.	R	34	24	0	0
	P	14	14	0	0
	I	16	16	0	0
	I	9	9	0	0
COMPENDEX	R	22	22	4	4
	R	25	25	5	5
	P	16	6	2	2
	I	10	10	1	1
	I	7	7	0	0
	R	14	13	2	2
CAD/CAM	R	15	14	2	2
	P	3	2	0	0
	I	3	3	0	0
	I	2	2	0	0
	R	8	4	2	2
NTIS	R	9	5	2	2
	P	3	3	1	1
	I	2	2	2	2
	I	1	1	0	0
EI Meetings	R	14	2	0	0
	R	8	6	1	1
	P	9	9	1	1
	I	15	15	7	7
	I	47	47	7	7
	R	14	14	3	3
Japan Technology	R	16	16	4	4
	P	5	5	2	2
	I	4	4	1	1
	I	2	2	0	0
	R	17	16	3	3
Supertech	R	22	21	3	3
	P	11	11	0	0
	I	14	14	0	0
	I	9	9	0	0

R = Relevant
P = Partially Relevant
I = Irrelevant

TABLE 20.2 *Distribution by document type*

Document type	No. of articles	Cum. articles	%
Journal	208	208	54.3
Conference, symposium, seminars, colloquium	151	359	34.4
Technical report	45	404	10.2
Books	28	432	6.3
Dissertion/thesis	6	438	1.3

concept of the economic justification of APMS was viewed as a frame in a frame-based system. A set of concepts in which each concept is represented with at least synonymous terms, broader concepts, narrower concepts, and related concepts formed a thesaurus of terms or a set of controlled vocabulary. This thesaurus formed a basis of our search strategy.

20.4 Thesaurus

Assembly automation economics
Benefit-cost analysis
Capital budgeting
Capital investment analysis
Computer-aided manufacturing (CAM)
Computer-integrated manufacturing (CIM)
Cost analysis
Cost effectiveness
Cost estimating
Costing methods
Decision analysis
Economic comparisons
Economic justification
Economic models for advanced manufacturing
Economic models for planning systems
Economic valuation
Economics of robots
Engineering-economic analysis
Evaluation
Expert systems in economic analysis
Expert systems in equipment selection
Factory automation economic justification
Financial methods
Flexible manufacturing systems (FMS)
Investment decision
Investments
Long-range planning
Managerial economics
Manufacturing costs
Manufacturing flexibility
Manufacturing systems
Mathematical modelling of FMS
Non-traditional justification
Production systems for computer-aided manufacturing
Productivity measurement weighted evaluation
Quantitative economic analysis
Risk analysis
Selection of manufacturing systems
System justification

20.5 The algorithm

For on-line searching of the databases, we use DIALOG. The following algorithmic logic was used to conduct the search.

FMS/TI, DE, ID
Flexible () Manufact? () System?
CAM/TI, DE, ID
Computer () Aided () Manufact?
Computer () Integrated () Manufact?
Economic? () Justification
Economic? () FMS
Factory () Automation () Economic?
Robotic? () Economic? () Justification
Engineering () Economic? () Analysis
Costing () Method? () ((FMS or Flexible () Manufacturing () System?))
Expert () system? () Economic () Analysis
Flexible () Automation () Economic () Justification
Planning () ((Flexible () Manufacturing () System? () or FMS))
Cost () Estimating
Factory () Automation
Productivity () Improvements
Financial () Justification

20.6 Discipline impact factor (DIF)

First, a starting source or set of sources relevant to advanced production and manufacturing systems (APMSs) was selected. Call this source the citing set C. For each journal J cited by a source in C, we computed its DIF (the discipline impact factor) according to the formula (Hurst, 1978).

$$DIF = nc/ns$$

where: *nc* is the number of citations of J by sources in C over a time period *tc*; and *ns* is the number of citable items published by J over a time period *ts*.

Journals ranked according to their impact factor may offer a useful indication of the key sources for a specific field. To present a ranked list of sources specific to advanced production and manufacturing systems (APMSs), a two-step procedure was used. First, a pool of key sources of potential contributions was identified. This is known as a **selective journal set**. Second, a weighing score was computed for each journal in the **selective journal set** or **seed journals** (Table 20.3).

The ranking score of a source, based on citation influence of a selected discipline source set has been suggested by Hurst (1978). This source set

TABLE 20.3 *Seed journals*

Rank	Journal title	No. of articles	Cumulative articles
1	Harvard Business Review	16	16
2	Production Engineering	12	28
3	CIM Review	7	35
4	Industrial Engineering	7	42
5	American Metal Market	5	47
6	Engineering Economist	5	52
7	Production	5	57
8	Tooling and Production	5	62
9	Annals of CIRP	4	66
10	Journal of Manufacturing Systems	4	70
11	Robotics Today	4	74
12	Computing Canada	3	77
13	Industrial Robotics	3	80
14	International Journal of Production Research	3	83
15	International Labor Review	3	86
16	Modern Material Handling	3	89
17	Robotics & Computer Integrated Manufacturing	3	92
18	American Machinist	2	94
19	Chief Executive	2	96
20	Computer World	2	98
21	Economic Bulletin for Europe	2	100
22	Engineering Costs & Production Economics	2	102
23	Financial World	2	104
24	FMS Magazine	2	106
25	IIE Transactions on Engineering Management	2	108
26	Industrial Week	2	110
27	Journal of Post Keynesian Economics	2	112
28	Journal of System Management	2	114
29	Management Review	2	116
30	Manufacturing Systems	2	118
31	Mekhanika and Avtomatika Provizvid	2	120
32	National Productivity Review	2	122
33	New Scientist	2	124
34	Robotics	2	126
35	Robotics World	2	128

may be taken from a list of frequently used titles with high impact factors or titles selected by experts in the field. The discipline influence score (DIS) is computed as follows (He and Pao, 1986):

$$DIS_{\mathrm{A}} = \sum_{i=1}^{n} \frac{\text{number of times } J_i \text{ cited journal A}}{\text{total number of times } J_i \text{ cited all journals}}$$

which is the total sum of probabilities that journal A would be cited by a group of journals considered relevant to discipline, where: DIS_{A} is the

TABLE 20.4 *List of the top sources ranked by expert evaluation and Impact Factor (He and Pao, 1986)*

Journal title	Rank/expert evaluation	Impact factor	Discipline influence score
Harvard Business Review	1	*	1
Production Engineering	18	*	3
CIM Review	4	9	18
Industrial Engineering	7	*	8
American Metal Market	28	1	*
Engineering Economist	2	*	4
Production	27	*	*
Tooling and Production	17	*	*
Annals of CIRP	19	*	14
Journal of Manufacturing Systems	5	*	5
Robotics Today	16	13	*
Computing Canada	29	17	*
Industrial Robotics	20	*	*
International Journal of Production Research	6	6	*
International Labor Review	30	*	*
Modern Material Handling	15	3	*
Robotics & Computer Integrated Manufacturing	10	*	6
American Machinist	21	*	*
Chief Executive	26	*	*
Computer World	31	*	*
Economic Bulletin for Europe	32	18	*
Engineering Costs & Production Economics	3	8	7
Financial World	22	15	*
FMS Magazine	8	5	*
IIE Transactions on Engineering Management	9	*	*
Industrial Week	23	*	*
Journal of Post Keynesian Economics	33	4	*
Journal of System Management	11	*	*
Management Review	13	*	15
Manufacturing Systems	12	10	*
Mekhanika and Avtomatika Provizvid	35	7	*
National Productivity Review	14	*	*
New Scientist	34	*	*
Robotics	24	2	*
Robotics World	25	16	11

* Indicates rank greater than 20.

discipline influence source of journal A in the selective journal set; J_i is a member of the discipline journal set; and n is the total number of journals in the discipline journal set.

The discipline influence score (DIS) for the sources of information is given in Table 20.4.

20.7 Conclusion

The objective of this study was to demonstrate that for a specific discipline, an effective discipline-impact-factor selection algorithm could identify the key information sources in that field. These sources are often the basic science and engineering sources and multi-disciplinary sources. They are frequently cited titles. Authors of papers in these sources are the experts in the field. These sources frequently cite the publications in key journals of the discipline. The proposed algorithm incorporated both cited and citing sources, which in this instance was used to generate a balanced journal list for advanced production and manufacturing systems. Furthermore, this algorithm seems to be able to accommodate any discipline and to produce larger or smaller sets of sources depending on the needs. Experience has shown that this two-part algorithm is simple and easy to use.

20.8 References

Blair, D.C. (1986) Full text retrieval: evaluation and implementation. *International Classification*, **13** (1), 18–23.

Feng, S. (1989) A comparative study of indexing languages in single and multi-database searching. *Canadian Journal of Information Sciences*, **14** (2), 26–36.

He, C. and Pao, M.L. (1986) A discipline specific journal selection algorithm. *Information Processing and Management*, **22** (5), 405–16.

Hurst, G. (1978) Discipline impact factors: a method of determining core journal lists. *Journal of the American Soc. for Information Science*, **29** (4), 171–2.

Lancaster, W.F. (1979) *Information Retrieval Systems*, 2nd edn, Wiley.

Parsaei, H.R. and Wilhelm, M.R. (1989) A justification methodology for advanced manufacturing technologies. *Journal of Computers and Industrial Engineering*, **16** (6), 363–73.

Salton, G. and M.J. McGill (1983) *Introduction to Modern Information Retrieval*, McGraw-Hill.

Appendix

1. Abegglen, J.C. (1983) How to defend your business against Japan. *Business Week*, August 15.
2. Acquisitions and agreements: McDonnell-Douglas. (1986) *Tooling and Production*, March, 32.
3. Aggarwal, S.C. (1985) MRP, JIT, OPT, FMS? *Harvard Bus. Rev.*, **63**, Sept–Oct.
4. Airey, J. and Young, C. (1983) Economic justification: counting the strategic benefits. *Proc. 2nd Int. Conf. on Flexible Manufacturing Systems*, Oct. 26–8, London, UK, v. 2, 549–54.
5. Albus, J.S. (1976) *Peoples Capitalism: The Economics of the Robot Revolution*, New-World, Kensington Books, Maryland, USA.
6. Alexander, R. (1974) Computer-aided manufacturing. *Datamation*, **20**, 109.
7. Allen, J.J. and Hastings, W.F. (1982) Planning for CAD/CAM. *Inst. Cont. Syst.*, **55**, 34–40.
8. Anon. (1980) CAD/CAM policy called for. *Computing Canada*, **6**, 8.
9. Anon. (1980) *A Survey of Industrial Robots*. Productivity International, Dallas.
10. Anon. (1981) Industry has designs on computer techniques. *Eng. Mater. & Des.*, **25**, 28–30.
11. Anon. (1982) Chasing the robotics rainbow. *Financial World*, **15** (8), 40–1.
12. Anon. (1982) Invest in CAD/CAM to remain competitive. *Engineering*, **222**, 158–62.
13. Anon. (1982) Round two in robotics. *Financial World*, **151**, 37–41.
14. Ara, B.J. (1985) Economic justification of process control and information systems. (Chemical plants). *Proc. Controls West Conf. Long Beach, Cal.*, USA, Part of the Int. Indust. Controls Conf. and Exhibition/Controls, pp. 257–68.
15. Aradt, G. (1977) Integrated flexible manufacturing systems: towards automation in batch production. *NZ Eng.*, **32**, 150–5.
16. Arbel, A. and Seidmann, A. (1984) Selecting an FMS: a decision framework. *Proc. ORSA/TIMS FMS Conf.*, Ann Arbor, MI, USA, August, pp. 22–9.
17. Argote, L., Goodman, P.S. and Schkade, D. (1983) The human side of robotics: how workers react to a robot. *Sloan Man. Rev.*, **24**, 31–41.

18. Arnold, B.H. (1973) Cost-effectiveness analysis in the design of automated communications systems. *Proc. NATO Conf. Operational Research in the Design of Electronic Data Processing Systems*, pp. 229–30.
19. Arrigo, T.J. (1985) Planning for flexible manufacturing. *Proc. Flexible Manufacturing Systems '85 Conf.* Dallas, TX, USA, March, M585-149/1-21.
20. Autofact Four: (1983) *Proc. of Autofact 4th Conf. on Computer Integrated Manufacturing and the Automated Factory*, Philadelphia, PA, USA; Nov–Dec 1982. Elsevier, New York.
21. Autofact Five: (1983) *Proc. Autofact Five Conf. on Computer Integrated Manuf. and the Automated Factory*, Detroit, Mich., USA, Nov. 1983, SME, Dearborn Michigan.
22. Autofact Six: *Proceedings*, 1984, Dearborn, MI, USA, SME, 1280.
23. The Automated Factory: Opinions and Insights. (1985) *Mater. Handl. Eng.*, **40**, 64–82.
24. Averyanov, N.K. and Gusev, Yu.V. (1987) Economic machining operations using a flexible production system. *Mekh. & Avtom. Proizvid.* (USSR), **7**, 8–9.
25. Awadalla, E.S., Burns, A. and Jenkins, W.M. (1976) Economical Design of Trussed Beam Structures. *Proc. CAD 76 2nd Int. Conf. on Computers in Engineering and Building Design*, March 23–25, 1976, London, England, pp. 108–11.
26. Ayers, J.B. (1985) Indirect cost identification and control. *CIM Rev.*, **1**, 58–64.
27. Ayers, R.U. and Miller, S.M. (1982) *Social and Economic Aspects of Robotics*. Ballinger, Cambridge, Massachusetts.
28. Ayers, R.U. and Miller, S. (1982) Robotics, CAM and industrial productivity. *National Productivity Rev.*, Winter, 42–60.
29. Bajic, E. and Grasa, P. (1987) Production Analysis Methodology for Mechanical Pieces made in a Machining Cell. L'Automatique, La Productique Et Les Outils Informatiques. Convention Automatique Productique 1987 (Automation, Production and Information Tools. Automatic Production Conf., 1987). April. Paris, France.
30. Barash, M.M. (1980) *Computer Integrated Manufacturing Systems. The Factory of the Future* (Special Volume PED-1), Amer. Soc. Mech. Eng., Chicago.
31. Barash, M.M., Solberg, J.J. and Talavage, J.J. (1978) Optimal planning of computerized manufacturing systems. *Proc. Int. Mach. Tool Des. and Res. Conf.*, 18th, 1977, pp. 767–75. Macmillan, New York.
32. Barlow, S.E. and Freeman, D.R. (1986) *Guidelines for Automated Material Handling and Storage System Cost Justification. Automated Material Handling and Storage* (Vol. 1), Auerback, pp. 1–19.

33. Batzdorff, A. and Pennell, H.Y. (1980) Contemplating the acquisition of a CAD/CAM system pre-procurement economic analysis. Adv. in *Comput. Technol.*, Presented at Int. Comput. Technol. Conf., ASME Century 2 – Emerging Technol. ASME, New York, Vol. 1. pp. 460–8.
34. Baudin, M. (1985) Experience curve theory: a technique for quantifying CIM benefits. *CIM Rev.*, **1**, 51–8.
35. Baxter, R. (1985) Why a 'gut feeling' can be essential in justifying CAE? *Product. Eng.*, **64**, May, 47–51.
36. Beavans, J.P. (1982) First choose an FMS simulator. *American Machinist*, **126**, 143–5.
37. Behuniak, J.A. (1979) *Economic Analysis of Robot Applications*. SME TECH. PAP. SER. MS No. 79–777.
38. Bell, R. and Souza, R.B.R. (1987) The management of tool flows in highly automated flexible machining installations. *Proc. Second Conf. on Computer-Aided Production Engineering*, Edinburgh, Scotland. April.
39. Benedetii, M. (1977) Economic Rate of Return Calculations for Robots in Industrial Applications. ECE Seminar Ind. Robots and Programmable Logical Controllers. Economic Commission for Europe Working Party, Party on Automation.
40. Bennett, J.W. (1973) Discrete manufacturing industries workshop. *Proc. 1973 Autom. Res. Counc. Rep.* N 3, 1974, Univ. of Mich., Ann Arbor, May.
41. Bennett, R.G. (1985) What are companies spending on CIM and how are they justifying these expenditures. *Proc. CIMCON '85*, 15–18 APRIL, Dearborn, Michigan, p. 4.
42. Bernard, P. (1986) Structured project methodology provides support for informed business decisions. *Ind. Eng.*, March, 52–7.
43. Bich, R.E., Kozyrev, Yu.G. and Tarasevich, I.V. (1978) Determining the economic effectiveness of industrial robots. *Mach. & Tool*, **49**, 11–14.
44. Biegel, J.E. and Soloman, D.P. (1984) Some models for the economic justification. *Proc. 1984 Fall Indust. Eng. Conf.*, October 12–31, Atlanta, Georgia, pp. 53–9.
45. Bingham, B. (1983) Instruments to measure electricity: industry's productivity growth rises. *Monthly Labor Rev.*, **6**, 11–17.
46. Bjorke, O. (1978) The use of robots within manufacturing cells. *Proc. 2nd Int. Conf. Comput. Aided Manufact.* National Engineering Laboratory, East Kilbride, UK, June.
47. Blank, L. (1985) The changing scene of economic analysis for the evaluation of manufacturing system design and operation. *Eng. Econ.*, **30**, 227–44.
48. Blank, L. and Carrasco, H. (1985) The economics of new tech-

nology: system design and development methodology. *Proc. 1985 Ann. Int. Indust. Eng. Conf.*, pp. 161–8.

49. Boczany, W.J. (1983) Justifying office automation. *J. Syst. Manage.*, **34**, 15–19.
50. Boehm, B.W. (1981) *Software Engineering Economics*, Prentice-Hall, Englewood Cliffs, New Jersey.
51. Bonczeck, R.H., Holsaple, C.W. and Whinston, A.B. (1981) *Foundations of Decision Support Systems*, Academic Press, New York.
52. Bonine, K.D. (1982) Economic considerations of CAD/CAM. *Proc. Autofact, III Conf. Nov. 1981*, Detroit, Michichan, SME, Dearborn, Michigan, pp. 43–71.
53. Boothroyd, G. (1982) Economics of assembly systems. *J. Manuf. Syst.*, **1**, 111–25.
54. Boothroyd, G. (1984) Economics of general-purpose assembly robots. *Ann. CIRP*, **33**, 8.
55. Boothroyd, G. (1984) Use of robotics in assembly automation. *Ann. CIRP*, **33**, 20.
56. Boothroyd, G. and Dewhurst, P. *Design for Assembly Handbook, Automatic Assembly Program*, ME Dept., University of Massachusetts, Amhurst.
57. Boucher, T.O. and Muckstadt, J.A. (1984) *Cost Estimating Methods for Evaluating the Conversion from a Functional Manufacturing Layout to Group Technology*. Working Paper 84–125, Dept. Indust. Eng., Rutgers Univ.
58. Bowers, E.A. (1987) Defining your system needs to select the proper material handling system. *Ind. Eng.*, Feb., **19**, 34–36, **38**, 40–41.
59. Branam, J.W. (1984) Flexible manufacturing systems eliminate the need for work orders. *Proc. APICS Seminar on Zero Inventory Philosophy & Practices*, Oct.
60. Brock, T. (1978) *UK Industry Accepts Robots*. Report on Industrial Robot Use. British Robot Assoc., Kempston, Beds, UK.
61. Browne, J. and Rathmill, K. (1983) The use of simulation modeling as a design tool for FMS. *Proc. Second Int. Conf. on FMS*, London, p. 197.
62. Browne, J. *et al.* (1984) Classification of flexible manufacturing systems. *FMS Magazine*, 114.
63. Bruijel, N.J. (1981) Management problems in manufacturing. *Ind. Rob.*, **8**, 162–5.
64. Bryan, L. (1982) The Japanese and the American first-line supervisor. *Training and Dev. J.*, January.
65. Buffa, E.S. (1984) *Meeting the Competitive Challenge*. Dow Jones-Irwin.

66. Bultel, J. (1983) Flexibilité de production et rentabilité des investissements. L'exemple de la robotization de l'assemblage tolerie en soudage par points (With English summary). *Revue D'Econ. Indust.*, 4th Trimester, **26**, 1–13.
67. Burstein, M.C. and Talbi, M. (1984) Economic justification for the introduction of flexible manufacturing technology: traditional procedures versus a dynamics based approach. *Proc. ORRSA/TIMS FMS Conf.*, Ann Arbor, Michigan, pp. 100–6.
68. Burstein, M.C. and Ezzekmi, A. (1985) Market based planning to support the time-phased introduction of programmable automation. *Proc. Ann. Int. Indust. Eng. Conf.*, pp. 169–75.
69. Bylinsky, E. (1983) The race to the automatic factory. Feb. 21, *Fortune*, 52–64.
70. CAD/CAM as a basis for development of technology in developing nations, *Proc. IFIP WG5.2 Working Conf.*, Encarnacao, J.L., Torres, O.F.F. and Warmen, E.A. (eds.) (1982) Sao Paulo, Brazil, Oct. 21–23, 1981, North-Holland, New York.
71. CAD/CAM: brave new world creeps slowly in. (1986) *Computer Weekly*, April 3, 21, 22.
72. Canada, J.R. (1986) Non-traditional method for evaluating CIM opportunities. Assigns weights to intangibles. *Ind. Eng.*, March, **18**, 66–71.
73. Canada, J.R. (1986) Annotated bibliography on justification of computer-integrated manufacturing systems. *Eng. Econ.*, **31**, 137–50.
74. Canuto, E., Menga, G. and Bruno, G. (1983) Analysis of flexible manufacturing systems. In: *Efficiency of Manufacturing Systems* (ed. B. Wilson), (Nato Conf. Ser. II, v. 14). Plenum Press, New York, 189–201.
75. Carringer, R. (1984) Integrated decision support system: project 8205, USAF Integrated Computer-Aided Manufacturing Program, Wright-Patterson AFB, OH, USA.
76. Chakravarty, A.K. and Shtub, A. (1987) Capacity, cost and scheduling analysis for a multiproduct flexible manufacturing cell. *Int. J. Prod. Res. (GB)*, **25**, 1143–56.
77. Charles Stark Draper Lab. Inc. (1984) *FMS (Flexible Machining Systems) (Flexible Manufacturing System Handbook*, Vol. 6) Decision Support Software Case Studies. NTIS Report AD-A169 881/0/XAB, Dec.
78. Chasen, S.H. and Dow, J.W. (1980) CAD/CAM – justification and benefits. *Reprographics*, **18**, 12–14.
79. Chasen, S.H. and Dow, J.W. (1980) Considering CAD/CAM. *Computerworld*, **14** (1), 6–18.
80. Chasen, S.H. and Dow J.W. (1980) *The Guide for the Evaluation*

and Implementation of CAD/CAM Systems, CAD/CAM Decisions, Atlanta, Georgia.

81. Chen, P.H. and Talavage, J.J. (1982) Production decision support system for computerized manufacturing systems. *Manufact. Syst.*, **1**, 157–67.
82. Chitaley, A.D. (1986) CIM approach to robot justification. *Proc. Robots 10 Conf.* Chicago, Part. 2, 1–2, 17, April.
83. Choobineh, F. (1986) Economic justification of flexible manufacturing systems. Computers in Engineering 1986: *Proc. 1986 ASME Int. Computers in Engineering Conf. and Exhibition*, Chicago, July 20–24, v. **2**, pp. 169–75.
84. Choobineh, F. and Suri, R. (1971) Flexible manufacturing systems: current issues & models. *Trans. 25th Annual Technical Conf. Amer. Soc. Quality Control.* Chicago, May 1971, pp. 323–30.
85. Choobineh, F. and Suri, R. (1986) Flexible manufacturing systems: current issues & models. *Inst. Indus. Eng.*
86. Ciborra, C. and Romano, P. (1977) Industrial robots and industrial relations – an empirical research. *Proc. 7th Int. Symp. Ind. Robots.* Japan Industrial Robot Assoc., Tokyo, pp. 1–7.
87. Ciborra, C. and Romano, P. (1978) Economic evaluations of industrial robots – a proposal. *Proc. 8th Int. Symp. Indust. Robots.* Pub: Int. Fluidics Service Ltd., Bedford, UK, pp. 15–22.
88. Co. H.C. and Liu, J. (1984) Simulation and decision analysis in FMS justification. *Proc. 1984 Winter Simulation Conf.*, Dallas, November 28–30, 1984, pp. 407–12.
89. Cochran, J.K. and Viswanath N. (1985) Justifying the robotics replacement decision. *Robotics Age*, **7**, 8–16.
90. *Conference on computer graphics in CAD/CAM Systems, 1st, Proc.*, Cambridge, Massachusetts, April 9–11, 1979, MIT Press, Cambridge, Massachusetts.
91. Cook, N.H. (1979) Design and analysis of computerized manufacturing systems (CMS). *Ann. CIRP*, **28**, 377–80.
92. Cook, N. *et al.*, (1981) Design and analysis of computerized manufacturing systems for small parts with emphasis on non palletised parts of rotation. *Proc. National Science Foundation Grantees Conf.*
93. Cook, P. (1981) Considerations in purchasing CAD/CAM systems. *New Electron*, **14**, 66–9.
94. Cooke, J.W. (1981) Unemployment and automation. *Proc. 4th British Robot Assoc. Ann. Conf.*, British Robot Assoc., Kempston, Beds., England, pp. 1–10.
95. Cooke, P.N.C. (1986) Financial justification for industrial automation. ISATA 86, *Proc. 15th Int. Symp. on Automotive Tech. and Automation with Particular Reference to Computer Integrated Manufacture.* Flims, Switzerland, Oct. 1986, ISATA 86086/9, v. 2.

96. Cronin, M.J. (1975) Product cost reduction through interactive CAD/CAM systems. *Prof. Eng.*, **45**, 32–3.
97. Cross, R.E. (1980) The future of automotive manufacturing – evolution of revolution? *Automotive Eng.*, **88**, p. 35.
98. Crumpton, P.A. (1980) The economic justification of automatic assembly. *Proc. 1st Int. IFS Conf. on Assembly Automation*, Brighton, UK, Mar, p. 125.
99. Curtain, F.T. (1984) Evaluating investment in industrial automation systems. *Proc. 2nd Biennial Int. Machine Tool Tech. Conf.* Chicago, Sept 1984, Vol. 1, pp. 49–60.
100. Curtain, F.T. (1984) New costing methods needed for manufacturing technology. *Manage. Rev.*, **73**, 29–30.
101. Curtin, F.T. (1984) Planning and justifying factory automation systems. *Prod. Eng.*, **31**, 46–51.
102. D'Amore, R. (1984) Changing MFG's outdated accounting and control systems. *Proc. APICS Seminar on Zero Inventory Philosophy & Practices*, October.
103. Dallas, D.B. (1984) The impact of FMS (panel discussion of the impact and progress of flexible manufacturing systems). *Production*, 94, 33(6), Oct.
104. Davidson, P. (1985) Can effective demand and the movement toward further income equality be maintained in the face of robotics? An introduction. *J. Post Keynesian Econ.*, **7**, 422–5.
105. Davis, G.B. (1980) CAM: a key to improving productivity. *Mod. Mach. Shop*, **53**, 106–19.
106. Davis, J. (1986) Selling automation. *Prod. Eng.*, **33**, 49–51.
107. Davis, N. (1981) Cutting the cost of drafting and design. *Process Eng.*, **62**, 53–5.
108. Deane, J.N.S. (1980) Computer aided design and manufacture. The opportunities for British industry. *The Management of Automation Conference*, London.
109. Devaney, C.W. (1984) Building the bridge between CAD/CAM/MIS, *Proc. Fall Conf., Institute of Industrial Engineering*, Atlanta, Georgia.
110. Donahue, M. (1980) Pre-implementation considerations concerning a CAM facility. *Proc. Int. Conf. on Cybernetics and Society*, IEEE, New York, pp. 875–80.
111. Dooley, A. (1981) Grim future seen for firms shunning CAD/CAM seen reshaping US work habits. *Computerworld*, **15**, 24–5.
112. Dornan, S.B. (1987) Cells and systems: justifying the investment. (accounting for automated manufacturing processes). *Production*, **99**, 30(6), Feb.
113. Dornfield, D.A. (ed.) (1981) *Automation in Manufacturing, Systems, Processes, and Computer Aids*, (PED Series Vol. 4), ASME, New York.

114. Draper, Lab Inc. (1984) *Flexible Manufacturing Systems Handbook*, Noyes, Park Ridge, New Jersey.
115. Dronsek, M. (1979) Technical and economic problems in the production of aircraft. *Produktionstechnisches Kolloquium, Berlin PTK 79*, 107–115.
116. Drozda, T.J. (1979) Puma robots find cost effective jobs in batch manufacturing. *Production*, **84**, 86–9.
117. Drozda, T.J. (1983) Robotized multi-cells best for mid-value output. *Robotics Today*, **5**, 53–6.
118. Dunlap, G.J. (1979) Manufacturing modernization at Cincinnati Milacron. *Proc. Second Ann. Conf. on Manufacturing Management and Technology*, Amer. Inst. Aeronautics and Astronautics, Los Angeles.
119. Eaglen, R.L. (1980) Management overview of integrate CAD/CAM interactive graphics systems, *Proc. – AUTOFACT West, Vol. 2: Assemblex 7, Predict Maint. 2., PEMCON, Qualinspex 2, Mater. Flow 2, Robotics, SME*, Dearborn, Michigan, pp. 59–92.
120. Ebel, K.H. (1985) Social and labour implications of flexible manufacturing systems. *Int. Lab. Rev.*, **124**, 133–45.
121. Ebel, K.H. (1986) The impact of industrial robots on the world of work. *Int. Lab. Rev.*, **125**, 19–51.
122. Economic justification – how to cut the risks (1985) *Mod. Mater. Handl.*, **40**, 57–9.
123. Economic justification. III. Going beyond the basics (industrial project) (1986) *Mod. Mater. Handl.*, **41**, 68–71.
124. Edstrom, A. and Olhager, J. (1987) Production-economic aspects on set-up efficiency. *Eng. Costs & Prod. Econ.* (Netherlands)., **12**, 99–106.
125. Eiler, R.G., Muir, W.T. and Micheals, L.T. (1984) The relationship between technology and cost management. *Mater. Handl. Eng.*, **39**, Jan., Improving Cost-Benefits Analyses, **39**, Feb., Cost Benefit Tracking Procedures, **39**, March.
126. Eleventh Conf. on Production Research and Technology, Carnegie-Mellon Univ., Pittsburgh, Pennsylvania, SME, Dearborn, Michigan, 1984.
127. Encarnacao, J.L. and Messina, L.A. (1983) Systems simulation technique for the technical evaluation and economic justification of CAD Systems. *Proc. CAMP '83 Conf. on Computer Graphics Applications for Management and Productivity*, Berlin, Germany, March, pp. 943–60.
128. Engelberger, J.F. (1977) Robots make economic and social sense. *Atlanta Econ. Rev.*, **27**, 4–7.
129. Engelberger, J.F. (1980) *Robotics in Practice: Management and Applications of Industrial Robots*. AMACOM, New York.

130. Engwall, R.L. (1986) Flexible manufacturing system pays off for both Westinghouse and the Air Force. *Ind. Eng.*, 18, 41(7), Nov.
131. Eshleman, R.L. and Pagano, F.S. (1983) What managers need to know when choosing robots. *National Productivity Rev.*, Summer, 242–56.
132. Evans, D. and Schwab, P.C. (1984) Integrated manufacturing-financing: the backing to go forward. *Prod. Eng.*, Sept., 122–4.
133. Eversheim, W. and Westkamper, E. (1975) Computer controlled manufacturing systems-planning and organization. *Proc. NAMRAC III N. American Res. Conf.*, Dearborn, Michigan, 1975, p. 728.
134. Eversheim, W. and Witte, K.W. (1981) Economic production – aimed automation, *VDI Z.* (German), **123**, 367–74.
135. Eversheim, W. and Hennman, P. (1982) Recent trends in flexible automated manufacturing. *J. Manufact. Syst.*, **1**, 139–48.
136. Factory integration teams improve automation planning (1985) *Mod. Mater. Handl.*, **40**, 65–7.
137. Falkner, C.H. (1986) Multi-attribute decision analysis for justifying flexible automation. *Proc. 1986 IEEE Int. Conf. on Robotics and Automation*, San Francisco, IEEE, New York, pp. 1815–20.
138. Falkner, C.H. (1986) Multi-attribute decision analysis for justifying automation. *1986 IEEE Int. Conf. on Robotics and Automation*, 10 April, University of Wisconsin, Madison, Wisconsin, pp. 7–10.
139. Falkner, C.H. and Garlid, S. (1986) Simulation for the justification of an FMS. *Proc. 1986, Fall Indust. Eng. Conf.*, Boston, Dec. 1986, pp. 99–108.
140. Falter, B. and Choobineh, F. (1986) CAM economic justification – a case study in the electronic industry. *Proc. 1986 Int. Indust. Eng. Conf.*, May, 1986, Dallas, pp. 518–25.
141. Ferniani, D. and Ciborra, C. (1981) Economic consequences of the introduction of CAD/CAM in a small electronic products firm, CAD in medium sized and small industries. *Proc. First European Conf. on Computer Aided Design in Medium Sized and Small Industries*. MICAD 80. North-Holland, New York, pp. 223–31.
142. *First Step Towards FMS (Flexible Manufacturing Systems)*, NTIS Report MTIRA 84/01, Machine Tool Industry Research Association, Macclesfield, England, Nov. 1983.
143. Fishman, C.M. and Slovin, H.J. (1974) Life cycle cost impact on high reliability systems. *Proc. 1974 Annual Reliability and Maintainability Conf.* Jan. 29–31, 1974, Los Angeles, pp. 358–62.
144. Fleischer, G.A. (1985) Economic justification for automation. *Proc. 24th IEEE Conf. on Decision and Control Including the Synp. on Adaptive Processes*, Fort Lauderdale, Florida, Dec. 1985, IEEE, New York, pp. 1978–83.
145. Fleischer, G.A. (1982) A generalized methodology for assessing

the economic consequences of acquiring robots for repetitive operations. *Proc. May 1982 Annual Conf., Inst. Indust. Eng.*, Norcross, Georgia, pp. 130–9.

146. Flexible Assembly System is Moving Toward Reality. News Release, August 21, 1986, p. 1.
147. Flexible Manufacturing Systems 1: *Proc. Int. Conf. on Flexible Manufacturing Systems*, Brighton, England, Oct. 1982. Elsevier, New York, 1983.
148. *Flexible Manufacturing Systems*, Machine Tool Industry Research Association, Macclesfield, England. *Planning for FMS*, NTIS Report MTIRA 84/02, May 12, 1983.
149. Flexible systems grow slowly. *Cleveland Plain Dealer*, Feb. 16, 1986, SEC D, 1.
150. Flora, P.C. (1986) *International Computer-Aided Manufacturing (CAM) Directory* Tech Data TX 02/1986.
151. FMS strategy for small-lot tool production (1985) *Tooling and Production*, April, 114–18.
152. Fogarty, D.W. (1984) Scheduling manufacturing cells (MC) and flexible manufacturing systems (FMS). *Proc. APICS Seminar on Zero Inventory Philosophy & Practices*, October.
153. Fortin, P.E. (1985) *Economic Evaluation of Computerized Structural Analysis*. Report No. NAS 1.26:177078; NASA-CR-177078, August.
154. Foundyller, C.M. (1980) *Turnkey CAD/CAM Computer Graphics: A Survey and Buyer's Guide*, 3 vol., Daratech Associates, Cambridge, Massachusetts.
155. Foundyller, C.M. (1981) Buying a turnkey CAD/CAM system. *Mach. Dec.*, **53**, 77–8.
156. Fox, R.W. (1987) Strategic and financial justification for CIM. *Proc. Sixth Ann. Control Eng. Conf.* May 1987, Rosemount, Illinois, pp. 652–6.
157. Foyer, P. (1981) Justifying robots and automation. *Proc. 4th British Robot Ann. Conf.*, 1981, British Robot Assoc., Kempston, Beds., England, pp. 71–83.
158. Fraser, J.M. (1985) Justification of flexible manufacturing systems. *Proc. 1985 IEEE Int. conf. on Systems, Man and Cybernetics.*, IEEE, New York.
159. French, W.R. (1983) *Information Gaps and Cost-Benefit Uncertainties in the CIM Investment Decision*. Working paper by president of Lockheed-Georgia Company, Marietta, Georgia.
160. Funk, J.L. (1984) The potential societal benefits from developing flexible assembly technologies (economic, robotic, machine vision). PhD dissertation Carnegie-Mellon University.
161. Gelders, L. (1986) The justification of new business technologies. *PT/Elektrotech. Elektron.* (Netherlands), **41**, AK8–11.

162. Georgescu, D.M. (1986) Cybernetization and robotization of the operating processes in the wood and building materials industry. *Econ. Computation and Econ. Cybern. Studies and Res.*, **21**, 55–68.
163. Gerencser, P. and Veszi, A. (1986) Coordinated development of system and environment: a new CAD/CAM strategy. Information control problems in manufacturing technology 1986. *Proc. 5th IFAC/IFIP/IMACS/IFORS Conf.*, Suzdal, USSR., April.
164. Gerwin, D. (1982) Do's and don'ts of computerized manufacturing. *Harvard Bus. Rev.*, **60**, 107–16.
165. Giacomino, D.E. and Doney, L.D. (1986) The SAI movement in manufacturing. (manufacturing's moves to simplify, automate and integrate). *CPA J.*, 56, Sec. CTI, 64(6), Oct.
166. Goicoechea, A., Hansen, D.R. and Duckstein, L. (1982) *Multi-objective Decision Analysis with Engineering and Business Application*, Wiley, New York.
167. Gold, B. (1980) Revising managerial evaluations of computer-aided manufacturing systems. *Proc.-Autofact West, CAD/CAM 8 Conf.*, Anaheim, California, 1980, Vol. 1, pp. 13–24.
168. Gold, B. (1981) *Improving Managerial Evaluations of Computer-Aided Manufacturing.* Report for the US Air Force Systems Command, National Research Council Committee on Computer Aided Manufacturing, National Academy Press, Washington, D.C.
169. Gold, B. (1982) CAM sets new rules for production. *Harvard Bus. Rev.*, **60**, Nov.–Dec., 88–94.
170. Goldhar, J. and Jelinek, M. (1983) Plan for economics of scope. *Harvard Bus. Rev.*, **21**, Nov.–Dec., 141–8.
171. Goldhar, J.D. and Jelinek, M. (1986) Economics in the factory of the future. *CIM Rev.*, **2**, 21–8.
172. Golosovskii, S. (1985) Robotics and its efficiency. *Problems Econ.*, **28**, 72–87.
173. *Competitive Assessment of the United States.* Flexible Manufacturing Systems Industry Series, Government Printing Office, Washington, DC. 1985.
174. Graf, B. (1980) Flexible magazine stores for manipulative systems, *IND. – ANZ.* (German.), **102**, 29–30.
175. Granicki, J. (1980) Methodology of economic efficiency evaluation of industrial robots applications in Poland. *Proc. Int. Symp. on Ind. Robots; 10th Int. Conf. on Robot Technol.*, 1980, IFS, Kempston, Bedford, England, pp. 591–9.
176. Greguras, F.M. (1982) Robotics more attractive through leveraged lease financing. *Scott Rep.*, **1**, 6–9.
177. Grieve, R.J., Lowe, P.H. and Kelly, M.P. (1983) Robots: the economic justification. *Proc. Autofact Europe Conf.* Soc. Mech. Eng., Geneva, Switzerland, September 1983, pp. 2.1–12.

178. Grimm, J.J. (1984) Integrated manufacturing – technology's tomorrow just arrived. *Prod. Eng.* Sept., 72–6.
179. Grover, M.P. and Zimmers, E.W. (1984) *CAD/CAM: Computer-Aided Design and Manufacturing*, Prentice-Hall, Englewood Cliffs, New Jersey.
180. Groover, M.P. (1988) *Automation, Production Systems and Computer-Integrated Manufacturing*. Prentice-Hall, Engelwood Cliffs, New Jersey.
181. Gulati, V. (1984) Presenting project justification in a language top management. *Proc. Fall 1984, Industrial Eng. Conf.*, pp. 370–3.
182. Gunn, T.G. (1981) *Computer Applications in Manufacturing*, Industrial Press, New York.
183. Gupta, S.M. and Barash, M.M. (1977) *The Optimal Planning of Computerized Manufacturing Systems: Computer-Aided Selection of Machining Cycles and Cutting Conditions on Multi-Station Synchronous Machines*. NTIS Report PB81-113433, School of Industrial Engineering, Purdue Univ., Lafayette, Indiana.
184. Gustavson, R.E. (1981) Engineering economics applied to investments in automation. *Proc. Assembly Automation Conf.*, Brighton, UK, 1981, pp. 9–20.
185. Gustavson, R.E. (1983) Choosing manufacturing systems based on unit cost, *13th ISIR/ROBOTS 7 Conf.*, Chicago, April 17–21, 1983, (SME Paper MS 83-322), SME.
186. Gustavson, R.E. (1984) Computer-aided synthesis of least cost assembly systems. *14th Int. Symp. on Industrial Robots*, Gothenburg, Sweden, October 1984.
187. Haider, S.W. and Blank, L.T. (1983) A role for computer simulation in the economic analysis of manufacturing systems. *Proc. Winter Simulation Conf. 1983*, IEE, New York, pp. 199–206.
188. Hall, G. (1986) Robotic/intelligent machinery economic justification. *Robotics World*, **4**, 25–6.
189. Hamar, K., Molnar-Jobbagy, M. and Treer, R. (1975) Estimation of the accuracy of results of an ammonia synthesis model by means of sensitivity analysis. *Proc. Symp. on Computers in Design and Erection of Chemical Plants*, Karlovy, Czechoslovakia, Aug. 1975. Part II, pp. 575–8.
190. Harris, R.W., Cullinane, M.J. and Sun, P.T. (1982) *Process Design and Cost Estimating Algorithims for the Computer Assisted Procedure for Design and Evaluation of Wastewater Treatment Systems*, NTIS Report PB82-190455.
191. Harvey, D. (1979) When robots take over – what's left. *Chief Executive*, **47**, 43–4.
192. Hatvany, J. *et al.* (1983) *World Survey of CAM*, Butterworth, Sevenoaks, Kent, England.

193. Hayes, Richard H. and Weelwright, S.C. (1984) *Restoring our Competitive Edge: Competing through Manufacturing*, Wiley, New York.
194. Hayes, R.H. and Abernathy, W.J. (1980) Managing our way to economic decline. *Harvard Bus. Rev.*, **58**, July–Aug., 67–77.
195. Hayes, R.H. and Garvin, K. (1982) Managing as if tomorrow mattered. *Harvard Bus. Rev.*, **60**, May–June, 71–9.
196. Hee Man Bae and Devine, M. (1975) Economic optimisation models of windpower systems. *Bull. Oper. Res. Soc. Amer.*, **23**, Suppl. 2, B405.
197. Heginbotham, W.B. (1981) Present trends, applications and future prospects for the use of industrial robots. *Proc. Inst. Mech. Eng.*, **195**, 409–18.
198. Hegland, D.E. (1982) The automated factory: fitting the pieces together. *Prod. Eng.*, **29**, 56–64.
199. Hegland, D.E. (1982) Flexible manufacturing – a strategy for winners. *Prod. Eng.*, **29**, 40(7).
200. Henrici, S.B. (1981) How deadly is the productivity disease? *Harvard Bus. Rev.*, Nov/Dec, 123.
201. Herroelen, W., Degraeve, Z. and Lambrecht, M. (1986) Justifying CIM: quantitative analysis tool. *CIM Rev.*, **2**, 33–43.
202. Hill, N. and Dimnik, T. (1985) Cost justifying new technologies. (Special Supplement: Generating Profit From New Technology), *Bus. Quart.*, **50**, 91(6).
203. Hill, N.F. and Dimnijk, T. (1986) The accountant's role in cost-justifying new technologies. *CMA – The Man. Acc. Mag.*, **60**, 31(7), Nov.–Dec.
204. Hitz, K. (1987) Flexible integrated computer-aided manufacturing systems increase productivity. *Robotics & Comput.-Integrated Man.*, **3**, 123–8.
205. Hodder, J.E. and Riggs, H.E. (1985) Pitfalls in evaluating risky projects. *Harvard Bus. Rev.*, **63**, 128–35.
206. Holbrook, W. (1984) Accounting changes required for Just-in-Time production. *Proc. APICS Seminar on Zero Inventory Philosophy and Practices*, October 1984.
207. Hopkinson, R.J. (1986) Potential of low cost PC CAD workstations sharing design tasks in a mixed environment. *Conference Record – MIDCON 1986*, Dallas, TX, USA, Sept. 1986, Pap. 2.
208. Hughes, J. (1976) Flexible manufacturing systems for improved mid-volume productivity: *Proc. 3rd Ann. AIEE Systems Eng. Conf.*, November 1975, Reprinted in *Understanding Manufacturing Systems* (Vol. 1), Kearney and Trecker Corporation, Milwaukee, Wisconsin.
209. Hundy, B.B. (1984) Investment in new technology and corporate strategy. *Annals. CIRP 1984. Conf. on Manufacturing Technology.*

34th General Assembly of CIRP. Held at Madison, Wisconsin, v. 33, 341–4.

210. Hunt, W.D. (1983) *Industrial Robotics Handbook.* Industrial Press, New York, Ch. 9, 150–8.
211. Hunter, R.P. (1978) *Automated Process Control Systems: Concepts and Hardware*, Prentice-Hall, Englewood Cliffs, New Jersey.
212. Hunter, S. (1985) Cost justification: the overhead dilemma. *Proc. Robots 9 Conf.*, Detroit, June 2–6, 1985. v. 2, pp. 17.67–81, 1985.
213. Hutchinson, G.K. (1977) *The Control of Flexible Manufacturing Systems: Required Information and Algorithm in Information Control Problems in Manufacturing Technology.* Int. Federation of Automatic Control, Pergamon.
214. Hutchinson, G.K. (1977) *Flexible Manufacturing Systems in Japan.* Management Research Center, University of Wisconsin-Milwaukee, Wisconsin.
215. Hutchinson, G.K. (1977) *Flexible Manufacturing Systems in the Federal Republic of Germany (BRD).* Management Research Center, University of Wisconsin-Milwaukee, Wisconsin.
216. Hutchinson, G.K. (1979) *Flexible Manufacturing Systems in the United States.* Management Research Center, School of Business Administration, University of Wisconsin, Milwaukee, Wisconsin.
217. Hutchinson, G.K. and Holland, J.R. (1982) Economic value of flexible automation. *J. Manufact. Syst.*, **1**, 215–28.
218. Hutchinson, G.K. and Clementson, A.T. (1985) Manufacturing control systems: an approach to reducing software costs. Presented at *Int. Conf. on Manufacturing Science and Technology of the Future*, M.I.T., Cambridge, Massachusetts, October 9–12, 1985.
219. Hyer, N. and Wemmerlov, V. (1984) Group technology and productivity, *Harvard Bus. Rev.*, **62**, July–Aug., 140–9.
220. Hyer, N.L. (1984) *Group Technology at Work*, Dearborn, Michigan.
221. Infotech Limited (1980) *Factory Automation, State of the Art Report. 1. Analysis.* Infotech, Maidenhead, England.
222. International scene in Japan: FMS is simple and economical technology. *Amer. Metal Market.*, August 27, 44A, 45A.
223. Ito, Y. (1987) Evaluation of FMS: state of the art regarding how to evaluate system flexibility. *Robotics and Comp. Integrated. Manufact.* (GB), **3**, 127–34.
224. Ito, Y., Ohmi, T. and Shima, Y. (1985) Evaluation method of flexible manufacturing system – a concept of flexibility evaluation vector and its application. *Proc. 25th Int.; Machine Tool Design and Research Conf.*, Birmingham, England, April 22–24, 1985, pp. 89–95.
225. Iuan, Y.L. and Kiew, K.B. (1987) The economic evaluation of

flexible manufacturing system. *Proc. Int. Conf. on Optimization Techniques and Applications*, 8–10 April 1987, Singapore.

226. James, T.G. (1977) *Software Cost Estimating Methodology*, Air Force Avionics Lab. Wright-Patterson AFB, Ohio, Report No.: AFAL-TR-77-66. Aug., and NTIS Report AD-A048 192/9.
227. Jansson, S. (1982) Portfolio strategy: the robots are coming. *Institutional Investor*, **16**, 182–6.
228. Johnson, I.S.C. (1983) FMS operations management decision aids. *Proc. FMS West, Soc. Manufacturing Eng.*, Dearborn, Michigan.
229. Kafrissen, C. and Stephens, M. (1984) *Industrial Robots and Robotics*, Reston, Ch. 11, 303–15, Englewood Cliffs, New Jersey.
230. Kaplan, R.S. (1984) Yesterday's accounting undermines production. *Harvard Bus. Rev.*, **62**, 95–101.
231. Kaplan, R.S. (1986) Accounting lag: the obsolescence of cost accounting systems. *California Manage. Rev.*, **27**, 178–99.
232. Kaplan, R.S. (1986) Must CIM be justified by faith alone? *Harvard Bus. Rev.*, **64**, 87–95.
233. Kaschemekat, J. and Boeddeker, K.W. (1980) Parametric Study of Cost Factors in RO Desalination (GKSS Plate System), Report No. GKSS-80/E/34; September 1980, Int. Symp. on Fresh Water From the Sea and Exhibition of Desalination Equipment, Aquatech '80, Amsterdam, Netherlands.
234. Keeney, R. and Raiffa, H. (1976) *Decisions with Multiple Objectives: Preferences and Value Tradeoffs*, Wiley, New York.
235. Kellock, B.C. (1976) Industrial robot for investment foundry. *Mach. & Prod. Eng.*, **129**, 487–8.
236. Kennedy, W.J. (1987) Issues in the maintenance of flexible manufacturing systems. *Maint. Manage. Int.*, **7**, 43–52.
237. Kester, C.W. (1984) Today's options for tomorrow's growth. *Harvard Bus. Rev.*, **62**, Mar.–Apr., 153–9.
238. Klahorst, T.H. (1983) How to justify multimachine systems. *Amer. Machinist*, **127**, September, 67–70.
239. Knott, K., Bidanda, B. and Pennebauer, D. (1988) Economic analysis of robotic arc welding operations. *Int. J. Prod. Res.*, **26**, 107–17.
240. Komack, R.L. (1985) Business-oriented CAD system justification. *Proc. Conf. Autofact '85*, Detroit, Nov. 1985, pp. 2.1–23.
241. Konen, Y. (1983) *Computer Control of Manufacturing Systems*, McGraw-Hill, New York.
242. Konson, A.S. and Belodedov, V.I. (1979) Technical-economic problems of using automatic devices in the optical industry. *Sov. J. Opt. Tech.*, **46**, 551–4.

243. Kowalski, A. (1978) The industrial robots – the future of full automation of production. *Pol. Mach. Ind. Offers*, **12**, 16–18.
244. Kozar, Z. (1986) The concept of development of flexible automation in Czechoslovakia. Information control problems in manufacturing technology 1986. *Proc. 5th IFAC/IFIP/IMACS/IFORS Conf.* Suzdal, USSR, April 1986.
245. Krainz, G. (1987) A consultancy service in industrial robotics: methods and experience. *Robotics*, **3**, 59–63.
246. Krallmann, H. (1982) Capital expenditure planning for CAD/CAM systems. *ZWF Z. Wirtsch. Fertigung* (German.), **77**, 616.
247. Krolewski, S. and Gutowski, T. (1987) Effect of the automation of advanced composite fabrication processes on part cost. *Sampe J.*, **23**, 21–6.
248. Krouse, J.K. (1982) *What Every Engineer Should Know About Computer-Aided Design and Computer-Aided Manuf*: The CAD-CAM Revolution Series, Dekker, New York.
249. Kulatilaka, N. (1984) A managerial decision support system to evaluate investments in flexible manufacturing systems. *Proc. Orsa/TIMS Conf. on FMS*, Ann Arbor, Michigan, 1984, pp. 16–21.
250. Kulatilaka, N. (1985) *Capital Budgeting and Optimal Timing of Investments in Flexible Manufacturing Systems.* Working Paper – Boston University School of Management and Massachusetts Institute of Technology, (Forthcoming in *Annals of Operations Research*).
251. Kulatilaka, N. and Marks, S. (1985) The value of flexibility in a world of certainty. *Proc. INRIA Conf. on Production Systems, Scientific, Economic, Strategic Approach*, Paris, France, April 1985.
252. Kusiak, A. (1986) Application of operational research models and techniques in flexible manufacturing systems. *European J. Oper. Res.*, **24**, 336–45.
253. Labini, P.S. (1985) Valore e distribuzione in un'economia rototizzata. (With English summary.) *Economia Et Politica*, **2**, 359–63.
254. Labs, W. (1987) Flexible manufacturing: an investment that really pays off. *Chilton's I & CS*, **60**, 31–3, 40–1.
255. Lambrix, R.J. and Singhvi, S.S. (1984) Preapproval audits of capital projects. *Harvard Bus. Rev.*, **62**, 12–14.
256. Leather, G.M. (1972) An examination of the cost of installing and using a computer in an integrated design organization. *Proc. Int. Conf. on Computers in Architecture*, Sept. 1972, York, England, pp. 267–74.
257. Leaver, E.W. (1987) Process and manufacturing automation – Past, present, and future. *ISA Trans.*, **26**, 45–50.
258. Leclair, S.R. and Sullivan, W.G. (1985) Justification of advanced manufacturing technology using expert systems. *Proc. 1985 Ann. Int. Industrial Eng. Conf.*, Chicago, December 1985, pp. 362–70.

259. Leimkuhler, F. (1981) *The Optimal Planning of Computerized Manufacturing Systems. Economic Analysis of CMS.* NTIS Report PB81-193336, School of Industrial Engineering, Purdue Univ., Lafayette, Indiana.
260. Leimkuhler, F.F. (1979) Cost analysis of computerized manufacturing systems. *Proc. ASEE Annu. Conf. on Scarce Resource Allocation in the 80's*, Baton Rouge, Louisiana, June 1979, ASEE, Washington, DC, pp. 142–9.
261. Leimkuhler, F.F. (1984) *Economic Analysis of CIMS: Methods and Tools for Computer Integrated Manufacturing*. (Lecture Notes in Computer Science, 168), Springer-Verlag, Berlin and New York.
262. Leung, L.C. and Tanchoco, J.M.A. (1987) Multiple machine replacement within an integrated system framework. *Eng. Econ.*, **32**, 89–114.
263. Levine, S.J. and Yalowitz, M.S. (1983) Managing technology: the key to successful business growth. *Manage. Rev.*, **72**, 44–8.
264. Lindgren, R.K. (1984) Documenting productivity gains in CAD. *Comp. Graphics World*, **7**, 95–8.
265. Lipchin, L. (1981) The role of strategic planning for productivity improvement through computer-aided engineering. *Proc. 23rd IEEE Computer Soc. Int. Conf. Compcon Fall 1981 on Productivity an Urgent Priority*, IEEE, New York, pp. 203–12.
266. Llewelyn, A.I. and Freeman, G.C. (1973) Economics of CAD – a new approach. *Proc. AGARD Conf. on Computer Aided Design for Electronic Circuits*, May 1973, Lyngby, Denmark, pp. 1–10.
267. Longo, P.A. (1975) Design to cost – how much? How long? *Proc. Numerical Control Soc. of SME, 1975, CAD/CAM III Conf*; Chicago, Feb. 1975.
268. Losleben, P. (1972) The expensive myth of design automation and how to build an economically viable system. *Proc. Technical Program of Computer Systems Design, '72 West Conf.*, Feb. 1972, Anaheim, California, pp. 60–4.
269. *Low Cost Automation, Technical Services for Industry*. Department of Industry, UK, 1977. Obtainable from New English Library.
270. MacArthur, L. (1985) Traditional values can't justify CIM. *Automation News*, May 1, 8.
271. McCartney, J. (1982) Tooling economics in integrated manufacturing systems. *Int. Prod. Res.*, **20**, 493.
272. MacDonald, D.B. (1985) Financial alternatives for acquiring automated systems. *CIM Rev.*, Spring, 26–36.
273. McDonald, J.L. and Hastings, W.F. (1983) Selecting and justifying CAD/CAM. *Assembly Eng.*, April, 24–7.
274. McDonnell-Douglas to Invest $150 Million in C-17 Program. *Production*, **987**, 16(1), March 1986.

275. McGinnis, M.S., Gardnier, K.M. and Jesse, R. (1985) Capital equipment selection strategy under volatile economic conditions. *Proc. Autofact '85 Conf.*, Detroit, November 4–7, 1985, pp. 6.67–86.
276. Mai, O. and Liebler, U. (1975) Economic grinding of non-circular shapes by means of copying method. *Maschinenmarkt* (Germany) **81** (59), 1096–100.
277. Making unmanned machining pay. *Numerical Engineering*.
278. Maller, R.J. (1984) Integrated manufacturing – the structure: simplify and focus, or collapse. *Prod. Eng.* **31**, 62–5, Sept.
279. Mandelbaum, M. Flexibility in Decision Making Exploration and Unification. Ph.D. Thesis, University of Toronto, 1987.
280. Martins, G. (1987) Profitability and flexibility in an assembly automation. *Proc. 8th Int. Conf. on Assembly Automation*, Copenhagen, Denmark, March.
281. Mehlhope, K.D. (1976) Computer-aided management tools using group technology. *Proc. 12th IEEE Computers Society Int. Conf. on Computers: the Next 5 Years*, 24–26, Feb., 1976, San Francisco, pp. 171–4.
282. Merchant, M.E. (1976) Economic, social and technological challenges facing future manufacturing engineers. *Proc. Manufacturing Eng./Technology Education Forum*, Cleveland, Ohio, USA. Soc. Manufacturing Eng., Dearborn, Michigan.
283. Meredith, J. (1984) Economics of computer aided manufacturing. *Proc. 1984 Fall Indust. Eng. Conf.*, Oct. 28–31, 1984, Atlanta, pp. 125–9.
284. Meredith, J. (1987) New justification approach for CIM technologies. *CIM Rev.*, **3**, 37–42.
285. Meredith, J.R. and Hill, M.M. (1987) Justifying new manufacturing systems: a Managerial approach. *Sloan Manage. Rev.*, **28**, 49–61.
286. Meredith, J.R. and Suresh, N.C. (1986) Justification techniques for advanced manufacturing technologies. *Int. J. Prod. Res.*, **25**, 1043–57.
287. Messina, L.A. (1984) Justification of CAD systems. *Computers and Graphics*, **8**, 105–6.
288. Meyer, R.J. (1982) A cookbook approach to robotics and automation justification. *Proc. Robots VI Conf. Soc. Manufact. Eng.*, Detroit, Mar. 1982.
289. Michael, G.J. and Millen, Robert A. (1984) Economic justification of modern computer-based factory automation equipment: a status report. *Proc. First ORSA/TIMS Special Interest Conf. on FMS*, Ann Arbor, Michigan, pp. 30–5.
290. Michaels, L.T. (1986) New guidelines for selecting and justifying

factory automation projects. *Proc. 29th Ann. Int. Conf. Amer. Prod. & Inventory Control Soc.*, 22–24 October, 1986, St. Louis, Missouri, pp. 483–5.

291. Milacron Exec: Outsouring + FMS = Profits. (1985) *Amer. Metal Market*, June 17, 5, 29.
292. Milberg, J. and Lutz, P. (1987) Integration of autonomous mobile robots into the industrial production environment. *Proc. 1987 IEEE Int. Conf. on Robotics and Automation*, Raleigh, North Corolina, 31 March–3 April 1987.
293. Miller, D.M. (1971) Economic advantages of integrating standard DIP hardware with flexible software. *Western Electronic Show & Convention Meeting*, San Francisco, Aug. 1971, pp. 22–3.
294. Miller, R.K. (1983) The bottom line – justifying a robot installation. *Robotics World*, April 32–5.
295. Miller, R.K. (1986) *Strategic Planning for Computer Integrated Manufacturing*, SEAI Tech. Pubns, 11/1986.
296. Miltenburg, G.J. (1986) Economic evaluation and analysis of flexible manufacturing systems. eng. costs & prod. econ. (NETHERLANDS), 12, 79–92, July 1987. Also in *Proc. 4th Int. Work Seminar on Prod. Econ.*, Innsbruck, Austria, Feb. 17–21, 1986, pp. 79–82.
297. Miltenburg, G.J. and Krinsky, I. (1987) Evaluating flexible manufacturing systems. *IIE Trans.* (USA), **19**, 222–33.
298. Monahan, G.E. and Smunt, T.L. (1984) The flexible manufacturing system investment decision. Presentation to the Dallas ORSA/TIMS Conf., November 1984.
299. Monahan, G.E. and Smunt, T.L. (1985) A Markov decision process model of the automated flexible manufacturing investment decision. Graduate School of Business Administration, Washington University, St. Louis, Missouri, December 1985.
300. Muir, W.T. (1984) *Alternative for Evaluating CIM Investments – A Case Study*. Tech. Paper, Soc. Manuf. Eng. MS84-184.
301. Muir, W.T. and Michaels, L.T. (1985) Technology management and factory automation. *Proc. APICS 28th Annual Int. Conf.* 21–25 October, 1985, Toronto, Amer. Production and Inventory Control Soc., pp. 503–7.
302. Nelson, C.A. (1984) A scoring model for flexible manufacturing systems project selection. *Proc. ORSA/TIMS Conf. on FMS*, Ann Arbor, Michigan, pp. 43–8.
303. Norton, F.J. (1980) Cost justification for an interactive computer-aided design drafting/manufacturing system. *Proc. Ann. ADUA Meeting*, Denver, Sept., 1980, (UCRL Report 84966, Univ. California-Lawrence Livermore Nat. Lab).
304. O'Conner, M.J. and Botero, S.A. (1977) *Computer-Aided Final*

Design Cost Estimating System Overview. Report No.: CERL-IR-p-81, (Construction Eng. Res. Lab (Army), Champaign, Illinois) May, 1977, and NTIS Report AD-A040 119/0, May 1977.

305. Odeh, N., Najm, M. and Omurtag, Y. (1986) Justification of micro-computer-based CADD in a small to medium sized manufacturing industry. *Proc. 1st 1986 ICEM Int. Conf. in Engineering Management*, Arlington, Virginia, Sept. 1986, pp. 82–7.
306. Office of Technology Assessment, Washington, DC. *Computerized Manufacturing Automation: Employment, Education, and the Workplace*. NTIS Report PB 84196500.
307. Ogden, H. (1981) Justifying assembly automation. economic methods. *Proc. 2nd Int. Conf. on Assembly Automation*. May, 1981, IFS, Kempston, Bedford, England, pp. 1–8.
308. O'Grady, P.J. and Menon, U. (1985) A multiple criteria approach for production planning of automated manufacturing. *Eng. Optim.*, **8**, (3), 161.
309. Ouchi, W.G. (1981) *Theory Z: How American Business Can Meet the Japanese Challenge*, Addison-Wesley, Reading, Massachusetts.
310. Owen, T. (1985) *Strategic Issues in Automated Production: the Challenge of Robotics and Computer Integrated Manufacturing*, Cranfield Press, Cranfield, Bedford, England.
311. Palmer, G.D. (1985) Justification of a CAD/CAM system in a machine tool company: the theory and the reality. *NES Ann. Conf. on Advances in Manufacturing Tech*. May 1985, London, Numerical Eng. Soc., London, p. 10.
312. Park, C.S. and Son, Y.K. (1987) Computer-assisted estimating of nonconventional manufacturing costs. *Comput. Mech. Eng.*, **6**, 16–26.
313. Parkinson, S.T. and Avlonitis, G.J. (1982) Management attitudes to flexible manufacturing systems. *Proc. 1st Int. Conf. on Flexible Manufacturing Systems*, 1982, IFS, Kempston, Beds., England, pp. 405–12.
314. *Parts on Demand: Evaluation of Approaches to Achieve Flexible Manufacturing Systems for Navy parts on Demand* (2 Vols.), NTIS Report AD-A143 248/3, Feb. 1984, and NTIS Report AD A142 151/0. Feb. 1984.
315. Patsfall, R. (1984) *FMS Justification Strategy – General Electric's Approach. Flexible manufacturing Systems*. Soc. Manufact. Eng., Dearborn, Michigan, pp. 187–90.
316. Petkov, Kh.K. and Sirakov, Kh.I. (1981) Technical-economic approach to the automation of processes in instrument production by the use of industrial manipulators and robot (IMR). *ELKTRO PROM.-st* and *PRIBOROSTR*. (Bulgarian), **16**, 6–10.
317. Planning for Payback. (1986) *Plant Eng.*, March 27, pp. O–K.

318. Potter, R.D. (1983) Analyze indirect savings in justifying robots. *Ind. Eng.*, Nov., 28–30.
319. Primrose, P. and Leonard, R. (1987) Automation needs no justification (cost-benefit analysis). *New Sci.* (UK), **115**, No. 1576, 60–2.
320. Primrose P.L., Harrison, D.K. and Leonard, R. (1987) Obtaining the financial benefits of CIM today. *Proc. CAD/CAM March 1987 Conf.*, Birmingham, England.
321. Primrose, P.L. and Brown, C.C. (1987) FMS evaluation: a case study. *FMS Mag.* (GB), **5**, 131–2.
322. Primrose, P.L. and Leonard, R. (1987) Financial aspects of justifying FMS. *Proc. Second Int. Conf. on Computer-Aided Production Engineering*, Edinburgh, Scotland, April, 1987.
323. *Proc. Conference on Computer Aided Manufacturing and Productivity, COMPRO 81*, London, England, October 21–22, 1981: Inst. of Product Eng., London, England, 1981.
324. Raju, V. (1984) Management attitude to FMS. *Proc. Autofact 6 Conf.*, Anaheim, California, Oct. 1–4, 1984, pp. 12.1–9.
325. Randhawa, S.U. and West, T.M. (1986) A multi-attribute methodology for the evaluation of automated assembly systems. *Proc. Int. Indust. Eng. Conf.*, May 1986, Dallas, Inst. Indust. Eng. pp. 526–33.
326. Ranky, P.G. (1981) Increasing productivity with robots in flexible manufacturing systems. *Ind. Robot*, **8**, 234–47.
327. Ranky, P. (1984) *The Design and Operation of Flexible Manufacturing Systems*, Elsevier, New York.
328. Rathmill, K. (1983) Flexible manufacturing systems 2: *Proc. 2nd Int. Conf.*, London, Oct. Elsevier, New York.
329. Rathmill, K. (1987) Flexible manufacturing systems: *Proc. 5th Int. Conf.*, November 3–5, 1986; Stratford-upon-Avon, U.K., Springer-Verlag, New York.
330. Rathmill, K.M. (1988) *Computer Integrated Manufacturing*, Springer-Verlag, New York.
331. Regaining our competitive edge. *Chief Executive* (US), Wint., 32(6).
332. Rembold, U., Blume, C. and Dillman, R. (1987) *Computer-Integrated Manufacturing*. (Manuf. Engineering Ser.) Purdue Univ.
333. Rhodes, W.L. (1982) Manufacturing systems: tolerating change. *Infosystems*, **29** (1), 68–78.
334. Roberson, B.E. (1980) Test for justifying the introduction of CAD/ CAM, NC in a traditional apt environment. *Proc. 17th Ann. Meeting and Tech. Conf. of Numer. Control Soc. Pioneering in Technol . . . Build for the Future*, 1980, Numer. Control Soc., Glenview, Illinois, pp. 13–22.

335. Robinson, R. (1982) CAD – the financial aspects. *Des. Eng.*, Nov., 125.
336. Roch, A.J. (1986) Flexible automation holds key to competitive advantage for aerospace manufacturer. *Indust. Eng.* **18**, 52 (8), Nov.
337. Rohan, T.M. (1985) Getting it together – Flexibility; assembly's role in the factory of the future. *Ind. Week*, **226**, 39(5), July 8.
338. Rolland, W.C. (1984) Strategic justification of flexible automation. *Proc. 2nd Annual Int. Robot Conf.* October 1985, Long Beach, California, pp. 177–85.
339. Rose, D.W., Solberg, J.J. and Barash, M.M. (1977) *Optimal Planning of Computerized Manufacturing Systems*; Report no. 5, Unit Machining Operations. Part I. Conform – a Code for Machining. Volume 1, NTIS Report PB-274265/8SL, School of Industrial Engineering, Purdue Univ., Lafayette, Indiana.
340. Rose, L.M. (1976) *Engineering Investment Decisions*, Elsevier, Amsterdam.
341. Rosenberger, A. (1987) Achieving economics of scope (manufacturing automation) *Electron. Wkly.* (GB), **1356**, 30–1.
342. Rudy, T. (1983) Justifying flexible manufacturing systems. *SME FMS Symp.*, Southfield, Michigan, Mar, 1983, p. 83.
343. Runner, J.A. and Leimkuhler, F.F. (1978) *The Optimal Planning of the Computerized Manufacturing System. CAMSAM: A Simulation Analysis Model for Computer-Aided Manufacturing Systems*, NTIS Report PB81-137978, School of Industrial Engineering, Purdue Univ., Lafayette, Indiana.
344. Rutledge, A.L. (1986) Economics of computer integrated manufacturing pre 1986. *Int. Indust. Eng. Conf.* May 1986, Dallas, Inst. Indust. Eng., pp. 247–53.
345. Ryott, J.P. (1977) Industrial robots in production systems. *ECE Seminar Ind. Robots Programmable Logical Controllers.* Economic Commission for Europe Working Party on Automation, Sept. 1977.
346. Saadettan, H. and Shantz, H.M. (1981) CAD/CAM aids productivity/costs and return key questions. *Computing Canada*, **7** (18), 16–19.
347. Sales outlook better to USSR engineering. *Business Eastern Europe*, October 20, 1986, pp. 329, 330.
348. Salzman, R.M. (1981) Impact of automation on engineering manufacturing productivity: CAD in medium sized and small industries. *First European Conf. on Computer Aided Design in Medium Sized and Small Industries. MICAD 80*, 1980, North-Holland, New York, pp. 205–22.
349. Satanovskii, R.L. and Serebryanskay, L.L. (1987) Organizing production by robot units. *Mekh. & Avtom. Proizvod.* (USSR), **5**, 40–1.

350. Schmenner, R.W. (1983) Every factory has a life cycle. *Harvard Bus. Rev.*, **61**, 121–9.
351. Scrimgeour, J. (1976) CAD/CAM expected to yield significant improvement in productivity. *Eng. J.*, **59**, 12–13.
352. Seed, A.H. (1984) Cost accounting in the age of robotics. *Manage. Account.*, **66**, 39–43.
353. Selected Reports and Case Studies. Part III. Econ. Bull. For Europe, **37**, 237(70), Sept. 1985.
354. Seliger, G., Viehweger, B. and Wieneke, B. (1987) Decision support in design and optimization of flexible automated manufacturing and assembly. *Robotics & Comput. Integrated Manuf.* (GB), **3**, 221–7.
355. Sepheri, M. (1984) Cost justification before factory automation. *P & IM Rev. and APICS News*, Apr., 3.
356. Shah, R.R. and Yang, G. (1979) Practical technique for benefit-cost analysis of computer-aided design and drafting systems. *Proc. Design Automation Conf.*, Las Vegas, June, 1978, IEEE, New York, pp. 16–22, (AECL – Report – 6533, March 1979) (Atomic Energy of Canada).
357. Shewchuk, J. (1984) Justifying flexible automation. *Amer. Machinist*, **128**, October, 93–6.
358. Shu, H.H.H., Church, J.C. and Kornfeld, J.P. (1976) *Computerized Production Process Planning* (Vol. 2 *Benefit Analysis*), NTIS Report AD-A151 996/6/XAB.
359. Shunk, D.L. (1984) Integrated manufacturing – factory factors. *Prod. Eng.*, Sept., 50–3.
360. Simpson, J.A. (1984) *Investment Justification of Robotic Technology in Aerospace Manufacturing*, User's guide, NTIS Report – AD-A156 193/5/XAB.
361. Simpson, J.A. (1984) *Investment Justification of Robotic Technology in Aerospace Manufacturing*, NTIS Report – AD-A140 782/4.
362. Simpson, J.A. (1984) *Robotics Investment Decision Model User's Manual*, NTIS Report AD-A145 467/7.
363. Skinner, W. (1980) *The Factory of the Future – Always in the Future? The Factory of the Future* (Special Volume. PED-1), Amer. Soc. Mech. Eng., Chicago.
364. Slock, E. (1980) The prospect of a venture with the industrial robot. *Polytech. Tijdschr., Werktuigbouw* (Dutch.), **35**, 639–42.
365. Sloggy, J.E. (1984) How to Justify the cost of an FMS (flexible manufacturing system). *Tooling & Prod.*, **50**, 72(4), Dec.
366. SME. CAD/CAM, (1983) Dearborn, Michigan, SME.
367. Smith, A.C. (1985) Robotics: a strategic issue. *SAM Adv. Manage. J.*, **50**, 7(6), Spring.
368. Smith, R.D. (1983) Measuring the intangible benefits of computer-based information systems. *J. Syst. Manage.*, **34**, 22–7.

369. Solberg, J.J. (1978) Quantitative design tools for computerized manufacturing systems. manuf. eng. trans., 1978: *Proc. 6th Amer. North Amer. Metalwork Res. Conf.*, SME, Dearborn, Michigan, pp. 409–13.
370. Spiser's superstock solution: reply: can effective demand and the movement toward further income equity be maintained in the face of robotics? *J. Post Keynesian Econ.*, **8**, 642–6, 1986.
371. Spencer, M.S. (1984) Minimum cost manufacturing – a framework for planned change. *Proc. APICS Seminar on Zero Inventory Philosophy & Practices*. October 1984.
372. Stamberger, A. (1980) Social impacts of the computer. *Elektroniker* (German), **19**, EL11–13.
373. Stauffer, R.N. (1978) A new concept in flexible automation. *Manufact. Eng.*, Jan.
374. Stauffer, R.N. (1983) Equipment acquisition for the automatic factory. *Robotics Today*, **5**, 37–40.
375. Stauffer, R.N. (1986) Justification of robotic systems. *Robotics Today*, **8**, 35–40, 43.
376. Stecke, K., Solberg, J.J. and Barash, M.M. (1981) *The Optimal Planning of Computerized Manufacturing Systems.* NTIS Report PB82-110644, School of Industrial Engineering, Purdue Univ., Lafayette, Indiana.
377. Stecke, K.E. and Suri, R. (1986) *Proc. Second ORSA TIMS Conf. on Flexible Manufacturing Systems: Operations Research Models & Applications*. Elsevier, New York.
378. Steffy, W.R., Barvol, L.L. and Polacsek, D. (1967) *Numerical Control Justification: A methodology*, Univ. Michigan. Inst. Science and Technology, Ann Arbor, Michigan.
379. Steinbart, P. (1987) The construction of a rule-based expert system as a method for studying materiality judgments. *Account. Rev.*, **62**, 97–116.
380. Steudel, H.J. and Berg, L.E. (1986) Evaluating the impact of flexible manufacturing cells via computer simulations: *Large Scale Syst.-Theory & Appl.* (Netherlands), **11**, 121–30.
381. Stevenson, D.O. (1985) Purpose behind the projects: progress, not profits. *Proc. Robot 9, Conf.* Detroit, June 1985, v. 2, pp. 16–17.
382. Stiefel, M.L. (1981) CAD/CAM spells productivity. *Mini-Micro Syst.*, **14**, 115–22.
383. Stokes, P.G. (1985) Justifying CIM investments. robotic trends – applications, research, education and safety. *Proc. 8th Ann. British Robot Assoc. Conf.* Birmingham, England, May 1985, pp. 221–30.
384. Strategic planning for computer integrated manufacturing. TBC Inc. striving for technical excellence in manufacturing through

communication, planning, and professionalism: *Proc. 18th Int. Technical Conf. of Numerical Control Soc.*, Numerical Control Soc., Glenview, Illinois, 1981.

385. Suardo, G.M.S. (1979) Workland optimization in a FMS modelled as a closed network of queues. *Ann. CIRP*, **28**.
386. Sullivan, W.A. (1985) Computer integrated manufacturing program justification. *Proc. Autofact 1985 Conf.*, Detroit, Nov. 1985, Part 2, SME, Dearborn, Michigan, pp. 34–42.
387. Sullivan, W.G. (1986) Justifying new technology: models IES can use to include strategic, non-monetary factors in automation decision. *Ind. Eng.*, **18**, 42–50.
388. Sullivan, W.G. (1984) Project justification in high technology ventures. Presented to 92nd ASEE Ann. Conf., Salt Lake City, Utah.
389. Suresh, N.C. Meredith, J.R. (1984) A generic approach to justifying flexible manufacturing systems. *Proc. ORSA/TIMS Conf. on FMS*, Ann Arbor, Michigan, pp. 36–42.
390. Suresh, N.C. (1985) Financial justification of robotics in multimachine systems. *Proc. Robot 9, Conf.* Detroit, June 1985, pp. 1–17.
391. Suri, R. (1981) New techniques for modeling and control of FMS. *Proc. IFAC. Conf.*, Tokyo, 1981.
392. Suri, R. and Whitney, C.K. (1984) *Designing a Decision Support System for Flexible Manufacturing* (FMS Handbook, Vol. III, Section 6-0), NTIS Publ. No. AD/A127929.
393. Swindle, R. (1985) Financial justification of capital projects. *Proc. Autofact, 1985, Conf.* Detroit, Nov. 1985, part 5, 1–5, 17.
394. Talavage, H. (1987) *Flexible Manufacturing Systems: Designs, Analysis & Simulation Series*: Manuf. Engineering & Materials Process Ser., Dekker.
395. Talaysum, A.T., Hassan, M.Z. and Goldhar, J.D. (1987) Uncertainty reduction through flexible manufacturing. *IEEE Trans. Eng. Manage.* (USA), **EM-34**, 222–33.
396. Talaysum, A.T. (1987) Uncertainty reduction through flexible manufacturing. *IEEE Trans. Eng. Manage.*, **EM-34**, 85–91.
397. Tanski, K. (1977) An attempt to assess economic efficiency in plant-control systems of computer-operated automation. *BIUL. INF. Obiektowe Syst. Komputerowej. Austom.* (Polish), **16**, 51–63.
398. Tech update: adaptable assembly fixture devised by ROHR for AF flexible cell system. *Amer. Metal Market*, August 11, 1986, 9, 10.
399. Technology: automatic factories are slow to catch on. (1984) *New Scientist*, August 9, 27.
400. Temmes, J. and Ranta, J. (1987) Utilization of modern production automation: some technological, economic and social impacts. *Robotics* (Netherlands), **3**, 89–94.

401. Teresko, J. (1986) How should management assess today's advanced manufacturing options? (automation and the bottom line). *Ind. Week*, **229**, 41(19), May 26.
402. Thompson, H.B. (1982) CAD/CAM: the strategic opportunities. *Tool. & Prod.*, **47**, 66–9.
403. Thornton-Bryar, I.C.M. (1985) A practical approach to workstation cost-justification. *Comp. Aided Des.*, **17**, 125–9.
404. Tipnis, V.A. and Misal, Anant C. (1985) Economics of flexible manufacturing systems. *Proc. Spring 1985. Flexible Manufacturing Systems Conf.*, March 1985, Dallas, SME.
405. Total automation seen needing 'clean-sheet' top-down approach, tech-update. (1981) *Amer. Metal Market/Metal Working News*, Nov.
406. Troxler, J.W. (1987) An economic analysis of flexible automation in batch manufacturing with emphasis on fabrication. *Diss. Abst. Int.*, **48B**, 1778.
407. United Kingdom Dept. of Industry. *Low Cost Automation* (1977) New English Library.
408. US Dept. Air Force (1985) *Software Available for Calculating Investment in Robotics.* (NTIS Tech. Note).
409. Use of simulation for the technical and economic evaluation of flexible manufacturing system. (1984) Part 2: Results of the simulation, Report PNR-90233-Pt-2; TRANS-16391/TLT-00900, November, Weskaemper, E.
410. Vail. *Computer Integrated Manufacturing.* PWS-Kent Pub., 1988, Boston, Massachusetts.
411. Van Blois, J.P. (1983) Economic models: the future of robotic justification. *Proc. 13th Int. Symp. on Industrial Robots & Robots 7*, Chicago, April 1983, v. 1, pp. 24–34.
412. Van Blois, J.P. and Andrews, P.P. (1983) Robotic justification: the domino effect. *Prod. Eng.*, April, 52–4.
413. Van Blois, J.P. (1983) Strategic robot justification: a fresh approach. *Robotics Today*, **5**, 44–8.
414. Van Nostrand, R.C. (1984) A case study in successful CAD/CAM justification, *CIM Rev.*, Fall, 45–52.
415. Varney, M.S., Sullivan, W.G. and Cochran, J.K. (1985) Justification of flexible manufacturing systems with the analytical hierarchy process. *Proc. 1985 Ann. Int. Indus. Eng. Conf.*, pp. 181–90.
416. Vern, E. (1984) Robot justification: a lot more than dollars and cents. *Proc. RI/SME Robots 8th Conf.*, Detroit, June, 1984, v. 1, p. 2–1.
417. Von Voros, G. (1981) Using CAM methodology to increase productivity. *Tool & Prod.*, **46**, 72–3.

418. Wallace, W.J. and Thuesen, G.J. (1987) Annotated bibliography on investment in flexible automation. *Eng. Econ.*, **32**, 247–57.
419. Warnecke, H.J. and Vettin, G. (1977) Strategies for the organization and control of discontinuous conveyors in flexible manufacturing systems. *Manufact. Syst.* (German), **3**, 197–209.
420. Warnecke, H.J. and Vettin, G. (1982) Technical investment planning of flexible manufacturing system: the application of practice oriented methods. *J. Manufact. Syst.*, **1**, 89–98.
421. Warnecke, H.J. (1985) Flexible manufacturing systems: *Proc. of the 3rd Int. Conf.*, Boeblingen, Near Stuttgart, BRD, 11–13 September, 1984, Elsevier, New York.
422. Warnecke, H.J. and Steinhilper, R. (1985) *Flexible Manufacturing Systems* (Int. Trends in Manuf. Tech. Ser.), Springer–Verlag, New York.
423. Watanabe, S. (1986) Labour-saving versus work-amplifying effects of micro-electronics. *Int. Lab. Rev.*, **125**, 243–5.
424. Weaver, L. (1985) Investing in advanced manufacturing technology. *Prod. Eng.*, **64**, 19–20.
425. Weinstein, M.S., Leu, M.C. and Infelise, F.A. (1985) Design and analysis of robotic assembly for a printer compensation arm. *Amer. Soc. Mech. Eng.*, Production Engineering Division publication v. 15, pp. 243–9.
426. Weiss, B. (1985) Milacron exec: outscoring + FMS = profits (flexible manufacturing systems). *Amer. Metal Market*, **93** 5(2), June 17.
427. Wheelwright, S.C. (1978) Reflecting corporate strategy in manufacturing decisions. *Bus. Horizons*, Feb.
428. Wheelwright, S.C. (1981) Japan – where operations really are strategic. *Harvard Bus. Rev.*, **59**, 67–74, July–Aug.
429. Wildemann, H. (1986) Justification on strategic planning for new technologies. *Hum. Syst. Manage.* (Netherlands), **6**, 253–63.
430. Williams, D.F., El-Tamimi, A.M. and Suliman, S.M.A. Product selection and investment appraisal for flexible assembly. *Proc. Inst. Mech. Eng.* Part B, 201, pp. 35–40.
431. Winter, D.D. and Huber, Rf. (1981) Robot pays off for finishing Operation Zama: where Japan works wonders. *Production*, **87**, 84–7.
432. Wirtsc, E.D. and Westkaemper, E. (1975) *Use of Simulation for the Technical and Economic Evaluation of Flexible Manufacturing Systems. Part 2 – Results of the Simulation.* NTIS Report N85-18614/6/XAB. Translated into English from *Ind-Anzeiger*, **97**, 1302–5 (1975).
433. Yonemoto, E.K. (1981) The socio-economic impacts of industrial robots in Japan. *Indus. Robot.*, **8**, 238–41.
434. Yost, L. (1987) Justification of Automation. Electro/87 and Mini/

Micro Northeast: Forcusing on the Dem. Conf. Record, New York, NY, USA, April 1987, 13/5/1–5, New York.

435. Yost, L. (1987) CIM: the global equalizer. *Proc. 1987 IEEE Industrial Application Soc. Conf.* 18–23 October, 1987, Piscataway, New Jersey, pp. 1295–8.
436. Young, M.P. (1975) A computer method in economic design of conductor sizes of distribution lines. *J. Korean Inst. Elec. Eng.*, **24**, 107–20.
437. Zeldenberg, J. (1984) New manufacturing tech aid to cut costs by 40 per cent. *Computing Canada*, October.

Subject index

Author index